职业教育双语教材

Bilingual Textbooks of Vocational Education

# 数控加工技术

# NC Machining Technology


赵　慧　王　安　主编

Edited by Hui Zhao An Wang

刘万菊　副主编

Subeditor Wanju Liu

MOHAMED　AHMED　ALI　BAIOUMY　MOHAMED　参编

Co-edited by MOHAMED AHMED ALI BAIOUMY MOHAMED

李云梅　主审

Reviewed by Yunmei Li

· 北 京 ·

· Beijing ·

本书主要包括数控车床的编程与操作、数控铣床的编程与操作两部分内容。本书以埃及“鲁班工坊”为背景，以“鲁班工坊”数控加工设备为载体，以培养高级技能型人才为目标，在注重基础理论教育的同时，突出实践性教育环节，力图做到深入浅出，便于教学，突出高等职业教育的特点，适合高职高专工科学生使用，也适合埃及开罗高级维修技术学校使用。

**图书在版编目（CIP）数据**

数控加工技术：汉英对照/赵慧，王安主编．—北京：化学工业出版社，2019.12
ISBN 978-7-122-35870-7

Ⅰ.①数… Ⅱ.①赵…②王… Ⅲ.①数控机床-加工-双语教学-高等职业教育-教材-汉、英 Ⅳ.①TG659

中国版本图书馆CIP数据核字（2019）第278801号

责任编辑：张绪瑞　刘　哲　　装帧设计：韩　飞
责任校对：王　静

出版发行：化学工业出版社（北京市东城区青年湖南街13号　邮政编码100011）
印　　装：大厂聚鑫印刷有限责任公司
787mm×1092mm　1/16　印张16　字数433千字　2020年7月北京第1版第1次印刷

购书咨询：010-64518888　　售后服务：010-64518899
网　　址：http：//www.cip.com.cn
凡购买本书，如有缺损质量问题，本社销售中心负责调换。

定　　价：49.50元

# 前言

在扩大与一带一路沿线国家的职业教育合作，贯彻落实天津市启动实施的将优秀职业教育成果输出国门与世界分享计划的要求中，职业教育作为与制造业联系最紧密的一种教育形式，正在发挥着举足轻重的作用。为了配合埃及“鲁班工坊”的理论和实训教学，开展交流与合作，提高中国职业教育的国际影响力，创新职业院校国际合作模式，输出我国职业教育优秀资源，课题组编写了本教材。

数控机床是以数字运算为核心，执行程序代码，计算机统一各部件协调工作，来加工零件的冷加工设备，是智能制造的核心技术。主要应用于IT、汽车、轻工、医疗、航空等行业，是高速、高精、高可靠性的机械加工设备。

本书为埃及“鲁班工坊”的培训教材，采用中英文两种语言编写。以埃及“鲁班工坊”为背景，以“鲁班工坊”数控加工设备为载体，以培养高级技能型人才为目标，在注重基础理论教育的同时，突出实践性教育环节，力图做到深入浅出，便于教学，突出高等职业教育的特点，适合高职高专工科学生使用，也适合埃及开罗高级维修技术学校使用。

本书共分两个项目，项目一介绍了数控车床的编程与操作；项目二介绍了数控铣床的编程与操作。

项目一由赵慧编写，项目二由王安编写，刘万菊教授对全书编写进行了悉心指导，埃及老师 MOHAMED AHMED ALI BAIOUMY MOHAMED 就教材内容联系埃及的实际情况，参与对内容进行合理化调整。天津轻工职业技术学院王娟、王丹阳、郭世杰、李晓彤参与了本书的翻译工作。

限于编者水平有限，书中定有不少疏漏，恳请读者批评指正。

编者

2019年10月

# 目 录

# 项目一　数控车加工

**【项目描述】**

数控车床是一种高精度、高效率的机电一体化自动机床，也是使用数量最多的数控机床，大约占数控机床总数的25%。数控车床主要用于轴套类和盘类等回转体零件的加工，通过数控加工程序的运行，可自动完成内外圆柱面、圆锥面、成形表面、螺纹和端面等工序的切削加工，并能进行车槽、钻孔、扩孔、铰孔等工作。车削中心可在一次装夹中完成更多的加工工序，提高了加工精度和生产效率，特别适合于复杂形状回转类零件的加工。FANUC 0i系统是目前数控车床上应用较多的一种数控系统，它的编程方法和指令格式具有一定的代表性，本项目以山东辰榜数控装备有限公司数控车床（配FANUC 0iFD系统）展开教学。

**【能力目标】**

1. 了解数控车床的组成。
2. 了解数控车削的加工特点。
3. 掌握G代码、M代码常用指令。
4. 掌握坐标系建立原则。
5. 熟练掌握数控车G00、G01、G02、G03等指令的用法。
6. 熟练掌握G71、G70等循环指令的用法。
7. 掌握G32、G92车等车削螺纹指令。

## 任务一　掌握数控车床安全文明生产

### 一、文明生产要点

文明生产是企业管理中的一项十分重要的内容，它直接影响产品质量，影响设备和工、夹、量具的使用效果及寿命，还会影响操作工人技能的发挥。高职学校的学生是工厂的后备工人，从开始学习本课程时，就要重视培养文明生产的良好习惯。因此，要求操作者在整个操作的全过程中必须做到以下几点。

① 进入数控实习场地后，应服从安排、听从指挥，不得擅自启动或操作机床数控系统。

② 开车前，应该仔细检查机床各部分机构是否完好，各传动手柄、变速手柄（主要指

经济型数控车床）的位置是否正确。还应按要求认真检查数控系统及各电气附件的插头、插座是否连接可靠。

③ 对改造后的两用（自动控制与手动操作）数控机床，使用前必须检查其传动机构是否相互干涉，以保证设备不受损坏并能正常运转。

④ 对机床主体，应按普通机床的有关要求进行文明使用和养护。

⑤ 数控系统在不使用时，要用布罩套上，防止进入灰尘，并应在专业人员指导下，定期进行内部除尘或细微清理。

⑥ 操作数控系统前，应检查两侧的散热风机是否运转正常，以保证良好的散热效果。

⑦ 操作数控系统时，对各按键及开关的操作不得用力过猛，更不允许用扳手或其他工具进行操作。

⑧ 在不加工螺纹时，主轴脉冲发生器应与主轴脱开连接，以延长其使用寿命。

⑨ 当自动转位刀架未回转到位时，不得强行用外力使刀架非正常定位，以防止损坏刀架的内部结构。

⑩ 虽然数控加工过程是自动进行的，但并不属无人加工性质，仍需要操作者经常观察，不允许随意离开生产岗位。

⑪ 下班时，除了按规定关机外，还应认真做好交接班工作，并应做好文字记录（如加工程序及程序执行情况等）。

## 二、安全操作技术

数控机床具有高精度、高效率和高适应性的特点。其运行效率的高低、各附件的故障率、使用寿命的长短等，很大程度上取决于用户的正确使用与维护。好的工作环境、好的使用者和维护者，将大大延长无故障工作时间，提高生产率，同时减少机械部件的磨损，避免不必要的失误，从而减少维修人员的负担。操作时，必须提高执行纪律的自觉性，严格遵守以安全技术要求为主的各项规章制度，并认真做到以下几点。

① 进入数控车削实训场地后，应服从安排，不得擅自启动或操作车床数控系统。

② 按规定穿、戴好劳动保护用品。

③ 不能穿高跟鞋、拖鞋上岗，不允许戴手套和围巾进行操作。

④ 开机床前，应该仔细检查车床各部分机构是否完好，各传动手柄、变速手柄的位置是否正确，还应按要求认真对数控机床进行润滑保养。

⑤ 操作数控系统面板时，对各按键及开关的操作不得用力过猛，更不允许用扳手或其他工具进行操作。

⑥ 完成对刀后，要做模拟换刀试验，以防止正式操作时发生撞坏刀具、工件或设备等事故。

⑦ 在数控车削过程中，因观察加工过程的时间多于操作时间，所以一定要选择好操作者的观察位置，不允许随意离开实训岗位，以确保安全。

⑧ 操作数控系统面板及操作数控机床时，严禁两人同时操作。

⑨ 自动运行加工时，操作者应集中思想，左手手指应放在程序停止按钮上，眼睛观察刀尖运动情况，右手控制修调开关，控制机床拖板运行速率，发现问题及时按下程序停止按钮，以确保刀具和数控机床安全，防止各类事故发生。

⑩ 实训结束时，除了按规定保养数控机床外，还应认真做好交接班工作，必要时应做好文字记录。

⑪ 操作前要穿紧身防护服，袖口扣紧，上衣下摆不能敞开，严禁戴手套，不得在开动

的机床旁穿、脱、换衣服或围布于身上，防止机器绞伤。必须戴好安全帽，辫子应放入帽内，不得穿裙子、拖鞋。要戴好防护镜，以防铁屑飞溅伤眼。

⑫ 车床开动前，必须按照安全操作的要求，正确穿戴好劳动保护用品，必须认真仔细检查机床各部件和防护装置是否完好、安全可靠，加油润滑机床，并作低速空载运行2～3min，检查机床运转是否正常。

⑬ 装卸卡盘和大件时，要检查周围有无障碍物，垫好木板，以保护床面，并要卡住、顶牢、架好，车偏重物时要按轻重做好平衡，工件及工具的装夹要紧固，以防工件或工具从夹具中飞出，卡盘扳手、套筒扳手要拿下。

⑭ 机床运转时，严禁戴手套操作；严禁用手触摸机床的旋转部分；严禁在车床运转中隔着车床传送物件。装卸工件，安装刀具，加油以及打扫切屑，均应停车进行。清除铁屑应用刷子或钩子，禁止用手清理。

⑮ 机床运转时，不准测量工件，不准用手去刹转动的卡盘；用砂纸时，应放在锉刀上，严禁戴手套用砂纸操作，磨破的砂纸不准使用，不准使用无柄锉刀，不得用正反车电闸作刹车，应经中间刹车过程。

⑯ 加工工件按机床技术要求选择切削用量，以免机床过载造成意外事故。

⑰ 加工切削时，停车时应将刀退出。切削长轴类须使用中心架，防止工件弯曲变形伤人；伸入床头的棒料长度不超过床头立轴之外，并慢车加工，伸出时应注意防护。

⑱ 高速切削时，应有防护罩，工件、工具的固定要牢固，当铁屑飞溅严重时，应在机床周围安装挡板使之与操作区隔离。

⑲ 机床运转时，操作者不能离开机床，发现机床运转不正常时，应立即停车，请维修工检查修理。当突然停电时，要立即关闭机床，并将刀具退出工作部位。

⑳ 工作时必须侧身站在操作位置，禁止身体正面对着转动的工件。

㉑ 工作结束时，应切断机床电源或总电源，将刀具和工件从工作部位退出，清理安放好所使用的工、夹、量具，并清扫机床。

㉒ 每台机床上均应装设局部照明灯，机床上照明应使用安全电压（36V以下）。

## 三、 6S职业规程

根据6S的管理精神和目前学校实训工场间的实际使用情况，制订6S管理规范如下。

**1. 1S——整理**

（1）定义

① 将工作场所任何东西区分为有必要的与不必要的。

② 把必要的东西与不必要的东西明确地、严格地区分开来。

③ 不必要的东西要尽快处理掉。

（2）目的

① 腾出空间，空间活用。

② 防止误用、误送。

③ 塑造清爽的工作场所。

**2. 2S——整顿**

（1）定义

① 对整理之后留在现场的必要的物品分门别类放置，排列整齐。

② 明确数量，有效标识，并做好相应的登记。

（2）目的
① 工作场所一目了然。
② 整整齐齐的工作环境。
③ 消除找寻物品的时间。
④ 消除过多的积压物品。

### 3. 3S——清扫

（1）定义
① 将工作环境、工作设施、工作设备清扫干净。
② 保持工作场所干净、亮丽。
（2）目的
① 消除脏污，保持工场间内干净、明亮。
② 稳定品质。
③ 减少工业伤害。

### 4. 4S——清洁

（1）定义
将上面的3S实施的做法制度化、规范化。
（2）目的
维持上面3S的成果。

### 5. 5S——素养

（1）定义
通过学习、参观等手段，提高员工文明礼貌水准，增强团队意识，养成按规定行事的良好工作习惯。
（2）目的
提升教师、工作人员、学生的品质，使教师、工作人员、学生对任何工作都讲究认真。

### 6. 6S——安全

（1）定义
重视全员安全教育，每时每刻都有安全第一观念，防患于未然。
（2）目的
建立起安全生产的环境，所有的工作应建立在安全的前提下。

### 习题

1. 机床操作过程中，有哪些注意事项？
2. 机床的安全操作要做到哪几点？
3. 阐述6S职业规程要点及内容。
4. 如何识读CAK6140dj/1000中的字母和数字？
5. 数控车床由哪些部分组成？它们的作用又是什么？
6. 什么是模态指令？什么是非模态指令？举例说明。
7. 程序由哪些部分构成？
8. 操作面板上有哪些操作模式？

# 任务二　数控车床基本操作

## 一、数控车床的初步认识

**1. 数控车床的标记**

CAK6140dj/1000

C——机床类别代号（车床类）；

A——生产厂家代号（改进的批次）；

K——机床结构特性代号（数控）；

6——机床组代号（卧式车床组）；

1——机床系代号（卧式车床系）；

40——主参数代号（最大车削直径为400mm）；

d——系统型号（FANUC系统）；

j——防护形式（半封式）；

1000——主参数代号（最大工件长度1000mm）。

**2. 数控车床组成部分及功能**

① 安全防护门：用以安全防护作用，门上有窥视孔。

② 卡盘：用以夹持工件。

③ 数控操作面板：数控编程、控制运动部件、调节加工参数。

④ 刀塔：安装各类车削用刀具。

⑤ 尾座：用以安装顶尖、钻头等工具。

⑥ 导轨：主要起导向与支承作用。

## 二、机床操作说明

（1）控制器面板（见图1-2-1）

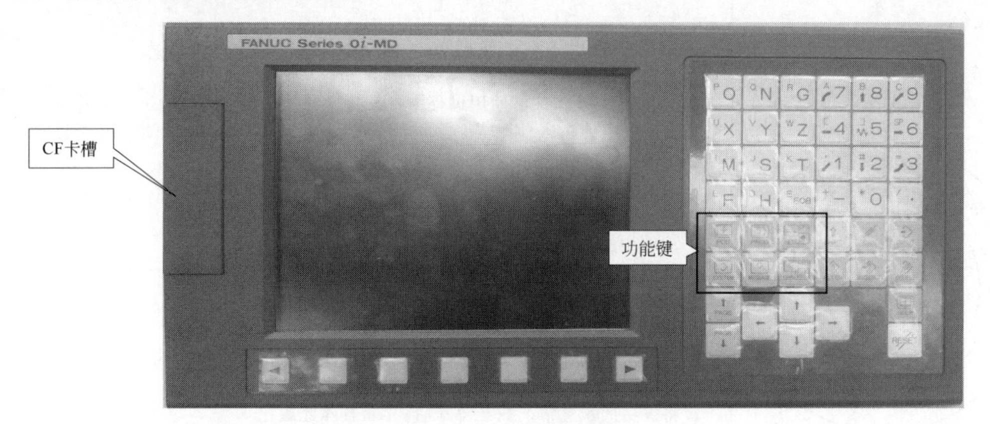


图1-2-1　控制器面板

（2）操作面板

本机配置方便、快捷操作的数控机床专用控制面板，面板按键等布局示意图如图1-2-2所示。

图 1-2-2　机床操作面板示意图

（3）系统电源开启及关闭开关（见表 1-2-1）

**表 1-2-1　系统电源开启及关闭开关**

| 图示 | 说明 |
| --- | --- |
|  | 名称：控制器开启按钮（绿色按钮）<br>操作面板上该按钮触发后，控制器电源开启，数控系统进入启动状态 |
|  | 名称：控制器关闭按钮（红色按钮）<br>操作面板上该按钮触发后，控制器电源关闭，数控系统进入关闭状态<br>程序运行中严禁触发此按钮 |

（4）紧急停止（见表 1-2-2）

**表 1-2-2　紧急停止**

| 图示 | 说明 |
| --- | --- |
| EMERGENCY STOP | 名称：紧急停止旋钮（红色旋钮）<br>紧急停止旋钮为红色，位于本机床操作面板左侧中间位置<br>① 本机床在操作或者使用过程中如发生紧急情况发生时，需要立即按下本按钮，按钮按下瞬间可使机床动作全面停止，同时输出到电动机的电流中断，但本机床不断电<br>② 按下紧急停止按钮现象<br>a. 如果伺服轴在运行中，则运行的轴停止移动（如机床配有第四轴，并且轴在运转中，第四轴将停止运转）<br>b. 旋转中的主轴停止旋转<br>c. 机床显示器报警信息画面，显示 ALARM 信息如下<br>1000 EMG STOP OR OVERTRAVEL<br>d. 如果刀盘旋转中按下紧急停止旋钮，则刀盘立即停止旋转<br>e. 如果换刀过程中按下紧急停止旋钮，则换刀动作停止，进入换刀异常中断情况<br>③ 需要注意以下条件<br>a. 将发生的紧急情况完全解除后，才能解除此旋钮<br>b. 在停止当时所有指令与机器状况均已被删除，因此需要重新核对加工程序，无异常再进行相关操作<br>c. 当执行自动换刀的中途，按下此旋钮所有动作将立即停止，因此刀盘可能位于不确定位置 |

（5）程序开关（见表 1-2-3）

**表 1-2-3　程序开关**

| 图示 | 说明 |
|---|---|
|   | 名称：程序启动按钮<br>在自动运转模式（手动输入、记忆、联机）下选中需要执行的加工程序，按下“程序启动”按钮后，程序即开始执行 |
|   | 名称：程序暂停按钮<br>① 在自动运转模式（手动输入、记忆、联机）下，按下“程序暂停”按钮后，各轴立即减速停止，进入运转休止状态<br>② 当再按下“程序启动”按钮后，加工程序将从当前暂停的单节继续执行 |

（6）程序保护开关功能（见表 1-2-4）

**表 1-2-4　程序保护开关功能**

| 图示 | 说明 |
|---|---|
| 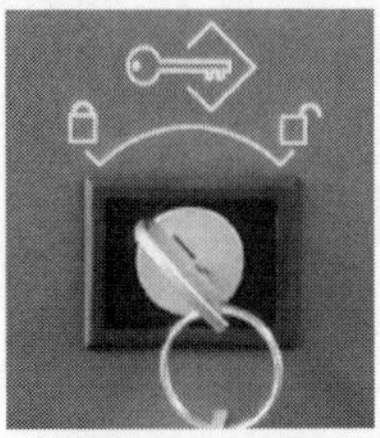  | 名称：程序保护开关<br>① 为防止本机床控制器中的程序被他人编辑、取消、修改、建立，钥匙应交由专人保管<br>② 一般情况，将此钥匙设定在“OFF”的位置，以确保程序不被修改或删除<br>③ 如果想对程序予以编辑、取消或修改时，应将本钥匙设定在“ON”的位置 |

（7）操作模式（见表 1-2-5）

**表 1-2-5　操作模式**

| 图示 | 说明 |
|---|---|
|  编辑方式 | 名称：编辑模式<br>本机床显示器左下角上显示为“编辑”<br>① 可以对程序予以编辑、修改、增加或删除时，将程序保护开关钥匙设定在 ON 状态，在此模式下才可编辑<br>② 本模式仅用于编辑，不能执行程序<br>③ 执行新编辑程序时，必须转至自动模式（手动输入、记忆），才会执行编制程序<br>④ 程序编辑完成控制器即自动存储，不必再执行存储动作<br>⑤ 本模式下，可由各个人计算机读入加工程序、NC 参数、刀具长度补正、刀具半径补正 |
|  自动方式 | 名称：自动模式<br>本机床显示器左下角上显示为“MEM”<br>① 在本模式下，按下“程序启动”按钮，即可执行当前已经选择的加工程序<br>② 本模式下，能执行 CNC 内存中的程序<br>③ 在本模式下的进给速率，参照进给倍率调整按键说明<br>④ 本模式下，程序在执行 M30 时即为程序结束 |

续表

| 图示 | 说明 |
| --- | --- |
|  | 名称：MDI 手动输入模式<br>本机床显示器左下角上显示为“MDI”<br>① 本模式下，在控制器 MDI 面板输入单节程序指令予以执行<br>② 本模式下，指令执行完成后，可通过设置参数来确定编制的程序是否需要消除<br>③ 本模式下，仅能输入部分程序段 |
|  | 名称：手摇模式<br>本机床显示器左下角上显示为“HAND”<br>① 本模式下，可用手持单元移动各轴<br>② 本模式下，轴向移动可由手持单元上的轴向旋钮选择轴向，从而控制选中轴向的移动<br>③ 本模式下，各轴手动移动速度可由手持单元上的进给倍率旋钮决定<br>④ 手摇脉冲发生器转动速度不能大于 5 圈/s |
|  | 名称：寸动模式<br>本机床显示器左下角上显示为“JOG”。<br>① 本模式下，欲移动各轴，可按各轴向键及选择慢速进给率<br>② 本模式下，移动进给速率，按照慢速进给率进行移动。速率调整范围为 0～1000mm/min<br>③ 本模式下按轴向键时，其指定轴向即可移动，松开按键，轴即停止<br>④ 配合快速按钮使用时，移动进给速率，按照快速进给率进行移动，按各轴向键，依指定轴向移动，松开按键，轴即停止 |
|  | 名称：原点复归模式<br>本机床显示器左下角上显示为“REF”<br>① 在手动方式下用于伺服轴机械回原点操作<br>② 机床每次开机后，需做一次原点复归操作。如果各轴位置在原点附近，则各轴需要手动移动远离原点位置后，再继续完成原点复归操作<br>③ 在本模式下，选择需机械原点复归的轴向键后，原点指示灯持续闪烁，直到原点复归动作完成后，原点指示灯不再闪烁处于常亮状态<br>④ 各轴的机械原点复归速率为：参数设定回零速度×进给倍率开关所在的倍率值(%) |

(8) 辅助功能(见表 1-2-6)

**表 1-2-6　辅助功能**

| 图示 | 说明 |
| --- | --- |
|  | 名称：单节执行<br>本功能仅在自动相关模式下有效<br>① 本按键内藏灯亮时，程序单节执行功能键有效<br>此功能打开后，程序将按单节执行，执行完当前单节后程序暂停，继续按程序启动按钮后方可执行下一单节程序，以后执行程序依此类推<br>② 当按键指示灯不亮时，程序单节执行功能无效。加工程序将一直被执行到程序终了 |
|  | 名称：空运行<br>本功能仅在自动相关模式下有效<br>① 本按键指示灯亮时，$Z$ 轴锁定功能键有效<br>此功能打开后，程序中所设定的 F 值(切削进给率)指令无效，其各轴移动速率依慢速位移速率所指定之速率位移<br>② 在功能有效时，若程序执行循环程序时，慢速进给率或切削进给率无法改变进给率，按照控制中的 F 值作固定的进给速率 |

续表

| 图示 | 说明 |
| --- | --- |
|   | 名称:选择跳过<br>本功能仅在自动相关模式下有效<br>① 本按键指示灯亮时,程序选择跳过功能有效<br>此功能打开后,自动运行中,当在程序段的开头指定了一个"/"(斜线)符号时,此程序段将略过不被执行<br>② 当按键内藏灯不亮时,程序选择跳过功能无效<br>此功能关闭后,即程序单节前有"/"(斜线)符号时,此程序段也可以正常执行 |
|   | 名称:选择停<br>本功能仅在自动模式有效<br>① 本按键指示灯亮时,程序选停功能键有效<br>此功能打开后,执行程序中,若有 M01 指令时,程序将停止于该单节。若欲继续执行程序时,按下程序启动键即可<br>② 当按键指示灯不亮时,程序选停功能无效<br>此功能关闭后,即使程序中有 M01 指令,程序也不会停止执行 |
|   | 名称:机械锁定<br>① 本按键指示灯亮时,所有轴机械锁定功能键有效<br>此功能打开后,无论在手动模式或自动模式中移动任意一个轴,CNC 均停止向该轴伺服电机输出脉冲(移动指令),但依然在进行指令分配,对应轴的绝对坐标和相对坐标也得到更新<br>② M、S、T、B 码会继续执行,不受机械锁定限制<br>③ 解除此功能后需要重新回归机械零点,在回零正确且完毕后,再进行其他相关操作<br>如果未回零而进行了相关操作则会造成坐标偏移,甚至出现撞机、程序乱跑等异常现象,从而导致危险 |
|   | 名称:F1<br>此按键依据机床实际配置<br>预备空格键,操作人员不能进行操作 |
|   | 名称:F2<br>此按键依据机床实际配置<br>预备空格键,操作人员不能进行操作 |
|   | 名称:F3<br>工作灯扩展<br>控制工作灯开启和关闭,不受任何操作模式限制 |

(9) 主轴功能（见表 1-2-7）

**表 1-2-7　主轴功能**

| 图示 | 说明 |
|---|---|
|  | 名称：主轴降速<br>① 此按钮位于本机床操作面板上，用于降低编程制定的主轴转速 S 速度，实际转速＝编程给定 S 指令值×主轴速度降低倍率值<br>② 与设定的主轴转速配合使用。 |
|  | 名称：主轴升速<br>① 此按钮位于本机床操作面板上，用于提高编程制定的主轴转速 S 速度，实际转速＝编程给定 S 指令值×主轴速度降低倍率值<br>② 编程设定速度超过主轴最高转速，转速达到 100%以上倍率时，主轴修调速度等于主轴最高转速<br>③ 与设定的主轴转速配合使用 |
|  | 名称：主轴正转<br>(1)在机床执行一次 S 代码后，选中手动操作模式，按主轴正转按键后，主轴进行顺时针旋转。主轴旋转速度＝先前执行的主轴速度 S 值×主轴修调旋钮所在的挡位<br>(2)使用条件<br>① 仅在“手动”模式、“快速”模式、“寸动”模式才能使用<br>② 在自动模式时，当程序中执行主轴正转 M03 指令后，本按键指示灯会亮<br>(3)“主轴停止”或“主轴反转”生效时，指示灯即熄灭<br>(4)需要进行主轴反向旋转时必须使主轴停止后，才可指定反向旋转操作 |
|  | 名称：主轴停止<br>(1)主轴无论处于正转或反转状态下，按此键均可以停止正在旋转中的主轴<br>(2)使用条件<br>① 本按键仅在“手动”模式、“快速”模式、“寸动”模式才能使用<br>② 在自动操作时无效<br>(3)主轴停止时本按键指示灯会亮，但如果“主轴正转”或“主轴反转”生效时，指示灯即熄灭 |
|  | 名称：主轴反转<br>(1)在本机床执行一次 S 代码后，选中手动操作模式，按主轴反转按键后，主轴进行逆时针旋转。主轴旋转速度＝先前执行的主轴速度 S 值×主轴修调旋钮所在的挡位<br>(2)使用条件<br>① 仅在“手动”模式、“快速”模式、“寸动”模式才能使用。<br>② 在自动模式时，当程序中执行主轴反转指令 M04 后，本按键指示灯会亮<br>(3)主轴反转时本按键内藏灯会亮，但如果“主轴正转”或“主轴停止”生效时，指示灯即熄灭<br>(4)需要进行主轴正向旋转时必须使主轴停止后，才可指定正向旋转操作 |

(10) 辅助功能（见表 1-2-8）

**表 1-2-8　辅助功能**

| 图示 | 说明 |
|---|---|
|  | 名称：冷却<br>① 在手动、快速、寸动模式下，按此键指示灯亮后，切削液开启<br>② 按“RESET”键，冷却液停止喷出，冷却液停止，指示灯灭<br>③ 冷却液开启时需注意冷却液喷嘴的朝向 |

续表

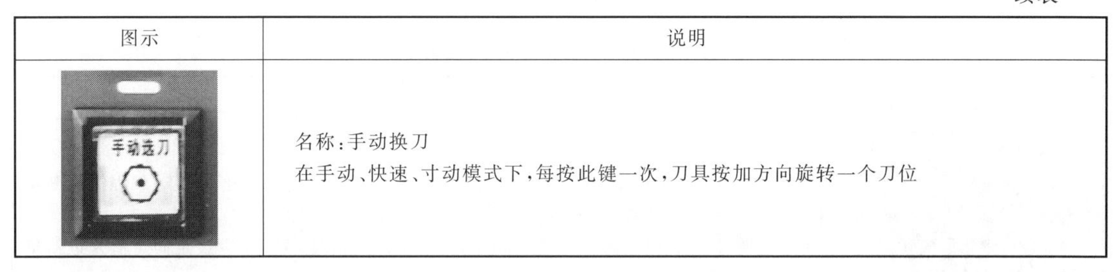

| 图示 | 说明 |
|---|---|
| | 名称：手动换刀<br>在手动、快速、寸动模式下，每按此键一次，刀具按加方向旋转一个刀位 |

（11）轴向选择键功能（见表 1-2-9）

**表 1-2-9　轴向选择键功能**

| 图示 | 说明 |
|---|---|
| | 名称：＋X 控制按键<br>＋X 按键：在 JOG 方式下，按住此键则 $X$ 轴依进给倍率/快速倍率之速度向机床 $X$ 轴“＋”方向（正方向）移动，同时按键指示灯点亮；当松开按键后，轴停止向“＋”方向移动，同时按键指示灯熄灭<br>此外，该按键也作为 $X$ 轴回零触发键 |
| | 名称：－X 控制按键<br>－X 按键：在 JOG 方式下，按住此键则 $X$ 轴依进给倍率/快速倍率之速度向机床 $X$ 轴“－”方向（负方向）移动，同时按键指示灯点亮；当松开按键后，轴停止向“－”方向移动，同时按键指示灯熄灭 |
| | 名称：＋Z 控制按键<br>＋Z 按键：在 JOG 方式下，按住此键则 $Z$ 轴依进给倍率/快速倍率之速度向机床 $Z$ 轴“＋”方向（正方向）移动，同时按键指示灯点亮；当松开按键后，轴停止向“＋”方向移动，同时按键指示灯熄灭<br>此外，该按键也作为 $Z$ 轴回零触发键 |
| | 名称：－Z 控制按键<br>－Z 按键：在 JOG 方式下，按住此键则 $Z$ 轴依进给倍率/快速倍率之速度向机床 $Z$ 轴“－”方向（负方向）移动，同时按键指示灯点亮；当松开按键后，轴停止向“－”方向移动，同时按键指示灯熄灭<br>另外，当程序中执行 $Z$ 轴负方向移动程序指令时，该键指示灯也将点亮，停止移动指令时，该按键指示灯熄灭 |
| | 名称：超程释放按键<br>① 当本机床的各轴行程超过硬限位时，机床会出现超程报警，机床的动作停止，这时按住这个按键，在手轮模式下用手持单元将机床超程的轴反方向移动<br>② 绝对式编码器机床超程无需按此键 |
| | 名称：手动快移<br>本功能仅在手动模式下有效<br>手动模式下按下此键指示灯点亮<br>在本模式下，实际快速进给速度＝参数设置 G00 最大速度值×快速倍率开关所在的倍率值（%） |

（12）进给倍率及进给修调（见表 1-2-10）

**表 1-2-10　进给倍率及进给修调**

| 图示 | 说明 |
| --- | --- |
| 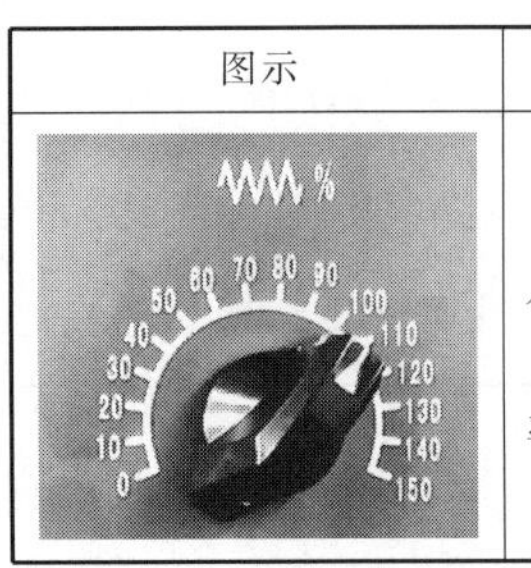 | 名称:进给倍率及进给修调<br>① 此旋钮位于本机床操作面板上,控制编程指定 G01 速度,实际进给速度＝编程给定 F 指令值×进给倍率开关所在倍率值(%)<br>② 寸动模式下,此时控制 JOG 进给倍率,实际 JOG 进给速度＝参数设定固定值×进给倍率开关所在倍率值(%)<br>③ 与设定的轴进给转速配合使用 |

（13）快速倍率（见表 1-2-11）

**表 1-2-11　快速倍率**

| 图示 | 说明 |
| --- | --- |
|  | 名称:快速倍率<br>① 此按钮位于本机床操作面板上,控制编程指定 G00 速度,实际进给速度＝参数设置 G00 最大速度值×快速进给倍率按钮所在倍率值(%)<br>② 快速模式下,此时控制手动快速进给倍率,实际快速进给速度＝参数设置 G00 最大速度值×快速进给倍率开关所在倍率值%。快移动倍率可以在 F0、25%、50%、100%四个挡位调整<br>③ 与设定的轴进给转速配合使用 |

（14）手摇脉冲（见表 1-2-12、表 1-2-13）

**表 1-2-12　轴选择、倍率选择与手轮每格的伺服轴移动量关系**

| 倍率选择 | ×1 | ×10 | ×100 |
| --- | --- | --- | --- |
| 公制移动量 | 0.001mm/每格 | 0.01mm/每格 | 0.1mm/每格 |
| 英制移动量 | 0.0001in/每格 | 0.001in/每格 | 0.01in/每格 |

注：1in＝25.4mm。

**表 1-2-13　手摇脉冲**

| 图示 | 说明 |
| --- | --- |
|  | 名称:轴向选择开关<br>① 本旋钮位于手持单元上,与手轮进给倍率×1、×10、×100 互相配合使用<br>② 本旋钮用于“手轮”模式<br>③ 本旋钮拨到“0”位不选择任何轴,拨到“X”位选择 $X$ 轴,拨到“Y”位选择 $Y$ 轴,拨到“Z”位选择 $Z$ 轴,拨到“4”位选择 4 轴 |
|  | 名称:倍率选择按钮<br>① 本旋钮位于手持单元上,与手轮进给倍率×1、×10、×100 互相配合使用<br>② 本旋钮用于“手轮”模式<br>③ 本旋钮拨到“×1”位选择手轮进给倍率为 0.001mm/每格,拨到“×10”选择手轮进给倍率为 0.01mm/每格,拨到“×100”选择手轮进给倍率为 0.1mm/每格 |

续表

| 图示 | 说明 |
| --- | --- |
|  | 名称：手轮 MPG<br>① 本码盘仅在“手轮”模式下有效，用于操作进给轴之方向与速度<br>② 手摇脉冲发生器的回转方向以顺时针方向为正(即正向旋转后伺服轴往正方向移动)，以逆时针方向为负(即负向旋转后伺服轴往负方向移动)<br>⚠警告<br>手轮旋转速度不得大于每秒 5 圈。如果手轮旋转速度超过了每秒 5 圈，刀具有可能在手轮停止旋转后还不能停止下来或者刀具移动的距离与手轮旋转的刻度不符<br>③ 手持单元使用应轻拿轻放、注意保护 |

## 三、程序输入练习

(1) 编辑程序

将操作面板中 MODE SELECT 旋钮切换到 EDIT 上，在 MDI 键盘上按PRGRM键，进入编辑页面，选定了一个数控程序后，此程序显示在 CRT 界面上，可对数控程序进行编辑操作。

① 移动光标　按 PAGE ↓或↑翻页，按 CURSOR ↓或↑移动光标。

② 插入字符　先将光标移到所需位置，点击 MDI 键盘上的数字/字母键，将代码输入到输入域中，按INSRT键，把输入域的内容插入到光标所在代码后面。

③ 删除输入域中的数据　按CAN键用于删除输入域中的数据。

④ 删除字符　先将光标移到所需删除字符的位置，按DELET键，删除光标所在的代码。

⑤ 查找　输入需要搜索的字母或代码；按 CURSOR ↓开始在当前数控程序中光标所在位置后搜索（代码可以是一个字母或一个完整的代码，例如“N0010”“M”等）。如果此数控程序中有所搜索的代码，则光标停留在找到的代码处；如果此数控程序中光标所在位置后没有所搜索的代码，则光标停留在原处。

⑥ 替换　先将光标移到所需替换字符的位置，将替换成的字符通过 MDI 键盘输入到输入域中，按ALTER键，把输入域的内容替代光标所在的代码。

(2) 显示数控程序目录

将操作面板中 MODE SELECT 旋钮切换到 EDIT 上，在 MDI 键盘上按PRGRM键，进入编辑页面，再按软键PRGRM。数控程序名显示在 CRT 界面上。

选择一个数控程序　将操作面板中 MODE SELECT 旋钮切换到 EDIT 或 AUTO 挡，在 MDI 键盘上按PRGRM键，进入编辑页面，按键入字母“O”；按数字键键入搜索的号码 XXXX（搜索号码为数控程序目录中显示的程序号），按 CURSOR ↓开始搜索。找到后，“OXXXX”显示在屏幕右上角程序号位置，NC 程序显示在屏幕上。

(3) 删除一个数控程序

将操作面板中 MODE SELECT 旋钮切换到 EDIT 上，在 MDI 键盘上按PRGRM键，进入编辑页面，按键入字母“O”；按数字键键入要删除的程序的号码 XXXX；按DELET键，程序即被删除。

新建一个 NC 程序　将操作面板中 MODE SELECT 旋钮切换到 EDIT 上，在 MDI 键盘上按PRGRM键，进入编辑页面，按键入字母“O”；按数字键键入程序号，但不可以与已有的程序号重复；按INSRT键，开始程序输入；每输入一个代码，按INSRT键，输入域中的内容显示在 CRT 界面上，用回车换行键EOB结束一行的输入后换行。

注：MDI 键盘上的数字/字母键，第一次按下时输入的是字母，以后再按下时均为数字。若要再次输入字母，须先将输入域中已有的内容显示在 CRT 界面上（按INSRT键，可将输入域中的内容显示在 CRT 界面上）。

（4）删除全部数控程序

将操作面板中 MODE SELECT 旋钮切换到 EDIT 上，在 MDI 键盘上按PRGRM键，进入编辑页面，按键入字母“O”；按键件入“-”；按键键入“9999”；按DELET键。

（5）程序模拟练习

NC 程序输入后，可检查运行轨迹。首先将光标移到程序头位置，同时按下MLK DRN AUX-LOCK这三个按键，再将操作面板中 MODE SELECT 旋钮切换到 AUTO 上，点击控制面板中AUX GRAPH命令，转入检查运行轨迹模式；再点击操作面板上的按钮START，即可观察数控程序的运行轨迹。通过程序模拟，可以效验输入的程序正确与否。程序模拟时，暂停运行、停止运行、单段执行等同样有效。

## 四、手动操作机床

当机床按照加工程序对工件进行自动加工时，机床的操作基本上是自动完成的，而其他情况下，要靠手动对机床操作。

（1）手动返回机床参考点

由于采用增量式测量系统，一旦机床断电后，其上的数控系统就失去了对参考点坐标的记忆。当再次接通数控系统的电源后，操作者必须首先进行返回参考点的操作。另外，机床在操作过程中遇到急停信号或超程报警信号，待故障排除后，恢复机床工作时，也必须进行返回参考点的操作。

具体操作步骤如下：将“MODE”开关置于 ZERO RETUEN 方式。提醒操作者注意：当滑板上的挡块距离参考点开关的距离不足 30mm 时，要首先用“JOG”按钮，使滑板向参考点的负方向移动，直到距离大于 30mm 停止点动，然后再返回参考点。

分别按下 $X$ 轴和 $Z$ 轴的“JOG”按钮，使滑板沿 $X$ 轴或 $Z$ 轴正向移向参考点。在此过程中，操作者应按住“JOG”按钮，直到参考点返回，指示灯亮，再松开按钮。在滑板移动到两轴参考点附近时，会自动减速移动。

（2）滑板的手动进给

当手动调整机床时，或是要求刀具快速移动接近或离开工件时，需要手动操作滑板进给。滑板进给的手动操作有两种，一种是用“JOG”按钮使滑板快速移动，另一种是用手摇轮移动滑板。

（3）快速移动

机床换刀或是手动操作时，要求刀具能快速移动接近或是离开工件，其操作如下：首先将“MODE”开关置于 RAPID 方式；用“APIDOVERRIDE”开关选择滑板快移的速度；按下“JOG”按钮，使刀架快速移动到预定位置。

（4）手摇轮进给

手动调整刀具时，要用手摇轮确定刀尖的正确位置，或是试切削时，一边用手摇轮微调进给速度，一边观察切削情况。其操作步骤是：将“MODE”开关转到“HANDLE”位置(可选择 3 个位置)，选择手摇轮每转动 1 格滑板的移动量。将“MODE”开关转至×1，手摇轮转 1 格滑板移动 0.001mm；若指向×10，手摇轮转 1 格滑板移动 0.01mm；若指向×100，手摇轮转 1 格滑板移动 0.1mm；使手摇轮左侧的 $X$、$Z$ 轴选择开关扳向滑板要移动的坐标轴；转动手摇脉冲发生器，使刀架按照指定的方向和速度移动。

（5）主轴的操作

主轴的操作主要包括主轴的启动、停止和主轴的点动。

① 主轴启动与停止　主轴的启动与停止是用来调整刀具或调试机床的。具体操作步骤是：将“MODE”开关置于手动方式（MANU）中任意一个位置。用主轴功能按钮中的“FWD-RVS”开关确定主轴旋转方向，在“FWD”位置，主轴正转；开关指向“RVS”位置，主轴反转。旋转主轴“SPEED”至低转速区，防止主轴突然加速。按下“START”按钮，主轴旋转，在主轴转动过程中，可以通过“SPEED”旋钮改变主轴的转速，且主轴的实际转速显示在 CRT 显示器上；按下主轴 STOP 按钮，主轴停止转动。

② 主轴的点动　主轴的点动是用于使主轴旋转到便于装卸卡爪或是便于检查工件的装夹的位置。其操作方法是：将“MODE”开关置于自动方式（AUTO）中的任意一个位置。将主轴“FWD-RVS”开关指向所需的旋转方向。压下“START”按钮，主轴转动；按钮抬起，主轴停止转动。

（6）刀架的转位

装卸刀具，测量切削刀具的位置以及对工件进行试切削时，都要在 MDI 状态下编程执行。其操作步骤是：将“MODE”开关置于“MDI”方式；按下功能键“PRGRM”输入 T10、T20、T30、T40 后再按下 START 键。

（7）手动尾座的操作

手动尾座的操作包括尾座体的移动和尾座套筒的移动。

① 尾座体的移动　手动尾座体使其前进或后退，主要用于轴类零件加工时调整尾座的位置；或是加工短轴和盘类零件时，将尾座退至某一合适的位置。其操作步骤是：将“MODE”开关置于“MANU”方式中的任一位置；压下“TALL STOCK INTERLOCK”按钮，松开尾座，其按钮上方指示灯亮；移动滑板带动尾座移动至预定位置；再次压下“TAIL STOK INTERLOCK”按钮，尾座被锁紧，且指示灯灭。

② 尾座套筒的移动　尾座套筒的伸出或退回是在加工轴类零件时，顶尖顶紧或松开工件。操作方法是：首先将“MODE”开关置于“MANU”方式中的任一位置，按下“QUILL”按钮，尾座套筒带着顶尖退回，指示灯灭。

（8）卡盘的夹紧与松开操作

机床在手动操作或自动运转时，卡盘的夹紧和松开是通过脚踏开关实现的，其操作步骤如下：扳动电箱内卡盘正、反卡开关，选择卡盘正卡或反卡；第一次踏下开关卡盘松开，第二次踏下开关卡盘夹紧。

## 五、机床的紧急停止

机床无论是在手动或自动运转状态下，遇有不正常情况，需要机床紧急停止时，可通过以下操作来实现。

（1）按下紧急停止按钮

按下“EMERG STOP”按钮后，除润滑油泵外，机床的动作及各种功能均被立即停止。同时CRT屏幕上出现CNC数控未准备好（NOT　READY）报警信号。待故障排除后，顺时针旋转按钮，被压下的按钮跳起，则急停状态解除，但此时要恢复机床的工作，必须进行返回机床参考点的操作。

（2）按下复位键（RESET）

机床在自动运转过程中，按下此键则机床全部操作均停止，因此可以用此键完成急停操作。按下NC装置电源断开键；按下NC的“OFF”键，机床停止工作。按下进给保持按钮（FEED HOLD）机床在自动运转状态下，按下“FEED HOLD”按钮，滑板停止运动，但机床的其他功能仍有效。当需要恢复机床运转时，按下“CYCLE START”按钮，机床从当前位置开始继续执行下面的程序。

（3）按下进给保持（FEED HOLD）

机床在自动运行状态下，按下“FEEK HOLD”按钮，滑板停止运动，但机床的其他功能仍有效。当需要机床运转时，按下“CYCLE START”按钮，机床从当前位置开始继续的程序。

## 六、对刀（刀具的几何补偿和磨损补偿）

（1）刀具几何补偿的方法

① 移动使 $X$、$Z$ 轴回归参考点，确认回归指示灯亮。

② 模式选择开关选择为手动进给。

③ 按“OFFST”键，CRT出现OFFSET/SETTING画面。

④ 选择开关选择所需刀具或在MDI方式下调用所需刀具。

⑤ 移动刀具使其靠近工件右端面，并平整端面，且不要移动 $Z$ 轴。

⑥ 移动光标至与之相对应的刀具几何补偿号的 $Z$ 轴，并输入 $Z0$，再按软键［测量］。

⑦ 移动刀具，车一刀工件外圆或内孔，见亮就可以，测量工件外圆直径或孔径，移动光标至与之相对应的刀具几何补偿号的 $X$ 轴，直接输入X D（工件外径值或内径值），再按软键［测量］，如图1-2-3所示。

⑧ 对其他所需用的刀具的补偿，与第一把刀具类似，但不要切削，只要轻轻地靠一下端面或工件外圆即可，其输入数据方法与第一把刀一样。

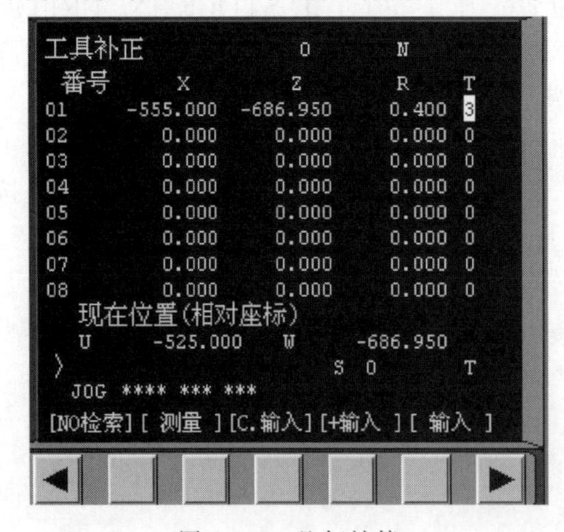


图1-2-3　几何补偿

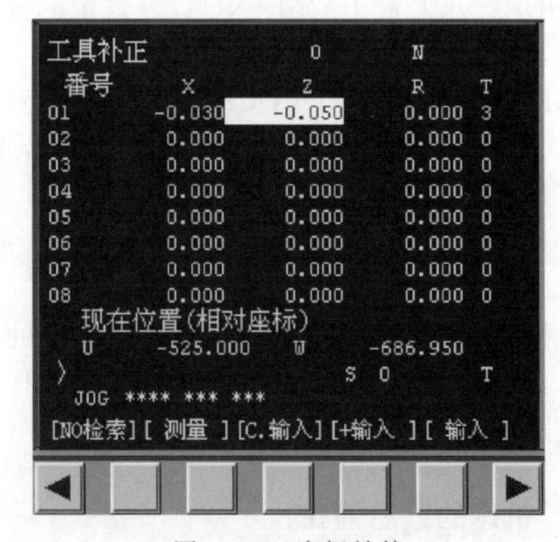


图1-2-4　磨损补偿

⑨ 确认各刀具的刀尖圆弧半径，并输入给数据库中相对应的刀具补偿号。

⑩ 确认刀具的刀尖圆弧假想位置编号，如右偏外圆刀是1、镗孔刀是3，并输入给刀具数据库中相对应的刀具补偿号。

（2）磨损补偿

① 按“OFFSET”键后，按［磨耗］软键，使CRT出现磨损补偿画面。

② 将光标移至所需进行磨损补偿的刀具补偿号位置。

如测量用T11刀具加工的工件外圆直径为$\phi$45.03mm，长度为20.05mm，而规定直径为$\phi$45mm，长度为20mm。实测值直径比要求值大0.03，长度大0.05mm，应进行磨损补偿：将光标移至X，键入－0.03后按［＋输入］软键；将光标移至Z，键入－0.05后按［＋输入］软键，X值变为在以前值的基础上加以－0.03mm，Z值变为在以前的基础上加以－0.05mm，如图1-2-4所示。

**习题**

1.数控车床机床原点的位置一般设在哪里？

2.什么是机床参考点？回参考点的目的是什么？

3.换刀点要注意什么？

4.机械加工为什么分为粗加工、半精加工和精加工？

5.开机操作的步骤是什么？

6.数控车床加工有何特点？

7.请介绍数控车床的编程特点。

8.遇到事故，机床应如何操作？

# 任务三　程序的基本构成

## 一、数控车床坐标轴与运动方向

为了简化编程和保证程序的通用性，对数控机床的坐标轴和方向命名制定了统一的标准，规定直线进给坐标轴，用 $X$、$Y$、$Z$ 表示基本坐标轴，其相互关系用右手定则决定。

**1. 数控车床坐标轴和方向命名原则**

① 用右手直角笛卡儿坐标定义原则（见图1-3-1）。

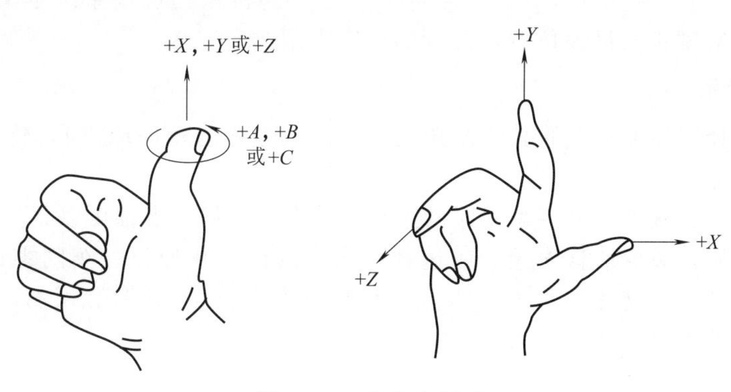


图1-3-1　直角坐标系

② 数控车床的 $Z$ 坐标轴规定为传递切削动力的主轴线方向；$X$ 坐标轴规定为水平方向，$X$ 坐标的方向是在工件的径向上，且平行于横向滑板，规定远离卡盘中心方向为正方向（见图 1-3-2）。

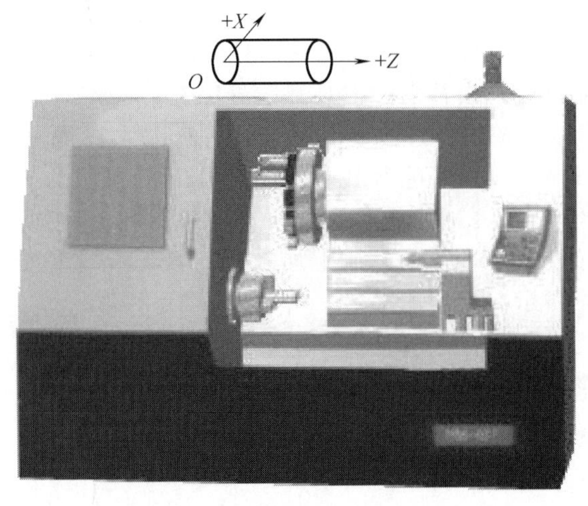


图 1-3-2　数控车床坐标轴及其方向

**2. 数控车床坐标系的确定与应用**

① 先确定 $Z$ 轴，$Z$ 轴是传递切削动力的主要轴，然后再确定其 $X$ 轴。

② 数控车床坐标系的原点一般定义在卡盘中心线与中间端面交点。

## 二、数控车床坐标系

### 1. 数控车床坐标系

（1）机床坐标系

数控车床生产厂家按照笛卡儿规则，在数控车床上建立的一个 $Z$ 轴与 $X$ 轴的直角坐标系，称为机床坐标系。

（2）机床原点

机床坐标系的零点称为机床原点，是机床上的一个固定点，一般定义在主轴旋转中心线与车头端面的交点或参考点上。

（3）参考点

参考点为机床上一固定点，由 $X$ 方向与 $Z$ 方向的机械挡块或系统定义的位置来确定，一般设定在 $Z$、$X$ 轴正向最大位置，位置的设定由制造商定义。

### 2. 编程坐标系

编程人员选择工件图样上的某一已知点为原点（也称为程序原点），建立一个新的坐标系，称为编程坐标系。

选择编程原点：从理论上讲编程原点选在零件上的任何一点都可以，但实际上，为了换算尺寸尽可能简便，减少计算误差，应选择一个合理的编程原点。车削零件编程原点的 $X$ 向零点应选在零件的回转中心。$Z$ 向零点一般应选在零件的右端面、设计基准或对称平面内。车削零件的编程原点选择见图 1-3-3。

### 3. 工件坐标系

操作者通过对刀等方式将编程坐标系的原点移到数控车床上，此时在数控车床上建立的坐标系称为工件坐标系。其原点一般选择在轴线与工件右端面、左端面或其他位置的交点上，工件坐标系的 $Z$ 轴一般与主轴轴线重合。车削零件的工件坐标原点选择见图 1-3-4。

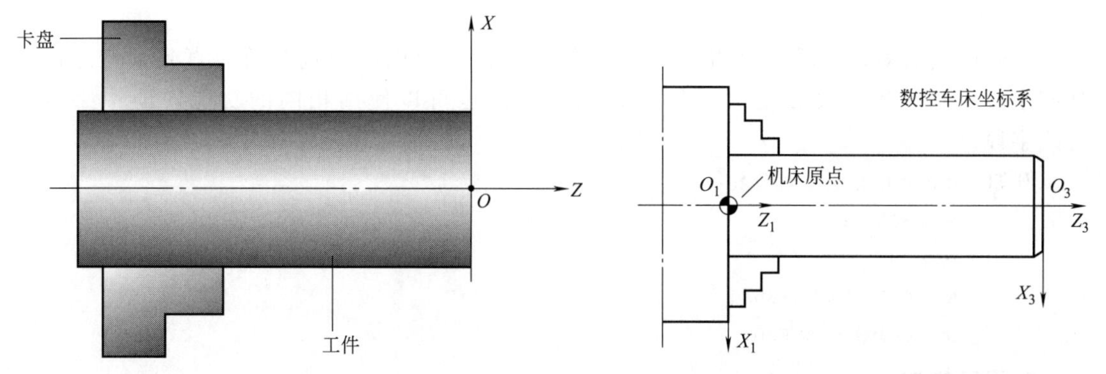


图 1-3-3　车削零件的编程坐标原点　　图 1-3-4　车削零件的工件坐标原点选择

**4. 对刀点、起刀点、换刀点**

将编程坐标系原点转换成机床坐标系的已知点并成为工件坐标系的原点，这个点就称为对刀点。起刀点是零件程序加工的起始点。在零件车削过程中需要自动换刀，为此必须设置一个换刀点，该点应离开工件有一定距离，以防止刀架回转换刀时刀具与工件发生碰撞。换刀点通常分为两种类型，即固定换刀点和自定义换刀点。

换刀点的位置通常要注意：①方便数学计算和简化编程；②容易找正对刀；③便于加工检查；④引起的加工误差小；⑤不要与机床、工件发生碰撞；⑥方便拆卸工件；⑦空行程不要太长。

## 三、编程坐标系

编程坐标分为绝对坐标（$X$，$Z$）、相对坐标（$U$，$W$）和混合坐标（$X/Z$，$U/W$）。

**1. 绝对坐标（*X*，*Z*）**

各点坐标参数以到坐标原点的距离作为参数值。

**2. 相对坐标（*U*，*W*）**

某点的坐标参数以到另外一点的距离作为参数值，即指令从前面一个位置到下一个位置的距离作为参数值。

**3. 混合坐标（*X/Z*，*U/W*）**

绝对坐标和相对坐标同时使用，即在同一个程序段中，可使用 $X$ 或 $U$，$Z$ 或 $W$。

**4 . 直径坐标**

$X$ 坐标参数值为直径。

**5. 半径编程**

$X$ 坐标参数值为半径值。

## 四、数控车床编程基础知识

**1. 初态、模态**

（1）初态

指运行加工程序之前的系统编程状态，即机器里面已设置好的，一开机就进入的状态，例如G98、G00。

（2）模态

一种连续有效的指令。指相应字段的值一经设置，以后一直有效，直到某程序段有对该字段重新设置。设置之后，如果是同一组的也可以使用相同的功能，而不必再输入该字段。

例如：
```
N30 G90 X32.0 Z0 F80;
N40 X30.0;
……
N… G02 X30.0 Z-30.0 R5.0 F50;
N… G01 Z-30.0 F30;
```

**2. 程序构成**

在数控车床上加工零件，首先要编制程序，然后用该程序控制机床的运动。数控指令的集合称为程序。在程序中根据机床的实际运动顺序书写这些指令。

（1）程序结构

一个完整的数控加工程序由程序开始部分、若干程序段、程序结束部分组成。一个程序段由程序段号和若干个“字”组成，一个“字”由地址符和数字组成。

下面是一个完整的数控加工程序，该程序由程序号开始，以M30结束。

| 程序 | 说明 |
|---|---|
| O1234 | 程序开始 |
| N10 T0101 G97 G99 M03 S500; | 程序段1 |
| N20 G00 X100.0 Z100.0; | 程序段2 |
| N30 G00 X26.0 Z0; | 程序段3 |
| N40 G01 X0 F0.1; | 程序段4 |
| N50 Z1.0; | 程序段5 |
| N60 G00 X100.0; | 程序段6 |
| N70 Z100.0; | 程序段7 |
| N80 M30; | 程序结束 |

（2）程序号

零件程序的起始部分一般由程序起始符号%（或O）后跟1～4位数字（0000～9999）组成，如%123，O1234等。

（3）程序段的格式和组成

程序段的格式可分为地址格式、分割地址格式、固定程序段格式和可变程序段格式等。其中以可变程序段格式应用最为广泛，所谓可变程序段格式就是程序段的长短是可变的。

例如：

| N10 | G01 | X40.0 Z-30.0 | F200 | ; |
|---|---|---|---|---|
| 程序段号 | 功能字 | 坐标字 | 给速度功能字 | 程序段结束 |

（4）“字”

一个“字”的组成如下所示：

```
Z                    -                    30.0
```

地址符　　符号（正、负号）　数据字（数字）

程序段号加上若干程序字就可组成一个程序段。在程序段中表示地址的英文字母可分为尺寸地址和非尺寸地址两种。表示尺寸地址的英文字母有X、Y、Z、U、V、W、P、Q、I、J、K、A、B、C、D、E、R、H共18个。表示非尺寸地址的英文字母有N、G、F、S、T、M、L、O共8个。

**3. 主轴转速功能字S**

主轴转速功能字的地址符是S，又称为S功能或S指令，用于指定主轴转速，单位为r/min。对于具有恒线速度功能的数控车床，程序中的S指令用来指定车削加工的线速度数。

有变速箱的：用S1（第一挡）、S2（第二挡）。

无变速箱的：直接输入转速，例如S100、S210、S500等。

**4. 进给功能字（切削速度）F**

进给功能字的地址符是F，又称为F功能或F指令，用于指定切削的进给速度。对于车床，F可分为每分钟进给和主轴每转进给两种；对于其他数控机床，一般只用每分钟进给。F指令在螺纹切削程序段中常用来指令螺纹的导程。

单位：G98为每分钟进给，mm/min；G99为每转进给，mm/r。

G00为快速定位，没有F值，速度由倍率控制快慢。

切削进给速度要有F值，F值的快慢也可在进给倍率中控制。

例如：
```
G00 X32.0 Z2.0;
G90 X24.0 Z-20.0 F50;
G90 X20.0 Z-15.0 F60;
```

**5. 刀具功能字T**

刀具功能字的地址符是T，又称为T功能或T指令，用于指定加工时所用刀具的编号。对于数控车床，其后的数字还兼作指定刀具长度补偿和刀尖半径补偿用。T后第一、二位是刀号，第三、四位是刀补号。

例：T0100 T0200 T0300 T0400无刀补，如T0200为2号刀无刀补。

T0101 T0202 T0303 T0404有刀补，如T0202为2号刀，执行2号刀补。

**6. 辅助功能字M**

辅助功能字的地址符是M，后续数字一般为1～3位正整数，又称为M功能或M指令，用于指定数控机床辅助装置的开关动作，见表1-3-1。

**表1-3-1　M功能字含义**

| M功能字 | 含义 | M功能字 | 含义 |
|---|---|---|---|
| M00 | 程序停止 | M07 | 2号冷却液开 |
| M01 | 计划停止 | M08 | 1号冷却液开 |
| M02 | 程序停止 | M09 | 冷却液关 |
| M03 | 主轴顺时针旋转 | M30 | 程序停止并返回开始处 |
| M04 | 主轴逆时针旋转 | M98 | 调用子程序 |
| M05 | 主轴旋转停止 | M99 | 返回子程序 |
| M06 | 换刀 | | |

### 7. 准备功能字 G

准备功能字的地址符是 G，又称为 G 功能或 G 指令，是用于建立机床或控制系统工作方式的一种指令。后续数字一般为 1～3 位正整数，见表 1-3-2。

表 1-3-2　G 功能字含义

| G 功能字 | 组别 | 功能 |
| --- | --- | --- |
| G00 |  | 快速移动点定位 |
| G01 |  | 直线插补(切削进给) |
| G02 | 01 | 顺时针圆弧插补 |
| G03 |  | 逆时针圆弧插补 |
| G04 | 00 | 暂停、准停 |
| G28 | 00 | 返回参考点(机械原点) |
| G32 | 01 | 螺纹切削 |
| G33 |  | $Z$ 轴攻螺纹循环 |
| G34 |  | 变螺距螺纹切削 |
| G50 | 00 | 坐标系设定 |
| G65 | 00 | 宏程序命令 |
| G70 |  | 精加工循环 |
| G71 |  | 外圆粗切循环 |
| G72 |  | 端面粗切循环 |
| G73 |  | 封闭切削循环 |
| G74 | 00 | 端面深孔钻循环 |
| G75 |  | 外圆、内圆切槽循环 |
| G76 |  | 复合螺纹切削循环 |
| G90 | 00 | 外圆、内圆车削循环 |
| G92 |  | 螺纹切削循环 |
| G94 | 00 | 端面车削循环 |
| G96 | 02 | 恒线速度 |
| G97 |  | 取消恒线速度 |
| G98 | 03 | 每分钟进给 |
| G99 |  | 每转进给 |

G 代码的使用方法如下。

一次性 G 代码：只有在被指令的程序段中有效的代码。例如：G04（暂停），G50（坐标设定），G70～G75（复合型车削固定循环）。

模态 G 代码：在同组其他代码指令前一直有效，即表中 01 组的 G 代码。例如：G00（定位），G01、G02、G03（插补），G90、G92、G94（单一型固定循环）。

初态 G 代码：即系统里面已经设置好的，一开机就进入的状态。初态也是模态。例如：G98、G00。

**习题**

1. G02、G03 指令的应用区别是什么？
2. M02 与 M30 的区别是什么？
3. 请写出 G70、G71、G73 的格式。
4. G01 与 G00 应用于程序的什么时候？
5. G98 与 G99 能一起使用吗？为什么？
6. 数控车床用于加工什么零件？
7. 切削三要素是什么？
8. 刀具有哪些重要角度？体现哪些作用？

# 任务四　光轴的编程与加工

## 一、图样与技术要求

如图 1-4-1 所示轴类特征零件，材料为 45 钢，规格为 $\phi$35mm 的圆柱棒料，正火处理，硬度 200HB。光轴加工的评分表见表 1-4-1。

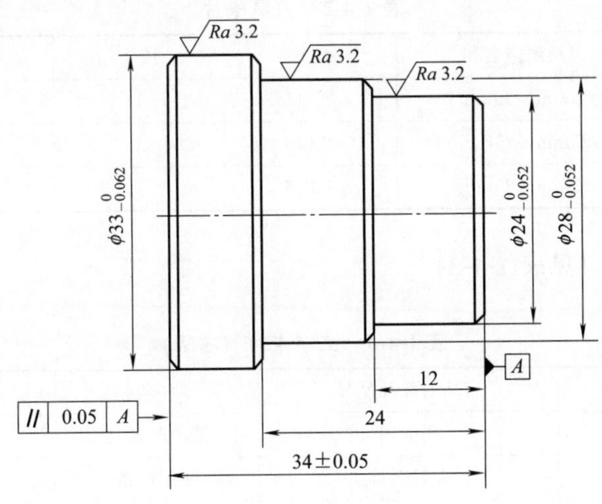


图 1-4-1　阶梯轴

**表 1-4-1　评分表**

| 序号 | 项目及技术要求 | 配分(IT/*Ra*) | 评分标准 | 检测结果 | 实得分 |
|---|---|---|---|---|---|
| 1 | 外径 $\phi 33_{-0.062}^{0}$,*Ra*3.2 | 18/4 | 超差全扣 | | |
| 2 | 外径 $\phi 28_{-0.052}^{0}$,*Ra*3.2 | 18/4 | 超差全扣 | | |
| 3 | 外径 $\phi 24_{-0.052}^{0}$,*Ra*3.2 | 18/4 | 超差全扣 | | |
| 4 | 长度 12、24 | 4 | 超差全扣 | | |
| 5 | 长度 36±0.05 | 6 | 超差全扣 | | |
| 6 | 倒角 1×45°(4 处) | 4 | 超差全扣 | | |
| 安全文明生产 | | 20 | | | |
| 加工工时 | | 60min | | | |

## 二、图纸分析

### 1. 零件图分析

零件的外轮廓是三个长度 12mm 的圆柱（$\phi$33mm、$\phi$28mm、$\phi$24mm）组成，属于简单的阶梯轴类零件。零件是根据训练、学习的前后顺序而设计的。

### 2. 工艺分析

① 结构分析：零件的结构都是简单的阶梯轴。

② 精度分析：零件的重点尺寸在外圆的精度等级都为 h9。对于中级工前期的训练是一个相对合理的公差尺寸。另外，长度、倒角等细节精度问题同样需要注意。

③ 定位及装夹分析：本零件采用三爪自定心卡盘进行定位和装夹。工件装夹时的夹紧力要适中，既要防止工件变形和夹伤，又要防止工件在加工时松动。工件装夹过程中应对工件进行找正，以保证各项形位公差。

④ 加工工艺分析：经过以上分析，本零件加工时总体安排顺序是，先加工零件的右端；切断工件后掉头找正后车削端面并倒角，保证工件总长尺寸。

### 3. 主要刀具选择（见表 1-4-2）

表 1-4-2　刀具卡片

| 刀具名称 | 刀具规格名称 | 材料 | 数量 | 刀尖半径 | 刀宽 |
|---|---|---|---|---|---|
| 90°外圆车刀 | 25mm×25mm | YT15 | 1 | 0 | |
| 45°外圆车刀 | 25mm×25mm | YT15 | 1 | 0 | |
| 切断刀 | 25mm×25mm | YT15 | 1 | 0 | 4.5mm |

### 4. 工艺规程安排（见表 1-4-3）

表 1-4-3　工序卡片（右端）

<table>
<tr><td colspan="2">单位</td><td colspan="2">产品名称及型号</td><td colspan="2">零件名称</td><td>零件图号</td></tr>
<tr><td colspan="2"></td><td colspan="2">任务四</td><td colspan="2">简单阶梯轴</td><td>图 1-4-1</td></tr>
<tr><td>工序</td><td>程序编号</td><td colspan="2">夹具名称</td><td colspan="2">使用设备</td><td>工件材料</td></tr>
<tr><td>001</td><td>O0001</td><td colspan="2">三爪卡盘</td><td colspan="2">SK50</td><td>45 钢</td></tr>
<tr><td colspan="7">阶梯轴零件</td></tr>
<tr><td>工步</td><td>工步内容</td><td>刀号</td><td>切削用量</td><td>备注</td><td colspan="2">工序简图</td></tr>
<tr><td>1</td><td>车端面</td><td>T11</td><td>$n$=600r/min<br>（手动进给倍率模式在×10 位置）</td><td>手动加工</td><td colspan="2">卡爪<br>(50～60)</td></tr>
</table>

续表

| 阶梯轴零件 | | | | | |
|---|---|---|---|---|---|
| 工步 | 工步内容 | 刀号 | 切削用量 | 备注 | 工序简图 |
| 2 | 粗车 φ24、φ28、φ33 外圆留 0.5mm 精加工余量 | T11 | $n$=500r/min<br>$f$=0.2mm/r<br>$a_p$=2.0mm | 自动加工 | |
| 3 | 精车 φ24、φ28、φ33 外圆 | | $n$=800r/min<br>$f$=0.1mm/r<br>$a_p$=0.5mm | | |

## 三、程序编制

### 1. 快速定位（G00）

（1）编程格式

```
N10  G00 X(U)__  Z(W)__;
```

式中 X，Z——快速点定位的终点坐标尺寸，是绝对值坐标编程；

U，W——快速点定位的终点坐标尺寸，是相对（增量）值坐标编程。

如图 1-4-2 中，从 $A$ 点到 $B$ 点快速移动的程序段为：

```
N10  G00 X20.0 Y30.0;
或是 N10  G00 U-20.0 W-10.0;
```

（2）G00 走刀路线

快速点定位指令控制刀具以点位控制的方式快速移动到目标位置，其移动速度由参数来设定。指令执行开始后，刀具沿着各个坐标方向同时按参数设定的速度移动，最后减速到达终点，如图 1-4-2 所示。注意：在各坐标方向上有可能不是同时到达终点。刀具移动轨迹是几条线段的组合，不是一条直线。例如，在 FANUC 系统中，运动总是先沿 45°角的直线移动，最后再在某一轴单向移动至目标点位置，如图 1-4-2（b）所示。编程人员应了解所使用的数控系统的刀具移动轨迹情况，以避免加工中可能出现的碰撞。

（3）G00 没有 F 值

快速移动速度由厂家设定，快速移动速度受快速倍率开关控制（F0、25%、50%、100%），用 F 值指定的进给速度无效。

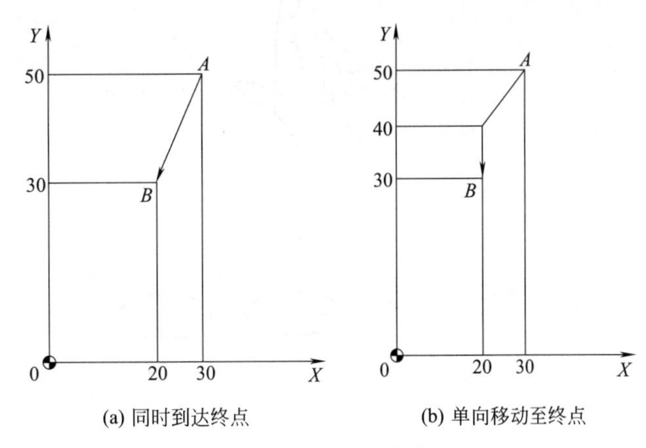

(a) 同时到达终点　(b) 单向移动至终点

图 1-4-2 快速点定位

### 2. 直线插补 G01

G01 是使刀具以指令的进给速度沿直线移动到目标点。

指令格式为：G01　X（U）__ Z（W）__ F __；

式中　X，Z——目标点绝对值坐标；

U，W——目标点相对前一点的增量坐标；

F——进给量，若在前面已经指定，可以省略。

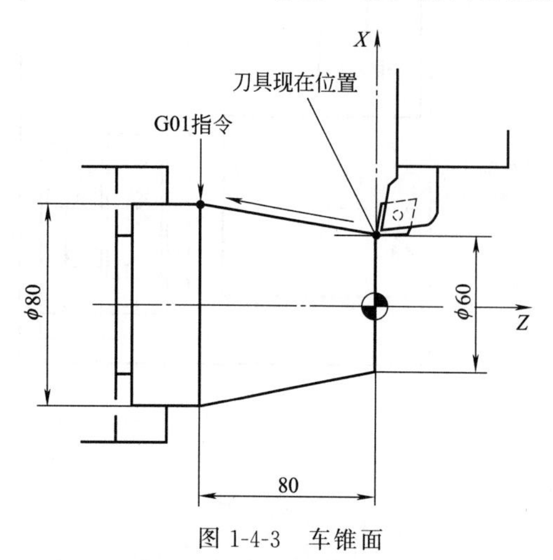


图 1-4-3　车锥面

通常，在车削端面、沟槽等与 $X$ 轴平行的加工时，只需单独指定 X（或 U）坐标；在车外圆、内孔等与 $Z$ 轴平行的加工时，只需单独指定 Z（或 W）值。图 1-4-3 为同时指令两轴移动车削锥面的情况，用 G01 编程为：

绝对坐标编程方式：G01 X80.0 Z-80.0F0.25

增量坐标编程方式：G01 U20.0 W-80.0F0.25

说明：

① G01 指令后的坐标值取绝对值编程还是取增量值编程，由尺寸字地址决定，有的数控车床由数控系统当时的状态决定。

② 进给速度由 F 指令决定。F 指令也是模态指令，它可以用 G00 指令取消。如果在 G01 程序段之前的程序段没有 F 指令，而现在的 G01 程序段中也没有 F 指令，则机床不运动。因此，G01 程序中必须含有 F 指令。

**3. 圆弧插补指令 G02、G03**

（1）指令格式

```
G02/G03  X(U)__ Z(W)__ I __ K __ F __;
G02/G03  X(U)__ Z(W)__ R __ F __;
```

圆弧顺逆的判断：圆弧插补指令分为顺时针圆弧插补指令 G02 和逆时针圆弧插补指令 G03。圆弧插补的顺逆可按图 1-4-4 给出的方向判断：沿圆弧所在平面（如 $XZ$ 平面）的垂直坐标轴的负方向（$-Y$）看去，顺时针方向为 G02，逆时针方向为 G03。

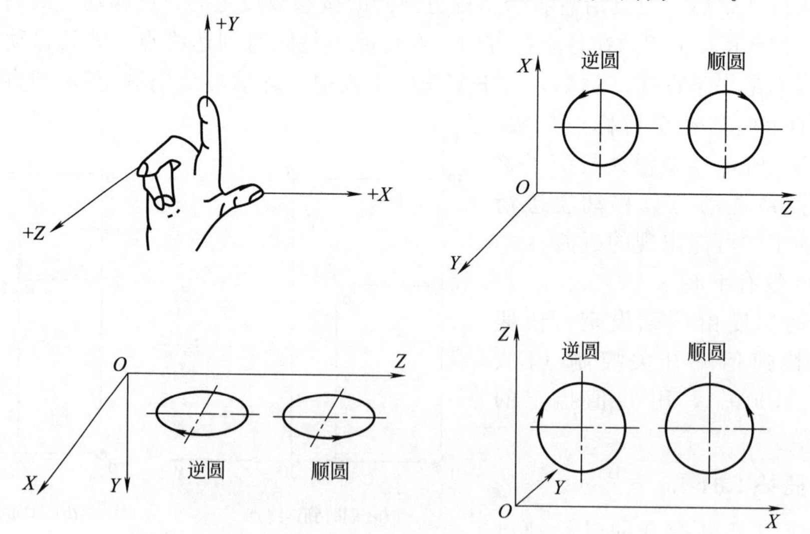


图 1-4-4　圆弧顺逆的判断

（2）说明

① 采用绝对值编程时，圆弧终点坐标为圆弧终点在工件坐标系中的坐标值，用 X、Z 表示。当采用增量值编程时，圆弧终点坐标为圆弧终点相对于圆弧起点的增量值，用 U、W 表示。

② 圆心坐标 I、K 为圆弧起点到圆弧中心所作矢量分别在 $X$、$Z$ 坐标轴方向上的分矢量（矢量方向指向圆心）。本系统 I、K 为增量值，并带有“±”号，当分矢量的方向与坐标轴的方向不一致时取“－”号。

③ 当用半径只指定圆心位置时，由于在同一半径的情况下，从圆弧的起点到终点有两个圆弧的可能性，为区别二者，规定圆心角≤180°时，用“＋R”表示。若圆弧圆心角＞180°时，用“－R”表示。

④ 用半径只指定圆心位置时，不能描述整圆。

（3）G02 应用实例

如图 1-4-5 所示。

① 用 I、K 表示圆心位置，绝对值编程：

```
N03 G00 X20.0 Z2.0;
N04 G01 Z-30.0 F80;
N05 G02 X40.0 Z-40.0 I0 K0 F60;
```

② 用 I、K 表示圆心位置，增量值编程：

```
N03 G00 U-80.0 W-98.0;
N04 G01 U0 W-32.0 F80;
N05 G02 U20.0 W-10.0 I0 K0 F60;
```

③ 用 R 表示圆心位置：

```
N04 G01 Z-30.0 F80;
N05 G02 X40.0 Z-40.0 R10.0 F60,
```

（4）G03 应用实例

如图 1-4-6 所示。

① 用 I、K 表示圆心位置，采用绝对值编程：

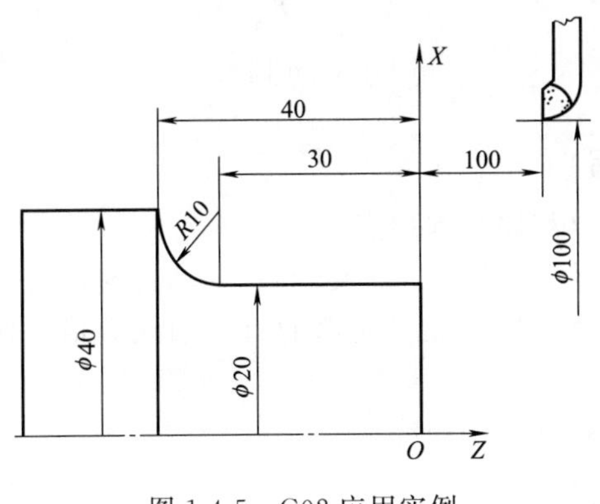


图 1-4-5　G02 应用实例

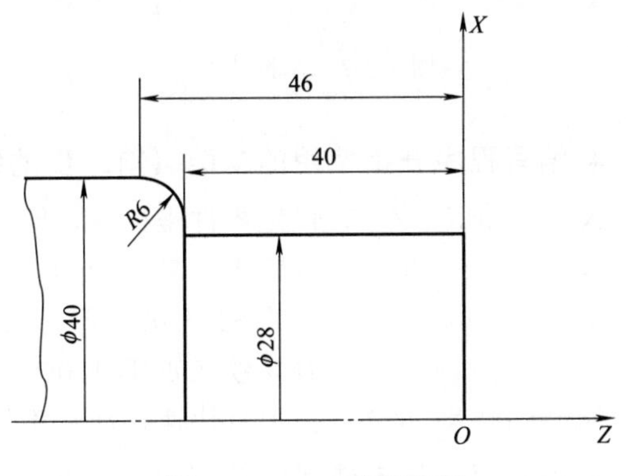


图 1-4-6　G03 应用实例

```
N04 G00 X28.0 Z2.0;
```

```
N05 G01 Z-40.0 F80;
N06 G03 K40.0 Z-46.0 I0 K-6.0 F60;
```

② 采用增量值编程：

```
N04 G00 U-150.0 W-98.0;
N05 G01 W-42.0 F80;
N06 G03 U12.0 W-6.0 I0 K-6.0 F60;
```

③ 用 R 表示圆心位置，采用绝对值编程：

```
N04 G00 X28.0 Z2.0;
N05 G01 Z-40.0 F80;
```

（5）G02/G03 车圆弧的方法

应用 G02（或 G03）指令车圆弧，若用一刀就把圆弧加工出来，这样吃刀量太大，容易打刀。所以，实际车圆弧时，需要多刀加工，先将大多余量切除，最后才车得所需圆弧。下面介绍车圆弧常用加工路线。

图 1-4-7 为车圆弧的车锥法切削路线。即先车一个圆锥，再车圆弧。但要注意，车锥时的起点和终点的确定，若确定不好，则可能损坏圆锥表面，也可能将余量留得过大。确定方法：连接 $OC$ 交圆弧于 $D$，过 $D$ 点作圆弧的切线 $AB$。

图 1-4-8 为车圆弧的同心圆弧切削路线。即用不同的半径圆来车削，最后将所需圆弧加工出来。此方法在确定了每次吃刀量 $a_p$ 后，对 90°圆弧的起点、终点坐标较易确定，数值计算简单，编程方便，常采用。但空行程时间较长。

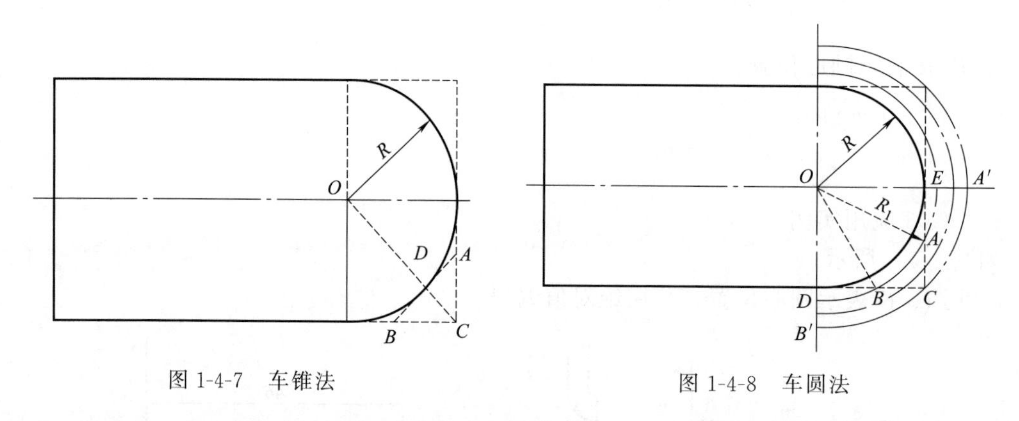


图 1-4-7　车锥法　　　图 1-4-8　车圆法

**4. 编写程序开始使用的功能（前三步的编写）**

N10 G50 X__Z__；设定零件坐标系，即刀具（程序）起始点

N20 M__S__T__；主轴正反转（M03 主轴正转，M04 主轴反转）；主轴转速，有变速箱的 S1 为第一挡，S2 为第二挡，无变速箱的直接输入数值；使用刀具号（如 T0100）

N30 G00 X__Z__；把刀具快速移动到工件准备加工的边缘

**5. 编写程序结束使用的功能（后三步的编写）**

N__G00 X__Z__；把刀具快速移动回到程序起点

N__M05 T__；主轴停止，换回基准刀

N__M30；程序结束，光标回到程序开始位置，为下一工件加工做准备

**6. 编写程序中间部分使用的功能**

根据加工图样的要求，选择加工工艺，编写程序的中间部分。

**7. 编程举例**

**【例 1-4-1】** 工件加工如图 1-4-9 所示，试编写数控加工程序。

编写程序：

```
O0401
N10 G50 X100.0 Z100.0;(刀具程序起始点)
N20 M03 S800 T0101;(主轴正转,转速为 800r/min,用一号基准刀)
N30 G00 X30.0 Z2.0;(快速定位到工件附近)
N40 ……
……
……        (切削加工部分)
……
……
N  G00 X100.0 Z100.0;(快速回到起始点)
N  M05 T0100;(主轴停止,换回基准刀)
N  M30;(程序结束,光标返回到程序开始)
```

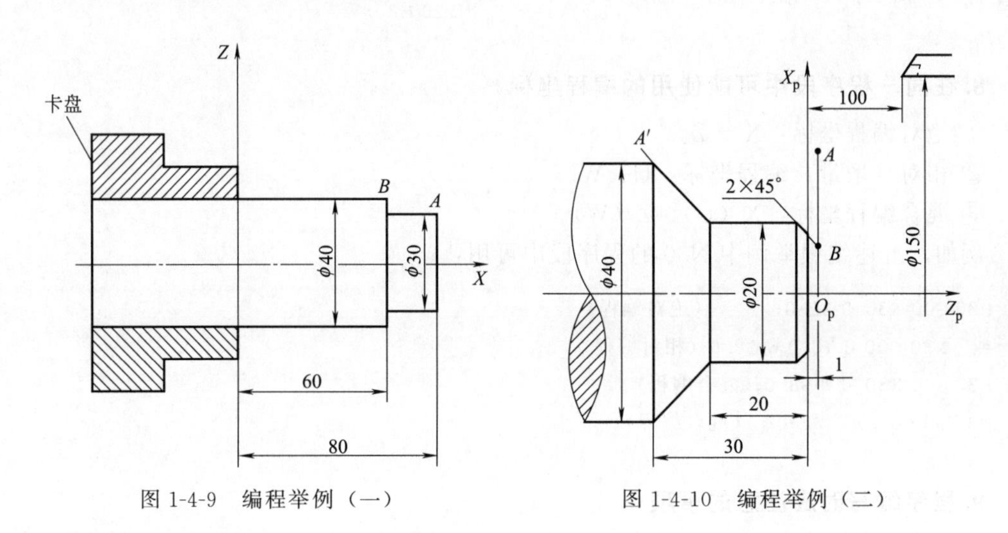


图 1-4-9　编程举例（一）　　图 1-4-10　编程举例（二）

**【例 1-4-2】** 如图 1-4-10 所示，工件已粗加工完毕，各位置留有余量 0.2mm，要求重新编写精加工程序，不切断。

编写加工程序

```
O0402
N10 G99 G50 X150.0 Z100.0;
N20 M03 S1000 T0101;
N30 G00 X16.0 Z2.0;
N40 G01 X16.0 Z0 F0.5;
N50 G01 X20.0 Z-2.0 F0.1;
N60 Z-20.0;
N70 X40.0 Z-30.0;
```

```
N80 G00 X150.0 Z100.0;
N90 M05 T0100;
N100 M30;
```

**【例 1-4-3】** 如图 1-4-11 所示，工件已粗加工完毕，各位置留有余量 0.2mm，要求重新编写精加工程序，不切断。

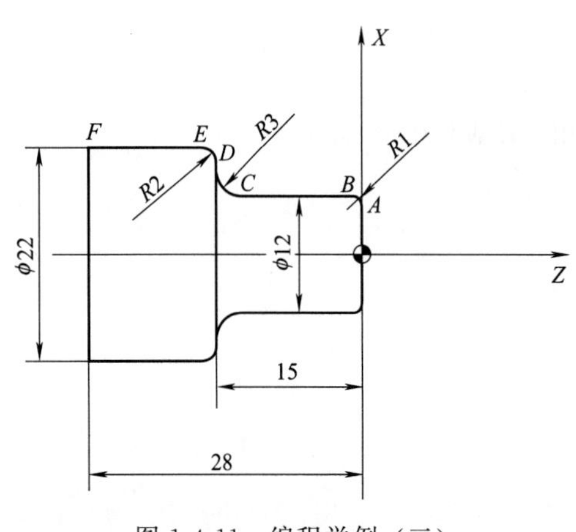


图 1-4-11　编程举例（三）

编写加工程序

```
O0403
N10 G99 G50 X100.0 Z100.0;
N20 M03 S1000 T0101;
N30 G00 X10.0 Z2.0;
N40 G01 Z0 F0.5;
N50 G03 X12.0 Z-1.0 R1.0 F0.2;
N60 G01 Z-12.0;
N70 G02 X18.0 Z-15.0 R3.0;
N80 G03 X22.0 W-2.0 R2.0;
N90 G01 Z-28.0;
N100 G00 X100.0 Z100.0;
N110 M05 T0100;
N120 M30;
```

**8. 在同一程序段中可能使用的编程坐标**

① 绝对编程坐标：X 、Z。

② 相对（增量）编程坐标：U、W。

③ 混合编程坐标：X（U）、Z（W）。

例如，上述编程举例中 N30 的程序段中可用：

N30 G00 X30.0 Z2.0;　　　（绝对编程）

或　N30 G00 U-70.0 W-98.0;（相对编程）

N30 G00 X30.0 W-98.0;（混合编程）

N30 G00 U-70.0 Z2.0;　（混合编程）

**9. 程序编写时应注意的事项**

① 坐标系的设定应根据加工工艺的要求，要尽可能靠近工件，只要换刀时不碰到工件就可以。

② 前三步和后三步的格式一样，但坐标系设定的范围不同，工件 G00 的定位根据加工要求确定。

③ G00 的程序段不能含有 F 值，有 F 值则无效。

**10. 加工程序**

略。

## 四、加工前准备

（1）机床准备（见表 1-4-4）

表 1-4-4　机床准备

| 项目 | 机械部分 | | | | 电气部分 | | 数控系统部分 | | | 辅助部分 | |
|---|---|---|---|---|---|---|---|---|---|---|---|
| 设备检查 | 主轴部分 | 进给部分 | 刀架部分 | 尾座 | 主电源 | 冷却风扇 | 电气元件 | 控制部分 | 驱动部分 | 冷却 | 润滑 |
| 检查情况 | | | | | | | | | | | |

注：经检查后该部分完好，在相应项目下打“√”；若出现问题及时报修。

（2）其他注意事项

① 安装外圆车刀时，注意控制刀杆伸出的长度及主偏角、副偏角的角度。

② 工件掉头装夹时注意控制夹紧力的大小，防止工件夹伤。

（3）参数设置

① 对刀的数值应输入在与程序中该刀具相对应的刀补号中；

② 在对刀的数值中应注意输入刀尖半径值和假想刀尖的位置序号。

## 五、实际零件加工

**1. 教师演示**

① 程序的输入及仿真校验。

② 对刀及刀具补偿的建立。

③ 外圆保证尺寸公差的方法。

**2. 学生加工训练**

训练中，指导教师巡回指导，及时纠正不正确的操作行为，解决学生练习中出现的各种问题。

## 六、零件测量

教学策略：讲授法、演示法。

重点讲授量具的选择及千分尺的使用方法及注意事项。由于学生在学习训练的初期阶段对数控加工了解得相对较少，加工每一个环节在初期都应细心地讲解给学生，为学生后期的成长打好基础。引入误差产生的因素及降低的方法。讲授和演示完毕后可以分组进行实物测量以强化检测的熟练度，提高测量的准确、稳定性。

（1）检查零件的外圆尺寸 $\phi 24_{-0.052}^{\ 0}$ mm 及表面粗糙度 $Ra3.2\mu m$

使用 0～25mm 的外径千分尺直接测量读数。

检查表面粗糙度，用表面粗糙度比较样板进行比较验定。

（2）检查零件的外圆尺寸 $\phi 28_{-0.052}^{\ 0}$ mm、$\phi 30_{-0.062}^{\ 0}$ mm 及表面粗糙度 $Ra3.2\mu m$

使用 25～50mm 的外径千分尺直接测量读数。

检查表面粗糙度，用表面粗糙度比较样板进行比较验定。

（3）检查长度尺寸

使用游标卡尺检测 36mm±0.05mm 长度尺寸。

（4）倒角尺寸

使用游标卡尺或目测进行倒角的检测。

**习题**

1. G90 与 G94 的区别是什么？

2. G40、G41 与 G42 有什么关系？

3. 请介绍 G04 的格式与含义。
4. 使用数控车床应掌握哪些基本测量工具?
5. 机床上如何删除单个程序和全部程序?
6. 主轴正转 M03 与主轴反转 M04 用于什么时候?
7. 如何计算螺纹的小径?
8. 请写出回参考点的程序。

# 任务五　连接轴的车削加工

## 一、图样与技术要求

如图 1-5-1 所示轴类特征零件，材料为 45 钢，规格为 $\phi$40mm 的圆柱棒料，正火处理，硬度 200HB。

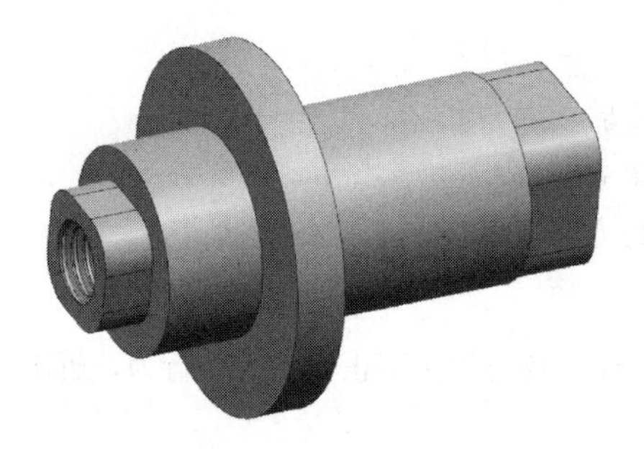

图 1-5-1　连接轴实物

## 二、图纸分析

### 1. 零件图分析

零件的外轮廓是三个不同长度的圆柱（$\phi$20mm、$\phi$35mm、$\phi$20mm）组成，属于简单的阶梯轴类零件，如图 1-5-2 所示。零件是根据训练、学习的前后顺序而设计的。

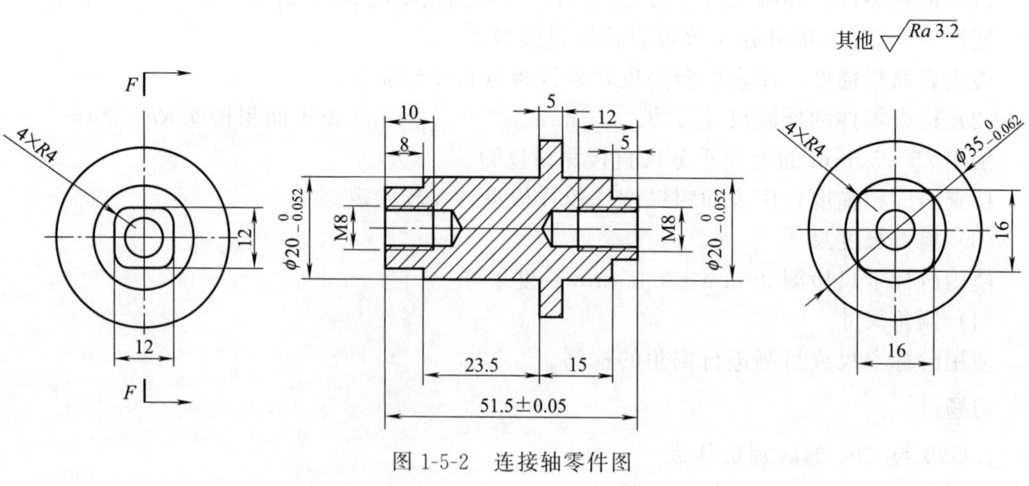

其他 $\sqrt{Ra\,3.2}$

F　F　4×R4　12　12　10　8　5　12　5　$\phi20^{\ 0}_{-0.052}$　M8　M8　$\phi20^{\ 0}_{-0.052}$　23.5　15　51.5±0.05　4×R4　$\phi35^{\ 0}_{-0.062}$　16　16

图 1-5-2　连接轴零件图

**2. 工艺分析**

① 结构分析：零件的结构都是简单的阶梯轴。

② 精度分析：零件的重点尺寸在外圆的精度等级都为 h9。对于中级工前期的训练是一个相对合理的公差尺寸。另外，长度、倒角等细节精度问题同样需要注意。

③ 定位及装夹分析：本零件采用三爪自定心卡盘进行定位和装夹。工件装夹时的夹紧力要适中，既要防止工件的变形和夹伤，又要防止工件在加工时的松动。工件装夹过程中应对工件进行找正，以保证各项形位公差。

④ 加工工艺分析：经过以上分析，本任务零件加工时总体安排顺序是，先加工零件的右端；切断工件后掉头找正后车削端面并倒角，保证工件总长尺寸。

**3. 主要刀具选择（见表 1-5-1）**

表 1-5-1　刀具卡片

| 刀具名称 | 刀具规格名称 | 材料 | 数量 | 刀尖半径 | 刀宽 |
| --- | --- | --- | --- | --- | --- |
| 90°外圆车刀 | 25mm×25mm | YT15 | 1 | 0 | |
| 45°外圆车刀 | 25mm×25mm | YT15 | 1 | 0 | |
| 切断刀 | 25mm×25mm | YT15 | 1 | 0 | 4mm |

**4. 工艺规程安排（见表 1-5-2）**

表 1-5-2　工序卡片（右端）

<table>
<tr><td colspan="2">单位</td><td colspan="2">产品名称及型号</td><td colspan="2">零件名称</td><td>零件图号</td></tr>
<tr><td colspan="2"></td><td colspan="2">任务五</td><td colspan="2">连接轴</td><td>图 1-5-2</td></tr>
<tr><td>工序</td><td>程序编号</td><td colspan="2">夹具名称</td><td colspan="2">使用设备</td><td>工件材料</td></tr>
<tr><td>001</td><td>O0001</td><td colspan="2">三爪卡盘</td><td colspan="2">SK50</td><td>45 钢</td></tr>
<tr><td colspan="7">阶梯轴零件</td></tr>
<tr><td>工步</td><td>工步内容</td><td>刀号</td><td colspan="2">切削用量</td><td colspan="2">备注</td></tr>
<tr><td>1</td><td>车端面</td><td>T11</td><td colspan="2">$n$=700r/min<br>（手动进给倍率模式在×10%位置）</td><td colspan="2">手动加工</td></tr>
<tr><td>2</td><td>粗车 $\phi$20、$\phi$30 外圆留 0.5mm 精加工余量</td><td>T11</td><td colspan="2">$n$=700r/min<br>$f$=0.2mm/r<br>$a_p$=3.0mm</td><td colspan="2" rowspan="2">自动加工</td></tr>
<tr><td>3</td><td>精车 $\phi$20、$\phi$35 外圆</td><td>T33</td><td colspan="2">$n$=900r/min<br>$f$=0.1mm/r<br>$a_p$=1.0mm</td></tr>
<tr><td>4</td><td>手动切断工件<br>总长留出 1.0mm<br>左右余量</td><td>T33</td><td colspan="2">$n$=500r/min<br>（进给倍率开关模式在×10%位置）</td><td colspan="2">$B$=4mm 切断工件，注意避免工件磕碰</td></tr>
<tr><td>5</td><td>掉头平端面保证<br>总长倒角</td><td>T11</td><td colspan="2">$n$=700r/min<br>（进给倍率开关模式在×10%位置）</td><td colspan="2">手动加工</td></tr>
</table>

## 三、程序编制

### 1. 外圆、内圆车削循环

功能：当零件的内、外圆柱面（圆锥面）上毛坯余量较大时，用 G90 可以去除大部分毛坯余量。

直线切削循环：

格式：G90 X（U）__ Z（W）__ F __；

式中　X，Z——终点绝对值坐标；

　U，W——相对（增量）值终点坐标尺寸；

　　F——切削进给速度。

其轨迹如图 1-5-3 所示，由 4 个步骤组成：

1（R）——第一步快速运动；

2（F）——第二步按进给速度切削；

3（F）——第三步按进给速度切削；

4（R）——第四步快速运动。

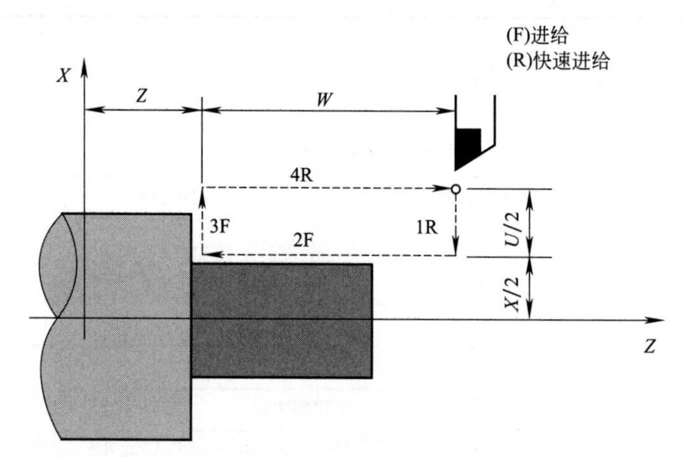


图 1-5-3　G90 粗车循环示意图

### 2. 锥体车削循环

（1）格式：G90 X（U）__ Z（W）__ R __　F __；

式中　X，Z——终点绝对值坐标；

　U，W——相对（增量）值终点坐标尺寸；

　　R——锥度尺寸 [$R=(D-d)/2$，$D$ 为锥度大端直径，$d$ 为锥度小端直径]，车削外圆锥度如是从小端车到大端时，切削锥度 R 为负值；车削内圆锥度如是从大端车到小端时，内圆锥度 R 为正值；

　　F——切削进给速度。

其轨迹如图 1-5-4 所示，R 值的正负与刀具轨迹有关。

（2）编程实例

G90 编程实例如图 1-5-5、图 1-5-6 所示。

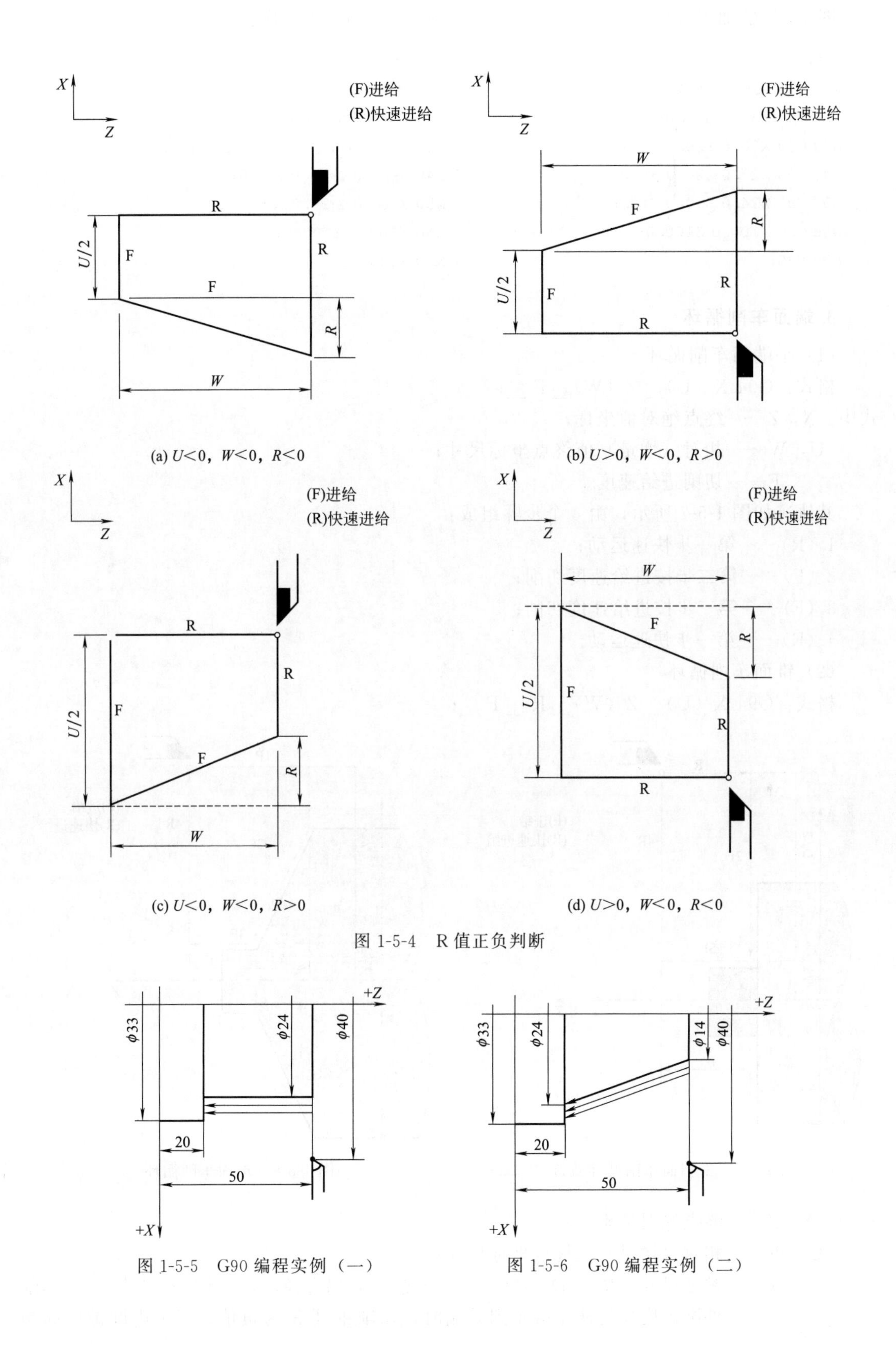


图 1-5-4　R 值正负判断

图 1-5-5　G90 编程实例（一）　　图 1-5-6　G90 编程实例（二）

图 1-5-5 的加工程序：

```
O0501
N10 T0101 M03 S800;
N20 G00 X35.0 Z51.0;
N30 G90 X30.0 Z20.0 F0.2;
N40 G90 X27.0 Z20.0 F0.2;
N50 G90 X24.0 Z20.0 F0.2;
N60 G00 X100.0 Z100.0;
N70 M30;
```

图 1-5-6 的加工程序：

```
O0502
N10 M03 S600 T0101;
N20 G00 X40.0 Z50.0;
N30 G90 X-10.0 Z-30.0 R-5.0 F0.1;
N40 X-13.0 Z-30.0 R-5.0;
N50 X-16.0 Z-30.0 R-5.0;
N60 X100.0 Z100.0;
N70 M30;
```

**3. 端面车削循环**

（1）平端面车削循环

格式：G94 X（U）__ Z（W）__ F __；

式中　X，Z——终点绝对值坐标；

U，W——相对（增量）值终点坐标尺寸；

F——切削进给速度。

其轨迹如图 1-5-7 所示，由 4 个步骤组成：

1（R）——第一步快速运动；

2（F）——第二步按进给速度切削；

3（F）——第三步按进给速度切削；

4（R）——第一步快速运动。

（2）锥面车削循环

格式：G94 X（U）__ Z（W）__ R __ F __；

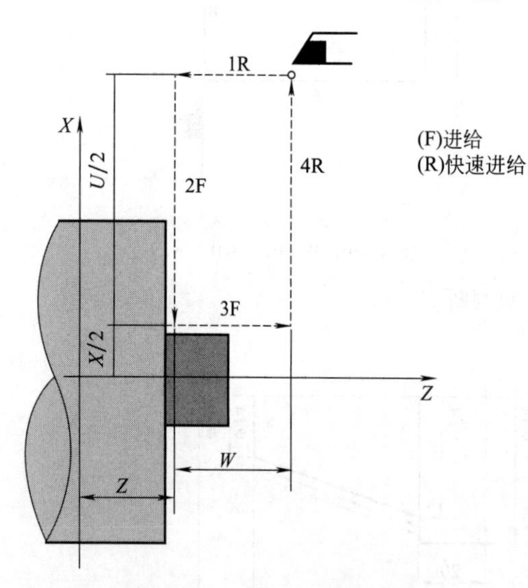


图 1-5-7　平端面车削循环轨迹

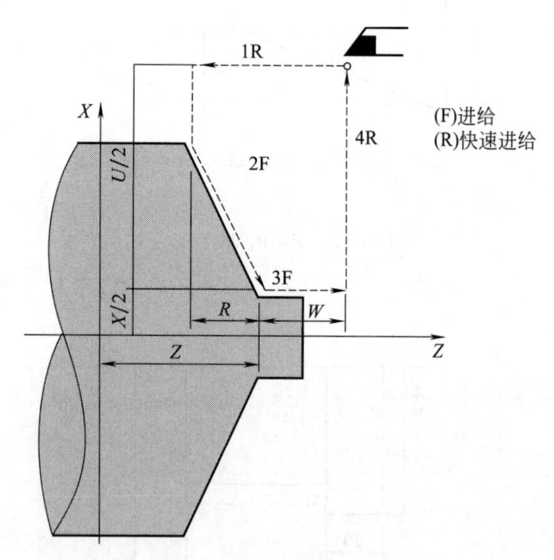


图 1-5-8　锥面车削循环

式中　X，Z——终点绝对值坐标；

U，W——相对（增量）值终点坐标尺寸；

R——锥度尺寸 [$R=(D-d)/2$，$D$ 为锥度大端直径，$d$ 为锥度小端直径]，车削外圆锥度如是从小端车到大端时，切削锥度 R 为负值；车削内圆锥度如是从大端车到小端时，内圆锥度 R 为正值；

F——切削进给速度。

其轨迹如图 1-5-8 所示，由 4 个步骤组成。

（3）G94 编程实例（见图 1-5-9、图 1-5-10）

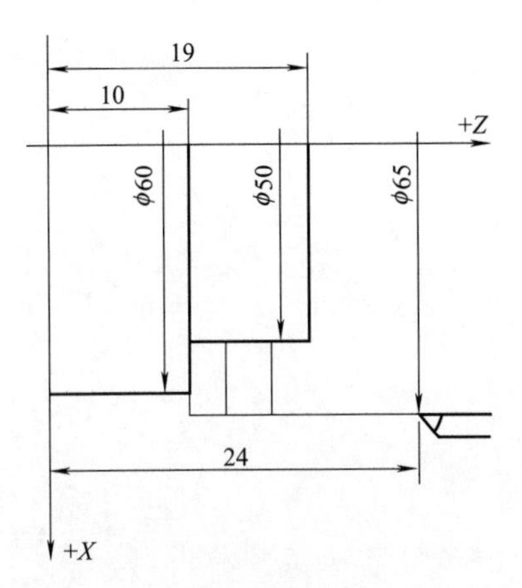

图 1-5-9　G94 编程实例（一）

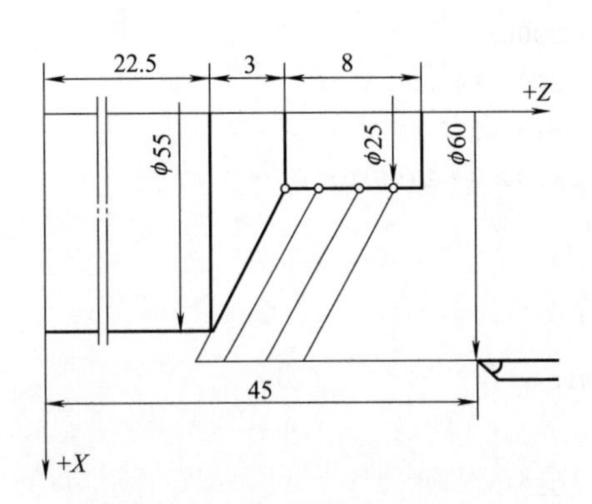

图 1-5-10　G94 编程实例（二）

图 1-5-9 的加工程序：

```
O0503
N10 M03 S600 T0202;
N20 G00 X65.0 Z24.0;
N30 G94 X-15.0 Z-8.0 F0.1;
N40 X-15.0 Z-11.0;
N50 X-15.0 Z-14.0;
N60 G00 X100.0 Z100.0;
N70 M30;
```

图 1-5-10 的加工程序：

```
O0504
N10 M03 S700 T0101;
N20 G00 X60.0 Z45.0;
N30 G94 X25.0 Z31.5 R-3.5 F0.15;
N40 X25.0 Z29.5 R-3.5;
N50 X25.0 Z27.5 R-3.5;
N60 X25.0 Z25.5 R-3.5;
N70 G00 X100.0 Z100.0;
N80 M30;
```

**4. 参考程序**

程序 1

```
O0001;
N1;
G97  G99;
T0101;
M03  S700;
G40  G00  X45.0  Z5.0;
G71  U1.5  R0.5;
G71  P10  Q20  U0.5  W0  F0.2;
N10  G00  X0;
G01  Z0;
X20.0,C0.5;
Z-15.0;
X35.0,C0.5;
```

```
W-6.0;
N20  X45.0;
G00  X100.0 Z100.0;
M05;
M00;
N2;
T0202;
M03  S900;
G42  G00  X45.0 Z5.0;
G70  P10  Q20  F0.1;
G00  X100.0  Z100.0;
M30;
```

程序 2

```
O0001;
N1;
G97  G99;
T0101;
M03  S700;
G40  G00  X45.0 Z5.0;
G71  U1.5  R0.5;
G71  P10  Q20  U0.5  W0  F0.2;
N10  G00  X0;
G01  Z0;
X20.0,C0.5;
Z-31.5;
X35.0,C0.5;
W-5.5;
N20  X45.0;
G00  X100.0 Z100.0;
M05;
M00;
N2;
T0202;
M03  S900;
G42  G00  X45.0 Z5.0;
G70  P10  Q20  F0.1;
G00  X100.0  Z100.0;
M30;
```

## 四、加工前准备

（1）机床准备（见表 1-5-3）

表 1-5-3 机床准备卡片

| 项目 | 机械部分 | | | | 电气部分 | | 数控系统部分 | | | 辅助部分 | |
|---|---|---|---|---|---|---|---|---|---|---|---|
| 设备检查 | 主轴部分 | 进给部分 | 刀架部分 | 尾座 | 主电源 | 冷却风扇 | 电气元件 | 控制部分 | 驱动部分 | 冷却 | 润滑 |
| 检查情况 | | | | | | | | | | | |

注：经检查后该部分完好，在相应项目下打“√”；若出现问题及时报修。

（2）其他注意事项

① 安装外圆车刀时，注意控制刀杆伸出的长度及主偏角、副偏角的角度。

② 工件掉头装夹时注意控制夹紧力的大小，防止工件夹伤。

（3）参数设置

① 对刀的数值应输入在与程序中该刀具相对应的刀补号中。

② 在对刀的数值中应注意输入刀尖半径值和假想刀尖的位置序号。

## 五、实际零件加工

**1. 教师演示**

① 程序的输入及仿真校验。

② 对刀及刀具补偿的建立。

③ 外圆保证尺寸公差的方法。

**2. 学生加工训练**

训练中，指导教师巡回指导，及时纠正不正确的操作行为，解决学生练习中出现的各种问题。

## 六、零件测量

教学策略：讲授法、演示法。

重点讲授量具的选择及千分尺的使用方法及注意事项。由于学生在学习训练的初期阶段对数控加工了解得相对较少，加工每一个环节在初期都应细心地讲解给学生，为学生后期的成长打好基础。引入误差产生的因素及降低的方法。讲授和演示完毕后可以分组进行实物测量以强化检测的熟练度，提高测量的准确、稳定性。

（1）检查零件的外圆尺寸 $\phi 20_{-0.052}^{\ 0}$ mm，检查表面粗糙度 $Ra3.2\mu m$

使用 0～25mm 的外径千分尺直接测量读数。

检查表面粗糙度，用表面粗糙度比较样板进行比较验定。

（2）检查零件的外圆尺寸 $\phi 35_{-0.062}^{\ 0}$ mm，检查表面粗糙度 $Ra3.2\mu m$

使用 25～50mm 的外径千分尺直接测量读数。

检查表面粗糙度，用表面粗糙度比较样板进行比较验定。

（3）检查长度尺寸

使用游标卡尺检测 51.5mm±0.05mm 长度尺寸。

（4）倒角尺寸

使用游标卡尺或目测进行倒角的检测。

## 七、加工注意事项

1. 刀具中心高度的高低对加工的影响。

2. 对刀所建立的工件坐标系要正确，并学会验证。

**习题**

1. 圆弧半径什么时候用正值？什么时候用负值？

2. 简述圆锥的基本参数。

3. 已知一外锥面，大端直径 $D$=80mm，小端直径 $d$=60mm，圆锥半角 30°，求其圆锥部分长度。

4. 编一段循环加工程序，已知：$d=40$mm，$d_1=30$mm，$d_2=20$mm，每段深度 10mm，刀具安全点为（$a$，$b$）。

5. 写出 M20×2 的螺纹程序，循环点自拟。

6. 验证螺纹的工具名称是什么？并做解释。

7. 切削液的作用有哪几点？

8. 画出前置刀架和后置刀架的刀位图。

# 任务六　螺纹轴的加工

## 一、图样与技术要求

图 1-6-1 所示为螺纹类特征零件，材料为 45 钢，规格为 $\phi$10mm 的圆柱棒料，正火处理，硬度 200HB。

## 二、图纸分析

教学策略：分组讨论、小组汇报、教师总结。

**1. 零件图分析**

零件的外轮廓是一个 $\phi$10mm 的圆柱面、M8×1.25 的外螺纹，属于简单的螺纹类零件，如图 1-6-2 所示。零件是根据训练、学习的前后顺序而设计的。评分表见表 1-6-1。

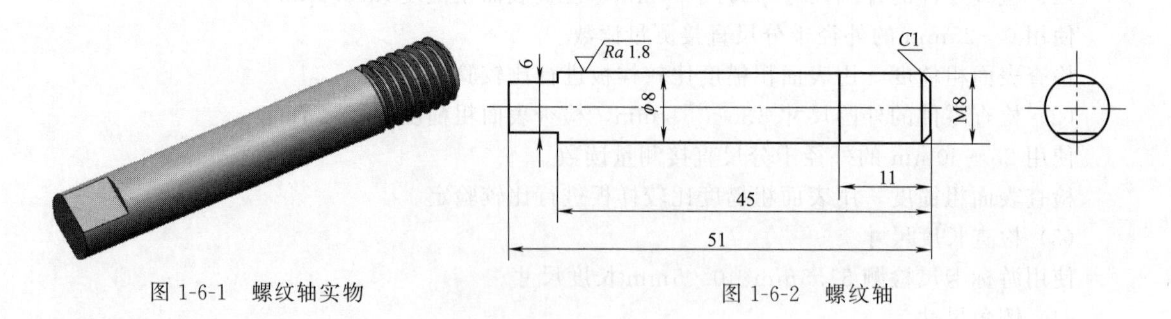


图 1-6-1　螺纹轴实物　　图 1-6-2　螺纹轴

**表 1-6-1　评分表**

| 序号 | 项目及技术要求 | 配分(IT/Ra) | 评分标准 | 检测结果 | 实得分 |
|---|---|---|---|---|---|
| 1 | 外径 $\phi 8_{-0.039}^{0}$,$Ra$1.6 | 16/2 | 超差全扣 | | |
| 2 | 圆扁 $6_{-0.052}^{0}$,$Ra$3.2 | 16/2 | 超差全扣 | | |
| 3 | 长度 51±0.1 | 16 | 超差全扣 | | |
| 4 | 长度 $45_{-0.06}^{0}$ | 16 | 超差全扣 | | |
| 5 | 长度 $11_{0}^{+0.1}$ | 16 | 超差全扣 | | |
| 6 | 倒角 $C_1$(共 4 处) | 8 | 超差全扣 | | |
| 7 | 形状轮廓完整 | 3 | 未完不得 | | |
| 安全文明生产 | | 5 | | | |
| 加工工时 | | 60min | | | |

## 2. 工艺分析

① 结构分析：零件的结构是简单螺纹轴。

② 精度分析：零件的重点尺寸在外圆的精度等级都为 h7，螺纹的中径公差等级为 6g。对于本模块的训练是一个相对合理的公差尺寸。特别是在编程时要注意螺纹的长度。另外，长度、倒角等细节精度问题同样需要注意。

③ 定位及装夹分析：本零件采用三爪自定心卡盘进行定位和装夹。工件装夹时的夹紧力要适中，既要防止工件的变形和夹伤，又要防止工件在加工时的松动。工件装夹过程中应对工件进行找正，以保证各项形位公差。

④ 加工工艺分析：经过以上分析，本任务零件加工时总体安排顺序是，先加工零件的右端；切断工件后掉头找正后车削端面并倒角，保证工件总长尺寸。

## 3. 主要刀具选择（见表 1-6-2）

**表 1-6-2　刀具卡片**

| 刀具名称 | 刀具规格名称 | 材料 | 数量 | 刀尖半径 | 刀宽 |
|---|---|---|---|---|---|
| 90°外圆车刀 | 25mm×25mm | YT15 | 1 | 0 | |
| 45°外圆车刀 | 25mm×25mm | YT15 | 1 | 0 | |
| 切断刀 | 25mm×25mm | YT15 | 1 | 0 | 4mm |
| 螺纹刀 | 25mm×25mm,刀尖角 60° | YT15 | 1 | 0 | |

## 4. 工艺规程安排（见表 1-6-3）

**表 1-6-3　工序卡片（右端）**

| 单位 | | 产品名称及型号 | | 零件名称 | 零件图号 |
|---|---|---|---|---|---|
| | | 任务六 | | 螺纹轴 | 图 1-6-2 |
| 工序 | 程序编号 | 夹具名称 | | 使用设备 | 工件材料 |
| 001 | O0001 | 三爪卡盘 | | SK50 | 45 钢 |
| 件 1 | | | | | |
| 工步 | 工步内容 | 刀号 | 切削用量 | | 备注 |
| 1 | 粗车 | T11 | $n$=700r/min<br>(进给倍率开关在×10%位置) | | 自动加工 |
| 2 | 精车 | T22 | $n$=900r/min<br>(进给倍率开关在×10%位置) | | 自动加工 |

续表

| 件 1 | | | | |
| --- | --- | --- | --- | --- |
| 工步 | 工步内容 | 刀号 | 切削用量 | 备注 |
| 3 | 粗车 M8×1.25 螺纹<br>外圆留 0.5mm<br>精加工余量 | T44 | $n=500$r/min<br>$f=1.25$mm/r<br>$a_p=0.2$mm | 自动加工 |
| 4 | 精车 M8×1.25<br>螺纹 | | $n=500$r/min<br>$f=1.25$mm/r<br>$a_p=0.1$mm | |
| 5 | 手动切断工件<br>总长留出 1mm 左右余量 | T33 | $n=500$r/min<br>(进给倍率开关在×10%位置) | 自动加工 $B=4$mm<br>切断工件,注意避免工件磕碰 |
| 6 | 掉头平端面保证总长、倒角 | T11 | $n=700$r/min<br>(进给倍率开关在×10%位置) | 自动加工 |

## 三、程序编制

### 1. 螺纹加工基础知识

加工螺纹时应注意以下两个问题。

① 车螺纹时一定要有切入段 $\delta_1$ 和切出段 $\delta_2$。见图 1-6-3。

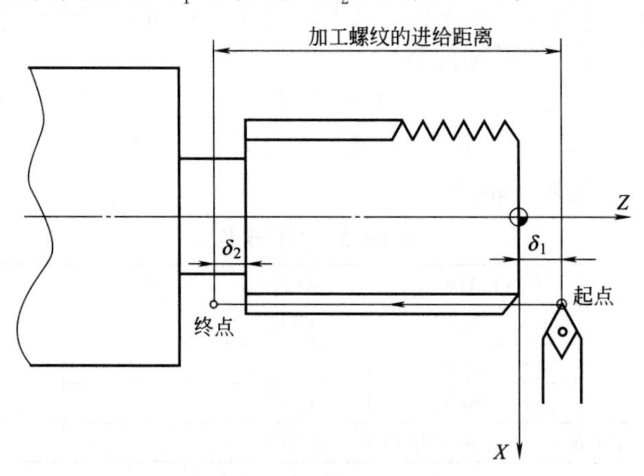


图 1-6-3　切入段 $\delta_1$ 和切出段 $\delta_2$

② 螺纹加工一般需要多次走刀，各次的切削深度应按递减规律分配，见图 1-6-4。

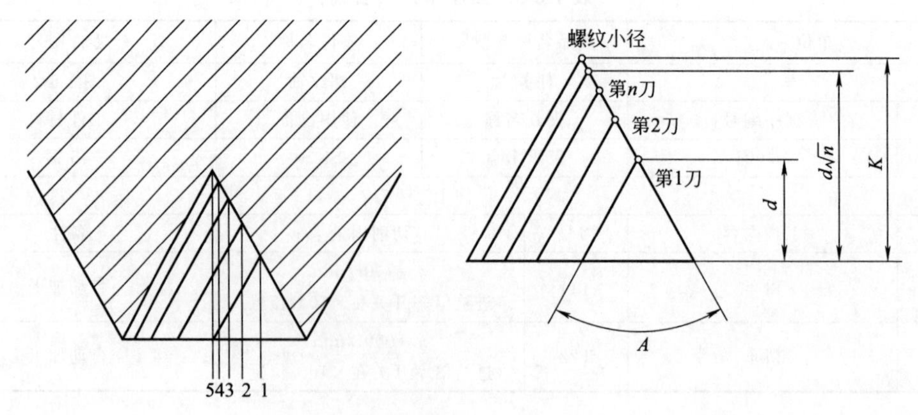


图 1-6-4　多次走刀路线

常用普通公制螺纹加工走刀次数与分层切削深度，见表 1-6-4。

表 1-6-4 普通公制螺纹走刀次数

| 普通公制螺纹 | | | | | | | | |
|---|---|---|---|---|---|---|---|---|
| 螺距/mm | | 1.0 | 1.5 | 2.0 | 2.5 | 3.0 | 3.5 | 4.0 |
| 牙型高度/mm | | 0.649 | 0.977 | 1.299 | 1.624 | 1.949 | 2.273 | 2.598 |
| 走刀次数及分层切削深度（直径值）/mm | 1次 | 0.7 | 0.8 | 0.9 | 1.0 | 1.2 | 1.5 | 1.5 |
| | 2次 | 0.4 | 0.6 | 0.6 | 0.7 | 0.7 | 0.7 | 0.8 |
| | 3次 | 0.2 | 0.4 | 0.6 | 0.6 | 0.6 | 0.6 | 0.6 |
| | 4次 | | 0.16 | 0.4 | 0.4 | 0.4 | 0.6 | 0.6 |
| | 5次 | | | 0.1 | 0.4 | 0.4 | 0.4 | 0.4 |
| | 6次 | | | | 0.15 | 0.4 | 0.4 | 0.4 |
| | 7次 | | | | | 0.2 | 0.2 | 0.4 |
| | 8次 | | | | | | 0.15 | 0.3 |
| | 9次 | | | | | | | 0.2 |

**2. 单一固定循环螺纹**

G92 X（U）_ Z（W）_ F _；圆柱面单一固定循环螺纹

式中 X，Z——车螺纹段的终点绝对坐标值；

U，W——车螺纹段的终点相对于循环起点的增量坐标值；

F——螺纹的导程（单头为螺距）。

单一固定循环车螺纹指令可以把一系列连续加工动作如切入→切削→退刀→返回，用一个循环指令完成，从而简化编程，见图 1-6-5。

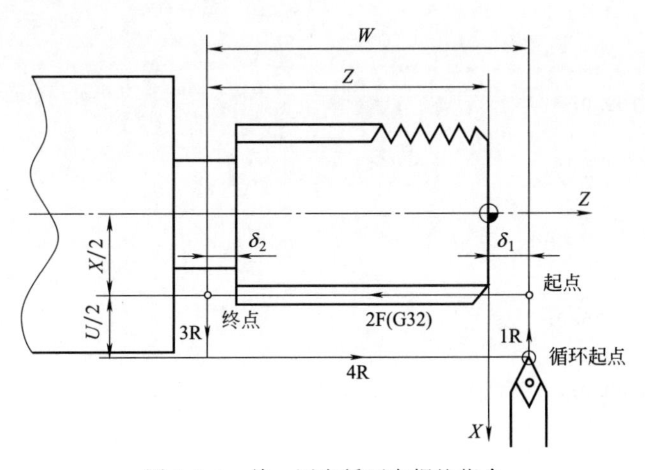


图 1-6-5 单一固定循环车螺纹指令

**3. 螺纹切削 G32 指令**

格式：G32 X（U）_ Z（W）_ F _；

式中 X，Z——绝对编程时，有效螺纹终点在工件坐标系中的坐标；

U，W——增量编程时，有效螺纹终点相对于螺纹切削起点的位移量；

F——螺纹导程，即主轴每转一圈，刀具相对于工件的进给值。

说明：

① 从螺纹粗加工到精加工，主轴的转速必须保持一常数；

② 在没有停止主轴的情况下，停止螺纹的切削将非常危险；因此螺纹切削时进给保持功能无效，如果按下进给保持按键，刀具在加工完螺纹后停止运动；

③ 在螺纹加工中不使用恒定线速度控制功能；

④ 在螺纹加工轨迹中应设置足够的升速进刀段 $\delta$ 和降速退刀段 $\delta'$，以消除伺服滞后造成的螺距误差。

**4. 螺纹切削循环 G92**

格式：G92 X（U）__ Z（W）__ F __；

式中 X，Z——绝对值编程时，螺纹终点在工件坐标系下的坐标；

U，W——增量值编程时，螺纹终点相对于循环起点的有向距离；

F——螺纹导程。

**5. 参考程序**

```
O0001;
N1;
G97  G99;
T0101;
M03  S700;
G40  G00  X35.0  Z5.0;
G71  U1.5  R0.5;
G71  P10  Q20  U0.5  W0  F0.2;
N10  G00  X0;
G01  Z0;
X7.8  ,  C1.0;
Z-54.0;
N20  X35.0;
G00  X100.0  Z100.0;
M05;
M00;
N2;
T0202;
M03  S900;
G42  G00  X35.0  Z5.0;
G70  P10  Q20  F0.1;
G00  X100.0  Z100.0;
M05;
M00;
N3;
T0404;
M03  S500;
G40  G00  X35.0  Z5.0;
G92  X7.4  Z-10.5  F1.25;
X7.;
X6.7;
```

```
X6.5;
X6.3;
X6.3;
G00  X100.0  Z100.0;
M30;
```

## 四、加工前准备

### 1. 机床准备（见表 1-6-5）

表 1-6-5　机床准备卡片

| 项目 | 机械部分 | | | | 电气部分 | | 数控系统部分 | | | 辅助部分 | |
|---|---|---|---|---|---|---|---|---|---|---|---|
| 设备检查 | 主轴部分 | 进给部分 | 刀架部分 | 尾座 | 主电源 | 冷却风扇 | 电气元件 | 控制部分 | 驱动部分 | 冷却 | 润滑 |
| 检查情况 | | | | | | | | | | | |

注：经检查后该部分完好，在相应项目下打“√”；若出现问题及时报修。

### 2. 其他注意事项

① 安装外圆车刀时，注意控制刀杆伸出的长度及主偏角、副偏角的角度。

② 工件掉头装夹时注意控制夹紧力的大小，防止工件夹伤。

### 3. 参数设置

① 对刀的数值应输入在与程序中该刀具相对应的刀补号中。

② 在对刀的数值中应注意输入刀尖半径值和假想刀尖的位置序号。

## 五、实际零件加工

### 1. 教师演示

① 程序的输入及仿真校验。

② 对刀及刀具补偿的建立。

③ 螺纹退刀槽的手动车削、螺纹加工及尺寸精度保证。

### 2. 学生加工训练

训练中，指导教师巡回指导，及时纠正不正确的操作姿势，解决学生练习中出现的各种问题。

## 六、零件测量

教学策略：讲授法、演示法。

重点讲授螺纹测量的方法及教学选择的测量方法。由于学生在学习训练的初期阶段对尺寸检测不熟悉，加工每一个环节在初期都应细心地讲解给学生，为学生后期的成长打好基础。引入误差产生的因素及降低的方法。讲授和演示完毕后可以分组进行实物测量以强化检测的熟练度，提高测量的准确、稳定性。

### 1. 参考检测工艺

（1）检查零件的外圆尺寸 $\phi$8mm，检查表面粗糙度 $Ra$3.2μm

使用 0～25mm 的外径千分尺直接测量读数。

检查表面粗糙度，用表面粗糙度比较样板进行比较验定。

(2) 检查零件的外螺纹 M8×1.25

使用 M8×1.25 螺纹环规的通规、止规测量螺纹是否合格。

(3) 检查长度尺寸

使用游标卡尺检测 51mm 长度尺寸。

(4) 倒角尺寸

使用游标卡尺或目测进行倒角的检测。

**2. 检测并填写记录表**

教学策略：小组互检、个人验证、教师抽验。

首先以组为单位，小组内进行互检，由检测同学按评分表给出一个互检成绩；然后个人对自己加工的工件进行自检并与互检成绩、检测结果进行比较，从中发现问题尺寸并找出检测出现不同结果的原因，更正出现失误的环节；最后由教师对学生的零件进行抽样检测，并针对出现的问题，集中解释出现测量误差的原因及提出改进的方法。

## 七、螺纹加工注意事项

① 螺纹加工过程中不能改变转速。

② 螺纹加工转速不能太高。

**习题**

1. 外螺纹刀装刀的注意事项是什么？
2. M10、M12、M16、M20、M24 螺纹的粗牙螺距分别是多少？
3. 内孔加工，程序余量正负值如何确定？有什么方法？
4. 自动运行和 NDI 有什么不同？
5. 如何使用电脑传输程序？
6. 如果刀具低于工件中心会有什么后果？
7. 加工螺纹时，进给速度 F 应如何定值？

# 任务七　手柄的加工

## 一、图样与技术要求

如图 1-7-1～图 1-7-3 所示，材料为 45 钢，规格为 $\phi$30mm 的圆柱棒料，正火处理，硬度 200HB。

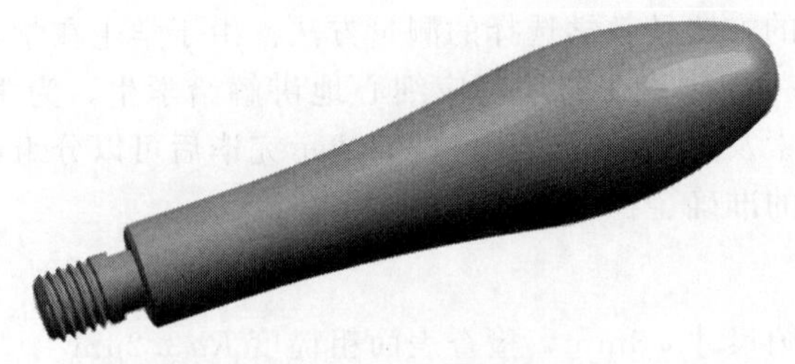

图 1-7-1　手柄实物渲染图

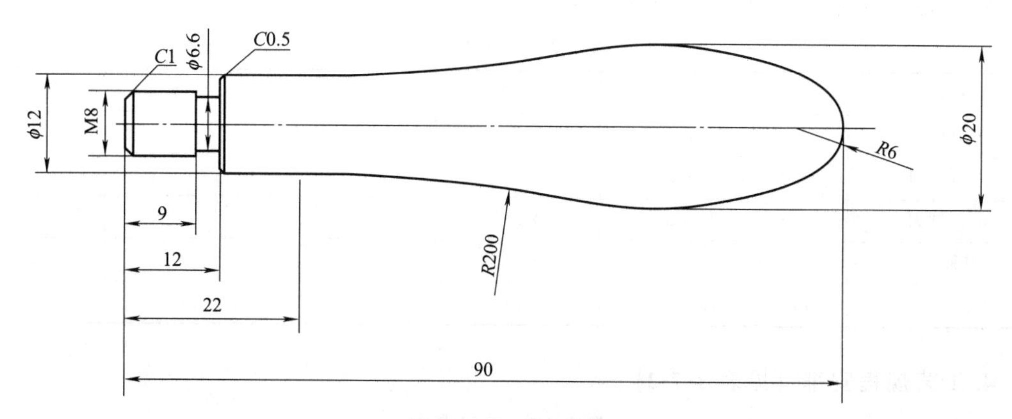


图 1-7-2　手柄零件

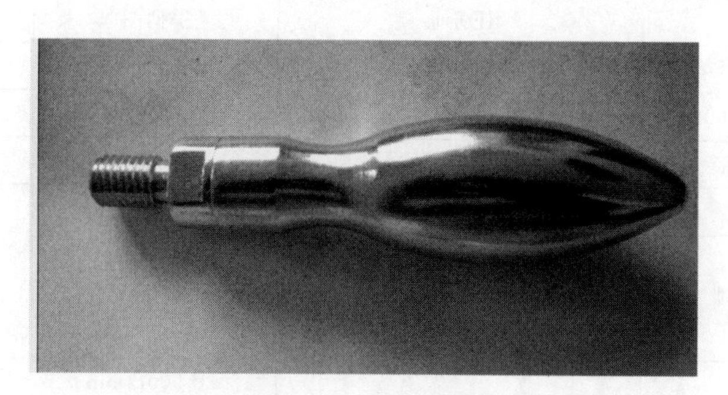

图 1-7-3　手柄实物

## 二、图纸分析

**1. 零件图分析**

如图 1-7-2 所示，手柄零件主要由外圆、宽槽及外螺纹组成，有关表面粗糙度为 $Ra$1. 6μm、$Ra$3. 2μm 和 $Ra$6. 3μm。

**2. 工艺分析**

① 结构分析：该零件结构相对比较复杂，在加工时应重点考虑刚性、编程指令、刀具工作角度、切削用量等问题。

② 精度分析：确保零件尺寸，还有同轴度、圆跳动、垂直度等形位公差要求；保证表面粗糙度要求，因此在加工时应注意工件的加工刚性、刀具刚性、加工工艺等问题。

③ 定位及装夹分析：本零件采用三爪自定心卡盘进行定位和装夹。特别是掉头加工，工件装夹时的夹紧力要适中，既要防止工件变形和夹伤，又要防止工件在加工时松动，工件装夹过程中应对工件进行找正，以保证各项形位公差。

④ 加工工艺分析：根据图纸分析，零件加工难点在宽槽的加工，因为切槽的径向力比较大，在第一次装夹加工零件右端时，把宽槽加工出来。又因为零件伸出来比较长，所以加工螺纹时，转速不能太高，否则容易产生锥螺纹。

在掉头加工零件时，对零件要找正，保证同轴度要求，确保长度尺寸 80mm±0. 03mm。

### 3. 主要刀具选择（见表 1-7-1）

表 1-7-1　刀具卡片

| 刀具名称 | 刀具规格名称 | 材料 | 数量 | 刀尖半径 | 刀宽 |
|---|---|---|---|---|---|
| 90°外圆刀 | 25mm×25mm | YT15 | 1 | 0 | |
| 45°外圆刀 | 25mm×25mm | YT15 | 1 | 0 | |
| 35°外圆刀 | 25mm×25mm | YT15 | 1 | 0.4mm | |
| 切断刀 | 25mm×25mm | YT15 | 1 | 0 | 3mm |
| 螺纹刀 | 25mm×25mm，刀尖角 60° | YT15 | 1 | 0 | |

### 4. 工艺规程安排（见表 1-7-2）

表 1-7-2　工序卡片

| 单位 | | 产品名称及型号 | 零件名称 | 零件图号 |
|---|---|---|---|---|
| | | 任务七 | 手柄图 | 图 1-7-2 |
| 工序 | 程序编号 | 夹具名称 | 使用设备 | 工序 |
| 001 | O0001，O0002 | 三爪卡盘 | SK50 | 001 |
| 工步 | 工步内容 | 刀号 | 切削用量 | 备注 |
| 1 | 车端面 | T11 | $n$=700r/min | 手动加工 |
| 2 | 粗车外圆 | T11 | $n$=700r/min<br>$f$=0.2mm/r<br>$a_p$=2mm | 自动加工 |
| 3 | 精车轮廓 | T22 | $n$=900r/min<br>$f$=0.1mm/r<br>$a_p$=0.5mm | 自动加工 |
| 4 | 切退刀槽 | T33 | $n$=500r/min<br>$f$=0.07mm/r | 自动加工 |
| 5 | 车螺纹 | T44 | $n$=500r/min<br>$P$=1.25mm | 自动加工 |
| 6 | 切断 | T33 | $n$=300r/min<br>（进给倍率开关在×10 位置） | 自动加工 |
| 7 | 掉头加工轮廓 | T11 | $n$=700r/min<br>$f$=0.15mm/r<br>$a_p$=1mm | 自动加工 |
| 8 | 精车轮廓 | T22 | $n$=900r/min<br>$f$=0.1mm/r<br>$a_p$=0.5mm | 自动加工 |

## 三、复合循环指令

### 1. G71 轴向粗车多重循环

（1）G71 指令格式

内、外圆粗车复合循环指令，适用于内、外圆柱面需要多次走刀才能完成的轴套类零件的粗加工，见图 1-7-4、图 1-7-5。

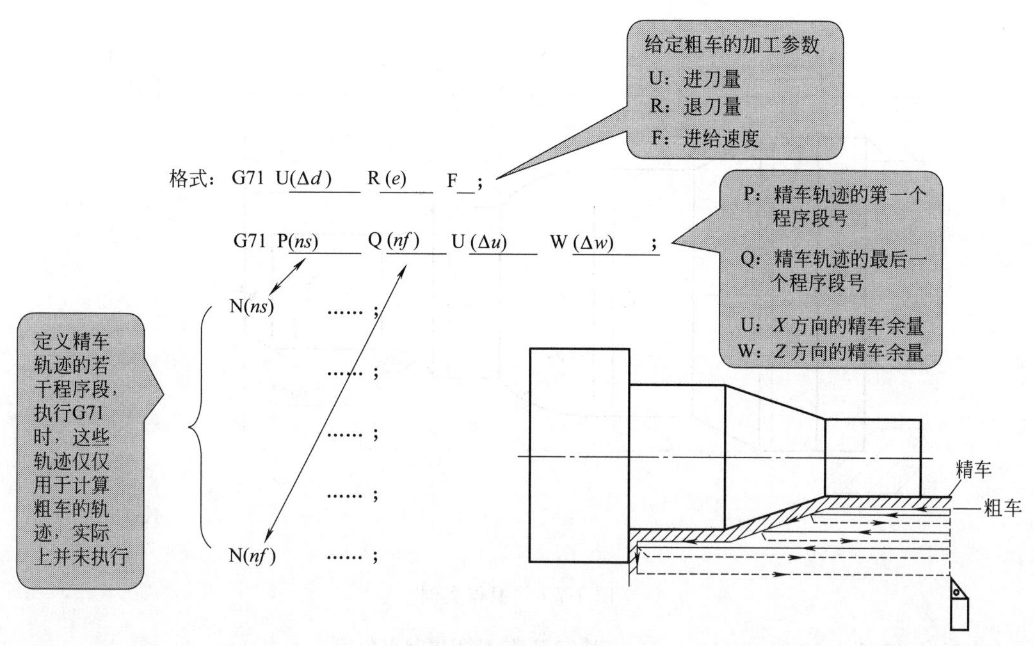


图 1-7-4　G71 指令格式

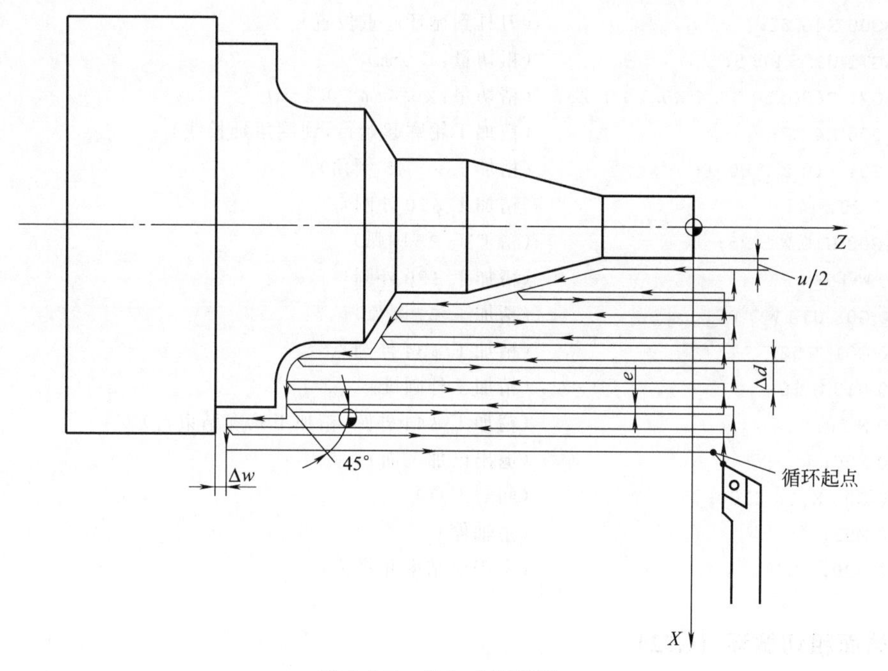


图 1-7-5　G71 走刀路线

（2）编程实例

**【例 1-7-1】** 编制图 1-7-6 所示零件的外径粗加工复合循环加工程序：要求循环起始点在（46，1），切削深度为 1.5mm（半径量）。退刀量为 1mm，$X$ 方向精加工余量为 0.4mm，$Z$ 方向精加工余量为 0.1mm，其中双点画线部分为工件毛坯。

O0701（见图 1-7-6）

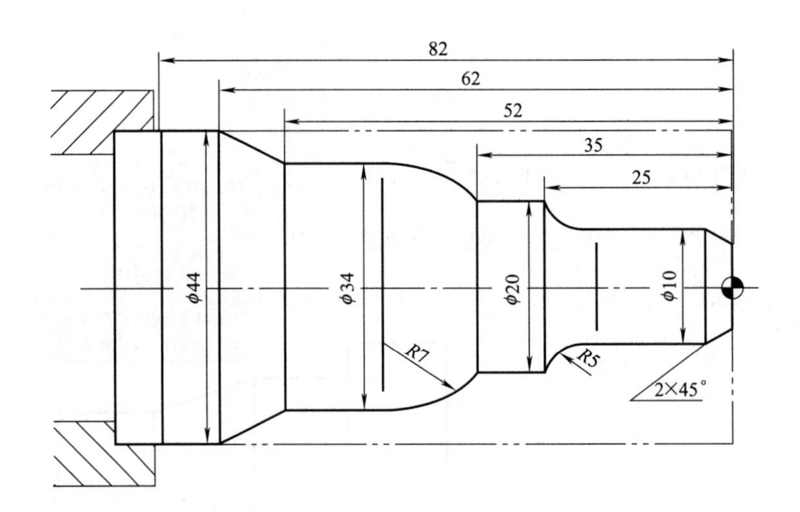

图 1-7-6　编程实例

| | |
|---|---|
| N10 T0101 G00 X100 Z100 G95; | (选定刀具,到程序起点位置) |
| N20 M03 S800; | (主轴以 800r/min 正转) |
| N30 G00 X46 Z2; | (刀具到循环起点位置) |
| N40 G71 U1.5 R0.5; | (粗切量:1.5mm) |
| N50 G71 P60Q150 U0.4 W0.1 F0.25; | (精切量:X0.4mm Z0.1mm) |
| N60 G00 X4 Z2; | (精加工轮廓起始行,到倒角延长线) |
| N70 G01 X10 Z-2 F0.1; | (精加工 2×45°倒角) |
| N80 Z-20; | (精加工 $\phi$10 外圆) |
| N90 G02 U10 W-5 R5; | (精加工 R5 圆弧) |
| N100 W-10; | (精加工 $\phi$20 外圆) |
| N110 G03 U14 W-7 R7; | (精加工 R7 圆弧) |
| N120 G01 Z-52; | (精加工 $\phi$34 外圆) |
| N130 U10 W-10; | (精加工外圆锥) |
| N140 W-20; | (精加工 $\phi$44 外圆,精加工轮廓结束行) |
| N150 X50; | (退出已加工面) |
| N160 G00 X100 Z100; | (回对刀点) |
| N170 M05; | (主轴停) |
| N180 M30; | (主程序结束并复位) |

### 2. 端面粗切循环（G72）

（1）编程格式

```
G72 U(Δd)R(e);
G72 P(ns)Q(nf)U(Δu)W(Δw)F(f)S(s)T(t);
```

式中　$\Delta d$——背吃刀量；

$e$——退刀量；

$ns$——精加工轮廓程序段中开始程序段的段号；

$nf$——精加工轮廓程序段中结束程序段的段号；

$\Delta u$——$X$ 轴向精加工余量；

$\Delta w$——$Z$ 轴向精加工余量；

$f$，$s$，$t$——F、S、T 对应的数值。

注意：

① $ns \to nf$ 程序段中的 F、S、T 功能，即使被指定对粗车循环也无效。

② 零件轮廓必须符合 $X$ 轴、$Z$ 轴方向同时单调增大或单调减少。

（2）循环路线

端面粗切循环是一种复合固定循环。端面粗切循环适于 $Z$ 向余量小、$X$ 向余量大的棒料粗加工，其循环路线如图 1-7-7 所示。

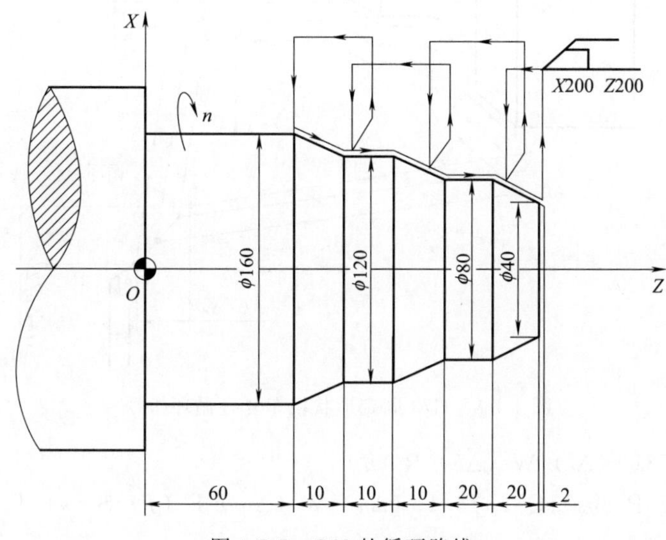


图 1-7-7 G72 的循环路线

（3）编程实例

**【例 1-7-2】** 按图 1-7-7 所示尺寸编写端面粗切、循环加工程序。

```
O0702
N10 G50 X200 Z200 T0101;
N20 M03 S800;
N30 G90 G00 G41 X176 Z2 M08;
N40 G96 S120;
N50 G72 U3 R0.5;
N60 G72 P70 Q120 U2 W0.5 F0.2;
N70 G00 X160 Z60;//ns
N80 G01 X120 Z70 F0.15;
N90 Z80;
N100 X80 Z90;
N110 Z110;
N120 X36 Z132;//nf
N130 G00 G40 X200 Z200;
N140 M30;
```

### 3. 固定形状粗车复合循环指令 G73

（1）指令格式

固定形状粗车复合固定循环指令，适用于铸件、锻件毛坯粗加工，如图 1-7-8 所示。

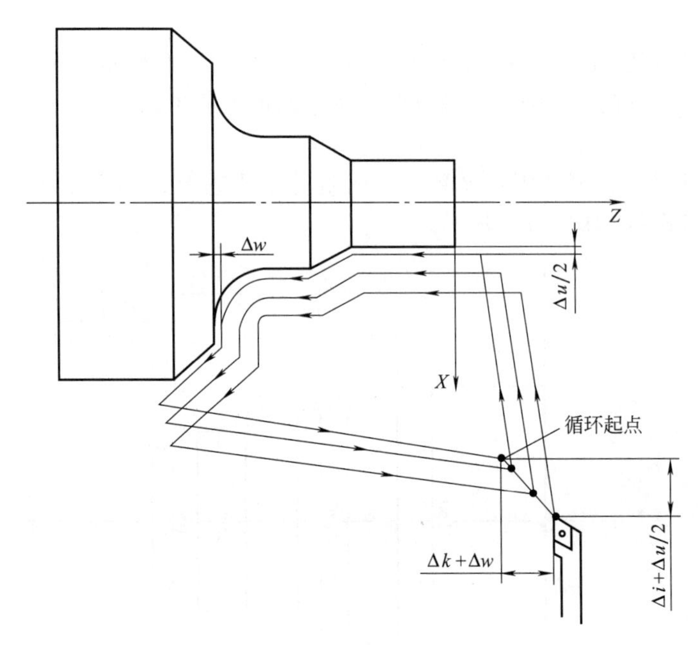


图 1-7-8　G73 固定形状粗车复合循环指令

编程格式：G73U（$\Delta i$）W（$\Delta k$）R（$d$）

G73P（$ns$）Q（$nf$）U（$\Delta u$）W（$\Delta w$）F（$f$）S（$s$）T（$t$）

N$ns$

……

N$nf$

式中　$ns$——精加工程序段的开始程序段号；

$nf$——精加工程序段的结束程序段号；

$\Delta i$——粗车时，径向（$X$ 方向）需要切除的总余量（半径值）；

$\Delta k$——粗车时，轴向（$Z$ 方向）需要切除的总余量；

$d$——粗车循环次数；

$\Delta u$——径向（$X$ 轴方向）给精加工留的余量；

$\Delta w$——轴向（$Z$ 轴方向）给精加工留的余量；

$f$——粗加工时的进给速度；

$s$——粗加工时的主轴转速；

$t$——粗加工时使用的刀具号。

说明：所谓封闭（或固定形状）粗车复合固定循环就是按照一定的切削形状逐渐地接近最终形状。所以，它适用于毛坯轮廓形状与零件轮廓形状基本形似的粗车加工。因此，这种加工方式对于铸造或锻造毛坯的粗车是一种效率很高的方法。

（2）编程实例

**【例 1-7-3】** 按图 1-7-9 所示尺寸编写封闭切削循环加工程序。

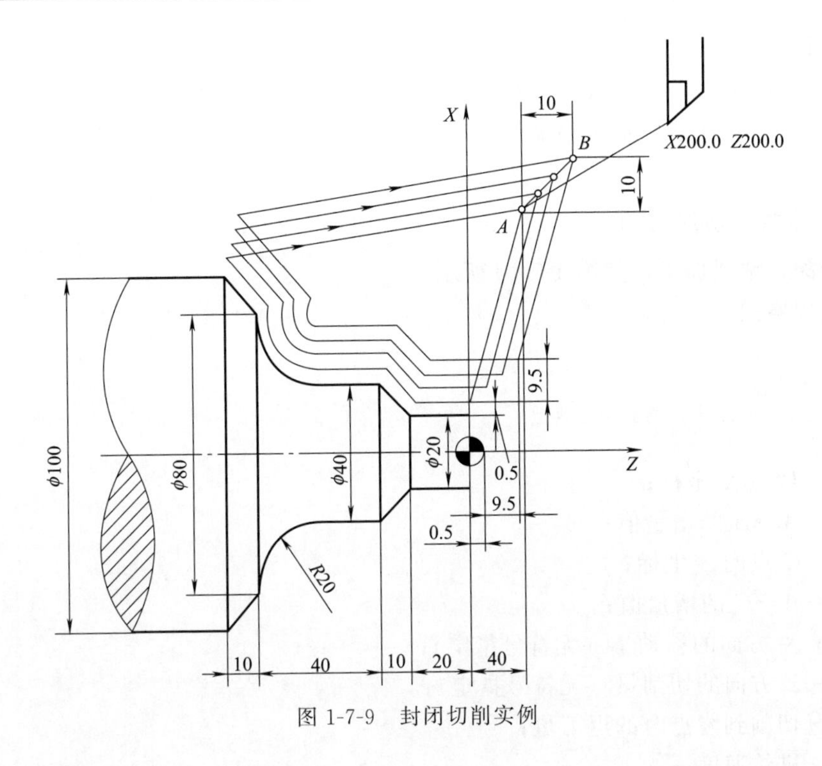


图 1-7-9　封闭切削实例

```
O0703
N10 G99 G50 X200 Z200 T0101;
N20 M03 S2000;
N30 G00 G42 X140 Z40 M08;
N40 G96 S150;
N50 G73 U9.5 W9.5 R3;
N60 G73 P70 Q130 U1 W0.5 F0.3;
N70 G00 X20 Z0;//ns
N80 G01 Z-20 F0.15;
N90 X40 Z-30;
N100 Z-50;
N110 G02 X80 Z-70 R20;
N120 G01 X100 Z-80;
N130 X105;//nf
N140 G00 X200 Z200 G40;
N150 M30;
```

#### 4. G70 精加工循环

由 G71、G72、G73 完成粗加工后，可以用 G70 进行精加工。

（1）编程格式

```
G70 P(ns )Q(nf )
```

式中　*ns*——精加工轮廓程序段中开始程序段的段号；

*nf*——精加工轮廓程序段中结束程序段的段号。

精加工时，G71、G72、G73 程序段中的 F、S、T 指令无效，只有在 *ns*～*nf* 程序段中的 F、S、T 才有效。

（2）编程实例

例：在 G71、G72、G73 程序应用实例中的 *nf* 程序段后再加上“G70 P*ns* Q*nf*”程序段，并在 *ns*～*nf* 程序段中加上精加工适用的 F、S、T，就可以完成从粗加工到精加工的全过程。

**5. 深孔钻循环功能 G74**

适用于深孔钻削加工，如图 1-7-10 所示。

（1）编程格式

```
G74 R(e )
G74 X(U)__ Z(W)__ I __ K __ D __ F __
```

式中 X——*B* 点 *X* 坐标；
U——*A*→*B* 的增量值；
Z——*C* 点的 *Z* 坐标；
W——*A*→*C* 的增量值；
I——*X* 方向的移动量（无符号指定）；
K——*Z* 方向的切削量（无符号指定）；
D——切削到终点时的退刀量；
F——进给速度。

如果程序段中 X(U)、I、D 为 0，则为深孔钻加工。

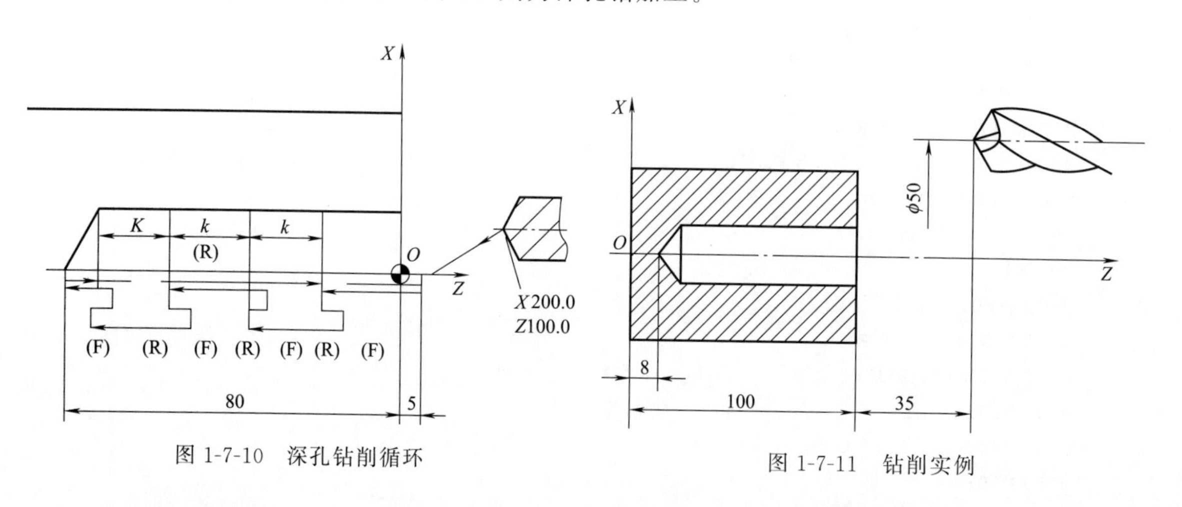


图 1-7-10 深孔钻削循环

图 1-7-11 钻削实例

（2）走刀路线

走刀路线如图 1-7-11 所示。

（3）编程实例

**【例 1-7-4】** 如图 1-7-11 所示，要在车床上钻削直径为 10mm、深为 100mm 的深孔，其程序为：

```
N01 G50 X50. 0 Z100. 0:          建立工件坐标系
N02 G00 X0 Z68. 0;               钻头快速趋近
N03 G74 Z 8. 0 K5. 0 F0. 1 S800; 用 G74 指令钻削循环
N04 G00 X50. 0 Z 100. 0;         刀具快速退至参考点
```

## 四、加工程序（参考）

程序 1

```
O0001;
N1;
G97  G99;
T0101;
M03  S700;
G40  G00  X35.0  Z5.0;
G71  U1.5  R0.5;
G71  P10  Q20  U0.5  W0  F0.2;
N10  G00  X0;
G01  Z0;
X7.8 ，C1.0;
Z-12.0;
X12.0,C0.5;
Z-13.0;
N20  X35.0;
G00  X100.0  Z100.0;
M05;
M00;
N2;
T0202;
M03  S900;
G42  G00  X35.0  Z5.0;
G70  P10  Q20  F0.1;
G00  X100.0  Z100.0;
M05;
M00;
N3;
T0303;
M03  S400;
G40  G00  X35.0  Z5.0 F0.07;
G00  Z-12.0;
G01  X6.6;
G04  X1.0;
G01  X35.0;
G00  X100.0  Z100.0;
M05;
M00;
N4;
T0404;
M03  S500;
G40  G00  X35.0  Z5.0;
G92  X7.4  Z-10.5  F1.25;
X7.0;
```

```
X6.7;
X6.5;
X6.3;
X6.3;
G00  X100.0  Z100.0;
M30;
```

程序 2

```
O0001;
N1;
G97  G99;
T0202;
M03  S700;
G40  G00  X35.0  Z5.0;
G73  U15.  W0  R15;
G73  P10  Q20  U0.5  W0  F0.15;
N10  G00  X0;
G01  Z0;
G03  X10.83  Z-3.42  R6.0;
G03  X18.47  W-28.73  R47.16;
G02  X12.0  W-35.85  R200.0;
G01  W-12.0;
N20  X35.0;
G00 X100.0  Z100.0;
M05;
M00;
N2;
T0202;
M03  S900;
G42  G00  X35.0  Z5.0;
G70  P10  Q20  F0.1;
G00  X100.0  Z100.0;
M30;
```

## 五、加工前准备

### 1. 机床准备（见表 1-7-3）

表 1-7-3　机床准备卡片

| 项目 | 机械部分 | | | | 电气部分 | | 数控系统部分 | | | 辅助部分 | |
|---|---|---|---|---|---|---|---|---|---|---|---|
| 设备检查 | 主轴部分 | 进给部分 | 刀架部分 | 尾座 | 主电源 | 冷却风扇 | 电气元件 | 控制部分 | 驱动部分 | 冷却 | 润滑 |
| 检查情况 | | | | | | | | | | | |

注：经检查后该部分完好，在相应项目下打“√”；若出现问题及时报修。

### 2. 其他注意事项

① 安装外圆刀时，主偏角为 90°～93°。

② 安装切断刀时，主切削刃要与主轴轴线平行。

**3. 参数设置**

① 对刀的数值应输入在与程序中该刀具相对应的刀补号中。

② 在对刀的数值中应注意输入刀尖半径值和假想刀尖的位置序号。

## 六、实际零件加工

**1. 教师演示**

① 工件的装夹、找正。

② 工件的测量方法。

**2. 学生加工训练**

训练中，指导教师巡回指导，及时纠正不正确的操作姿势，解决学生练习中出现的各种问题。

## 七、零件测量

（1）检查零件的外圆尺寸和长度尺寸，检查表面粗糙度 $Ra1.6\mu m$

用一级精度的外径千分尺对每个外圆尺寸进行测量，根据测量结果和被测外圆的公差要求判断被测外圆是否合格，再旋转主轴 90°重新测量一次。测量时注意千分尺的使用方法：应使千分尺的测量头轻轻接触被测外圆表面，旋转千分尺的微调棘轮响两三下，且旋转微调棘轮时同时沿外圆表面摆动千分尺的可活动测量头，找到被测外圆处的最大尺寸。

检查表面粗糙度，用表面粗糙度比较样本进行比较验定。

（2）检查宽槽的尺寸，检查表面粗糙度 $Ra3.2\mu m$

用内径千分尺对槽宽进行测量，尺寸 $3\pm0.03$ 可用公法线千分尺测量，根据测量结果和被测轴的公差要求判断被测轴是否合格。

检查表面粗糙度，用表面粗糙度比较样本进行比较验定。

（3）外螺纹检测

外锥面采用螺纹环规检测。

## 八、加工误差分析及后续处理

教学策略：学生反馈、讲授法、提问法。

针对学生出现加工误差并及时反馈的情况，教师进行集中汇总，针对出现的较多情况采用讲授的方法来指导学生了解出现的原因；对于出现概率不多或没有出现的情况，教师采用提问的方法引导学生自主分析加工误差产生的原因。

在数控车床上进行加工时经常遇到的加工误差有多种，其问题现象、产生的原因、预防和消除的措施见表 1-7-4。

**表 1-7-4 加工误差及后续处理**

| 问题现象 | 产生原因 | 预防和消除 |
|---|---|---|
| 切削过程出现振动 | ① 工件装夹不正确<br>② 刀具安装不正确<br>③ 切削参数不正确 | ① 检查工件安装，增加安装刚性<br>② 调整刀具安装位置<br>③ 提高或降低切削速度 |

续表

| 问题现象 | 产生原因 | 预防和消除 |
| --- | --- | --- |
| 表面质量差 | ① 切削速度不当<br>② 刀具中心过低<br>③ 切屑控制较差<br>④ 刀尖产生积屑瘤<br>⑤ 切削液选用不合理 | ① 调整主轴转速<br>② 调整刀具中心高度<br>③ 选择合理的刀具前角,进刀方式及切深<br>④ 选择合适的切削液并充分喷注 |
| 掉头加工轴向尺寸误差较大 | ① 主轴有间隙,轴向窜动<br>② 工件装卡不正 | ① 调整主轴间隙<br>② 找正工件 |

## 习题

1. 车孔的两个要解决的关键技术问题是什么?
2. 孔加工比车削外圆要困难得多，主要有哪些特点?
3. 在用 G71 进行内孔编程应该注意些什么?
4. 莫氏锥柄如何打开?
5. 麻花钻钻孔时，尾座应该如何操作?
6. 机床报警应如何寻找问题所在?
7. 如何给机床做好保养工作?

# 项目二　数控铣加工

【项目描述】

铣削加工是机械加工中最常用的加工方法之一，主要包括平面铣削和轮廓铣削，也可以对零件进行钻、扩、铰、镗、锪加工及螺纹加工等。

【能力目标】

1. 了解数控铣床的分类、功能、刀具，掌握数控镗铣削加工工艺分析。
2. 熟悉 FANUC 0i MC 数控铣系统操作面板及各按钮功能。
3. 熟悉数控铣床快速定位指令 G00 和直线插补指令 G01 的使用方法。
4. 熟悉数控铣床圆弧插补指令 G02、G03 的使用方法。
5. 掌握数控铣床对刀指令 G92 和定义坐标系 G54～G59 的使用方法。
6. 掌握数控铣床半径补偿和长度补偿的使用方法。
7. 掌握数控铣床中子程序调用的使用方法。
8. 掌握顺铣和逆铣的特点。
9. 掌握数控铣各个按键功能使用技能。
10. 掌握数控铣较复杂零件的加工技能。

## 任务一　支撑架的加工

### 一、图样与技术要求

见图 2-1-1、图 2-1-2、表 2-1-1、表 2-1-2。

图 2-1-1　支撑架实物

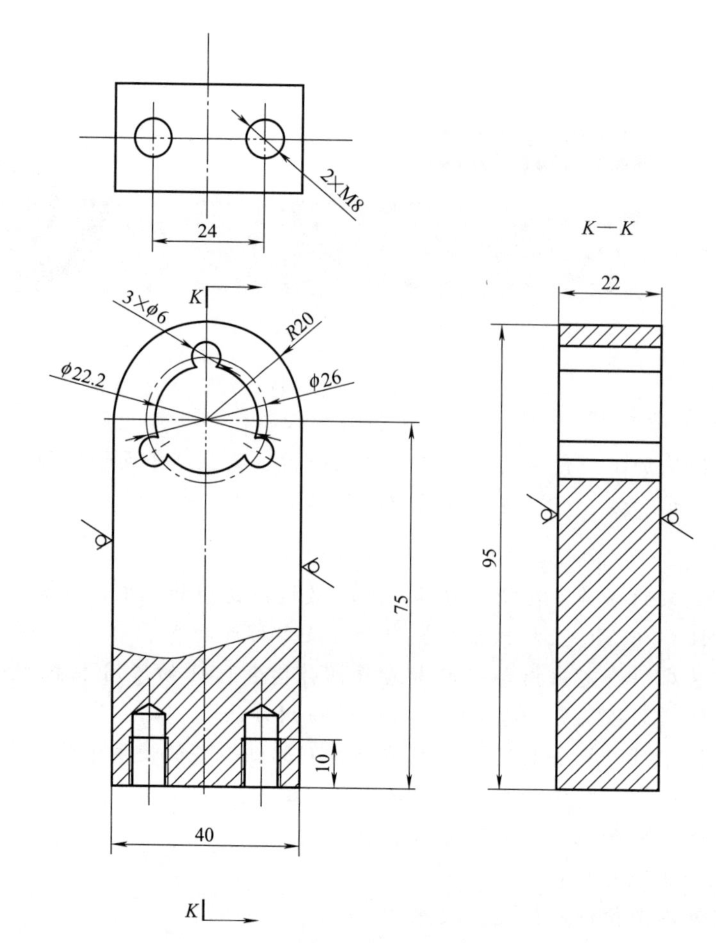


图 2-1-2　支撑架

**表 2-1-1　训练用具清单**

| 序号 | 类别 | 名称 | 规格 | 数量 | 备注 |
|---|---|---|---|---|---|
| 1 | 材料 | 6061 | 100mm×42mm×25mm | 2 | |
| 2 | 刀具 | 高速钢立铣刀 | $\phi$12mm | 1支 | |
| | | 麻花钻 | $\phi$6、$\phi$6.8mm | 各1支 | |
| | | 中心钻 | A3 | 1支 | |
| | | 丝锥 | M8 | 1支 | |
| 3 | 夹具 | 精密平口虎钳 | 0～300mm | 1套 | |
| 4 | 工具 | 铣夹头 | 0～13 | 1个 | |
| | | 攻螺纹扳手 | | 1个 | |
| | | 弹簧夹套 | $\phi$12mm | 各1个 | 与刀具配套 |
| | | 平行垫铁 | | 1副 | |
| | | 油石 | | 1支 | |

表 2-1-2　零件检测项目及评分表（配分 100 分）　　得分________

| 序号 | 考核项目 | 考核内容及精度要求 | 配 分 | 评分标准 | 实测结果 | 得分 |
|---|---|---|---|---|---|---|
| 1 | 轮廓尺寸 | 95±0.1 | 15 | 超差不得分 | | |
| 2 | | 40±0.05 | 15 | 超差不得分 | | |
| 3 | | 22±0.05 | 18 | 超差不得分 | | |
| 4 | | *R*20 | 12 | 超差不得分 | | |
| 5 | | $\phi$22.2 | 10 | 超差不得分 | | |
| 6 | | 3×$\phi$6 | 10 | 超差不得分 | | |
| 7 | 其他 | 表面粗糙度 3.2 | 5 | 超差面扣分 | | |
| 8 | | 棱边倒钝 | 2 | 超差全扣 | | |
| 9 | | 图形完整 | 5 | 不完整全扣 | | |
| 10 | | 文明生产 | 8 | 违规操作全扣 | | |
| 11 | | 工时 | | 每超 15 分钟扣 5 分 | | |

## 二、图纸分析

如图 2-1-1、图 2-1-2 所示，该零件主要外形尺寸为 95mm×40mm×22mm，上部为 $\phi22^{+0.03}_{0}$mm 的通孔；底部为 2 个 M8mm，深 10mm 的螺纹孔。

零件形状较简单，最高公差要求是 0.03mm。

### 1. 刀具选择

刀具材料的选择及合理应用是十分重要的，目前切削加工中常用的刀具材料主要有高速钢和硬质合金等材料。本例工件材料为硬铝，刀具选择刃口锋利、直线度好、精度高的高速钢整体立铣刀。

根据图纸，考虑零件的结构，选用切削加工刀具，见表 2-1-3。

表 2-1-3　刀具卡片

| 刀具名称 | 刀具规格 | 材料 | 数量 | 刀具用途 | 备注 |
|---|---|---|---|---|---|
| 立铣刀 | $\phi$12mm | 高速钢 | 1 | 平面加工，轮廓粗、精加工 | |
| 中心钻 | $\phi$3mm | 高速钢 | 1 | 钻中心孔 | |
| 麻花钻 | $\phi$6mm | 高速钢 | 1 | 钻孔 | |
| 麻花钻 | $\phi$6.8mm | 高速钢 | 1 | 钻螺纹底孔 | |

### 2. 切削参数选择

根据加工对象的材质、刀具的材质和规格，查找刀具切削速度、每齿进给量，确定选用刀具的转速、进给速度，参考切削参数见表 2-1-4。

表 2-1-4　切削参数卡片

| 刀具 | 切削速度 $v$/(m/min) | 每刃进给量 $f$/(mm/刃) | 主轴转速 $S$/(r/min) | 进给速度 $F$/(mm/min) | 备注 |
|---|---|---|---|---|---|
| $\phi$12m 立铣刀 | 50 | 0.04 | 1300 | 200 | 粗加工 |
| | 80 | 0.03 | 2100 | 240 | 精加工 |

续表

| 刀具 | 切削速度<br>$v$/(m/min) | 每刃进给量<br>$f$/(mm/刃) | 主轴转速<br>$S$/(r/min) | 进给速度<br>$F$/(mm/min) | 备注 |
|---|---|---|---|---|---|
| $\phi$3mm 中心钻 | 30 | 0.03 | 3200 | 200 | |
| $\phi$6mm 钻头 | 25 | 0.05 | 1400 | 140 | |
| $\phi$6.8mm 钻头 | 25 | 0.05 | 1400 | 140 | |

### 3. 切削深度 $a_p$

切削深度在粗加工时主要受机床和刀具刚度的限制，一般情况下，径向切削量较大时切削深度取（0.6～0.8）$D_{刀}$，否则切削深度可较大一些。

该零件轮廓加工量不大，每个轮廓加工深度按图纸标注尺寸加工即可，不需分层加工。

### 4. 工艺规程安排

零件各个轮廓加工工艺安排如表 2-1-5 所示。

表 2-1-5　零件工序卡片

| 单位 | | 产品名称及型号 | 零件名称 | 零件图号 |
|---|---|---|---|---|
| | | | 简单零件 | |
| 工序 | 程序编号 | 夹具名称 | 使用设备 | 工件材料 |
| 1 | O0001 | 精密平口钳 | VMC850 | LY12 |
| 工步 | 工步内容 | 刀号 | 刀具及切削用量 | 备注 |
| 1 | 铣上表面 | T01 | $\phi$12mm 立铣刀<br>$n$=2100r/min<br>$F$=240mm/min<br>$a_p$=0.3mm | 按精加工方式铣削 |
| 2 | 加工原点设定在工件上表面中心 | T01 | | 采用试切法 |
| 3 | 粗加工 95×40 为主要尺寸的外轮廓，留余量 0.2mm | T01 | $\phi$12mm 立铣刀<br>$n$=1300r/min<br>$F$=200mm/min<br>$a_p$=6.8mm | |
| 4 | 粗加工 $\phi$22 通孔，留余量 0.2mm | T01 | $\phi$12mm 立铣刀<br>$n$=1300r/min<br>$F$=200mm/min<br>$a_p$=14.8mm | |
| 5 | 精加工 95×40 为主要尺寸的外轮廓，至规定尺寸 | T01 | $\phi$12mm 立铣刀<br>$n$=1300r/min<br>$F$=200mm/min<br>$a_p$=4.8mm | |
| 6 | 精加工 $\phi$22 通孔，至规定尺寸 | T01 | $\phi$12mm 立铣刀<br>$n$=2100r/min<br>$F$=240mm/min<br>$a_p$=7mm | |
| 7 | 钻三个 $\phi$6mm 孔的定位中心孔 | T02 | $\phi$3mm 中心钻<br>$n$=3200r/min<br>$F$=200mm/min<br>$a_p$=3mm | |

续表

| 工步 | 工步内容 | 刀号 | 刀具及切削用量 | 备注 |
|---|---|---|---|---|
| 8 | 钻三个 $\phi$6mm，深 22mm 孔 | T03 | $\phi$5.8mm 钻头<br>$n$=1400r/min<br>$F$=140mm/min<br>$a_p$=2mm | |
| 9 | 钻 2 个 $\phi$6.8mm 孔的定位中心孔 | T02 | $\phi$3mm 中心钻<br>$n$=3200r/min<br>$F$=200mm/min<br>$a_p$=3mm | 底面朝上，重新装夹，对刀 |
| 10 | 钻 2 个 $\phi$6.8mm，深 14mm 螺纹底孔 | T03 | $\phi$5.8mm 钻头<br>$n$=1400r/min<br>$F$=140mm/min<br>$a_p$=2mm | |

### 5. 加工程序

```
O0001(主程序)
G69  G40;
G28  G91  Z0;
G54  G90  G00  X0  Y0;
Z100;
M03  S1500;
X0  Y0;
Z5;
G01  Z-5   F100;
G01  G41  D01  X-11.1  Y0;
G02  X-11.1  Y0  I11.1  J0;
G40  X0  Y0;
M98  P0002  L3;
G01  Z-22  F100;
G01  G41  D01  X-11.1  Y0;
G02  X-11.1  Y0;
G40  X0  Y0;
G00  Z100;
M05;
M30;
O0002(子程序)
Z5;
G01  Z-5  F100;
G01  G41  D01  X-11.1  Y0;
G02  X-11.1   Y0  I11.1  J0;
G40  X0  Y0;
Z5;
M99;
```

## 三、数控铣基础知识

### 1. 数控铣床的分类

数控铣床可以分为立式数控铣床、卧式数控铣床、复合式数控铣床和龙门式数控铣床。

(1) 立式数控铣床

立式数控铣床主轴轴线垂直于水平面，如图 2-1-3 所示。主要用于机械零件类的平面、内外轮廓、孔、攻螺纹等以及各类模具的加工。目前数控铣床中三坐标立式数控铣床占有很大的比例，一般可进行三坐标联动加工。

(2) 卧式数控铣床

卧式数控铣床主轴的轴线平行于水平面，如图 2-1-4 所示。为了扩大加工范围和扩充功能，卧式数控铣床通常采用增加数控转盘（万能数控转盘）来实现四、五坐标加工。这样既可以加工工件侧面的连续回转轮廓，又可以实现在一次安装中通过转盘改变工位，进行“四面加工”。卧式数控铣床主要适用于箱体类机械零件的加工。

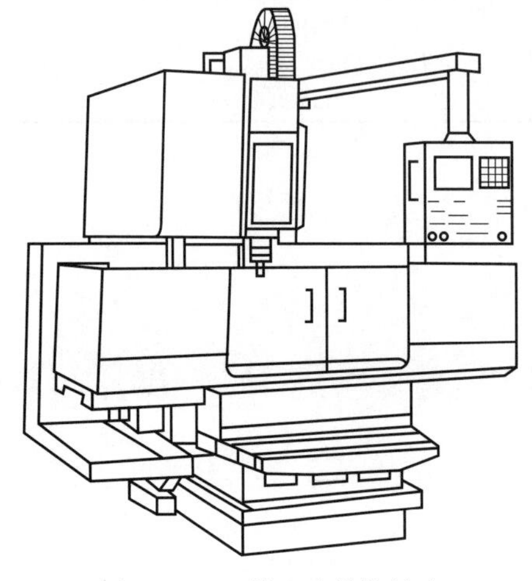

图 2-1-3　三轴立式数控铣床

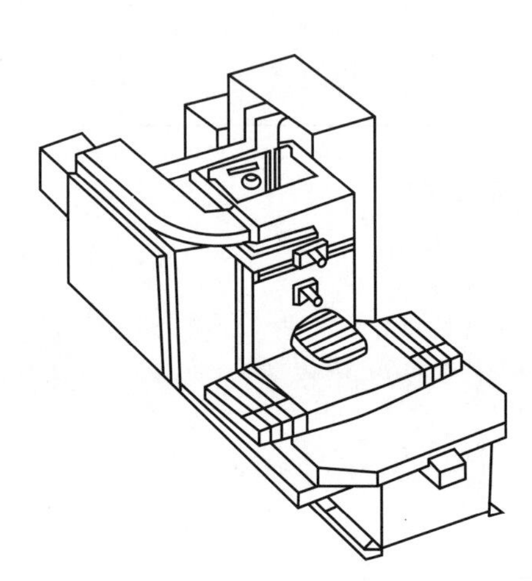

图 2-1-4　卧式数控铣床

(3) 复合式数控铣床

复合式数控铣床是指一台机床上有立式和卧式两个主轴，或者主轴可做 90°旋转的数控铣床，同时具备立、卧式铣床的功能。如图 2-1-5 所示为具有立式和卧式两个主轴的复合式数控铣床。

复合式数控铣床主要用于箱体类零件以及各类模具的加工。

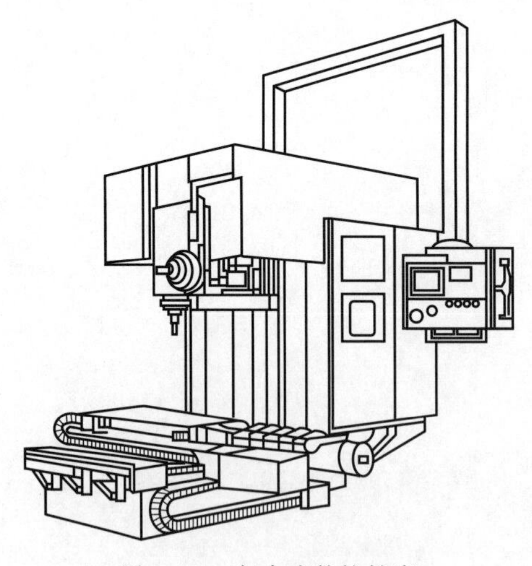

图 2-1-5　复合式数控铣床

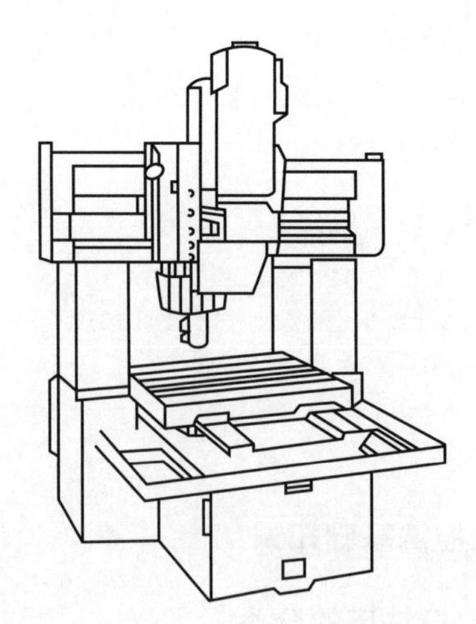

图 2-1-6　龙门式数控铣床

（4）龙门式数控铣床

龙门式数控铣床主轴固定于龙门架上，如图 2-1-6 所示。龙门式数控铣床主要用于大型机械零件及大型模具的加工。

**2. 数控铣床的主要功能**

各种类型数控铣床所配置的数控系统虽然各有不同，但各种数控系统的功能，除一些特殊功能不尽相同外，其主要功能基本相同。

① 点位控制功能　此功能可以实现对相互位置精度要求很高的孔系加工。

② 连续轮廓控制功能　此功能可以实现直线、圆弧的插补功能及非圆曲线的加工。

③ 刀具半径补偿功能　此功能可以根据零件图样的标注尺寸来编程，而不必考虑所用刀具的实际半径尺寸，从而减少编程时的复杂数值计算。

④ 刀具长度补偿功能　此功能可以自动补偿刀具的长短，以适应加工中对刀具长度尺寸调整的要求。

⑤ 比例及镜像加工功能　比例功能可将编好的加工程序按指定比例改变坐标值来执行。镜像加工又称轴对称加工，如果一个零件的形状对坐标轴对称，那么只要编出一个或两个象限的程序，而其余象限的轮廓就可以通过镜像加工来实现。

⑥ 旋转功能　该功能可将编好的加工程序在加工平面内旋转任意角度来执行。

⑦ 子程序调用功能　有些零件需要在不同的位置上重复加工同样的轮廓形状，将这一轮廓形状的加工程序作为子程序，在需要的位置上重复调用，就可以完成对该零件的加工。

⑧ 宏程序功能　该功能可用一个总指令代表实现某一功能的一系列指令，并能对变量进行运算，使程序更具灵活性和方便性。

**3. 刀具**

数控铣床上所采用的刀具要根据被加工零件的材料、几何形状、表面质量要求、热处理状态、切削性能及加工余量等，选择刚性好、耐用度高的刀具。常见刀具见图 2-1-7。

图 2-1-7　常见刀具

（1）铣刀类型选择

被加工零件的几何形状是选择刀具类型的主要依据。

① 加工曲面类零件时，为了保证刀具切削刃与加工轮廓在切削点相切，而避免刀刃与

工件轮廓发生干涉，一般采用球头刀，粗加工用两刃铣刀，半精加工和精加工用四刃铣刀，如图 2-1-8 所示。

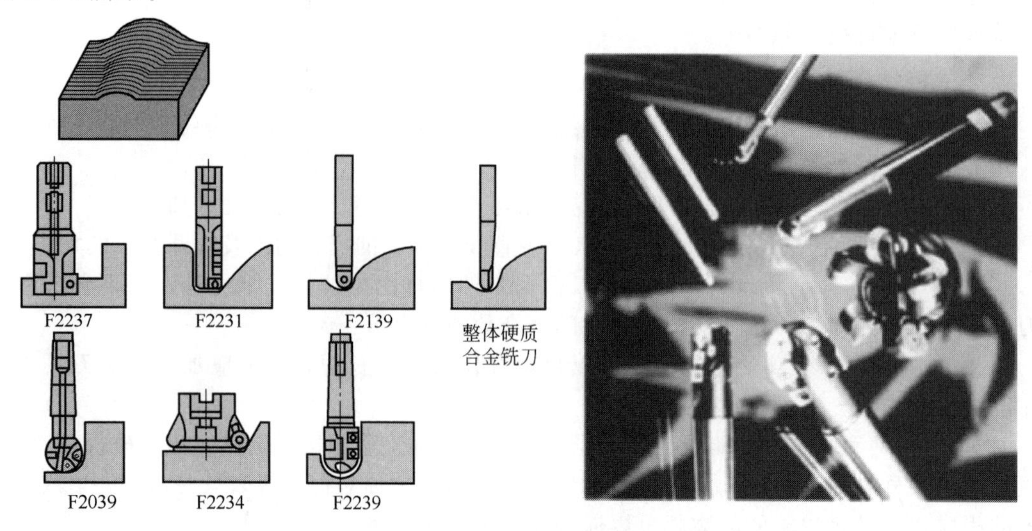


图 2-1-8　加工曲面类铣刀

② 铣较大平面时，为了提高生产效率和提高加工表面粗糙度，一般采用刀片镶嵌式盘形铣刀，如图 2-1-9 所示。

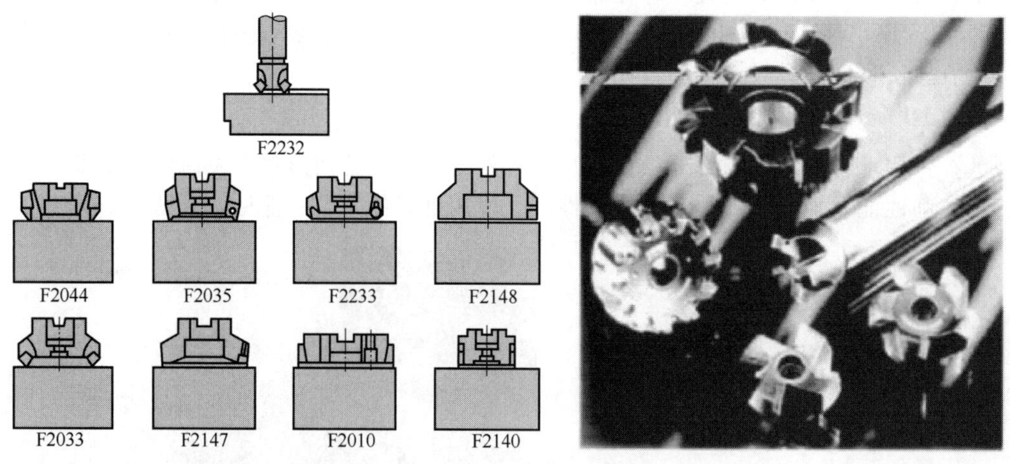


图 2-1-9　加工大平面铣刀

③ 铣小平面或台阶面时一般采用通用铣刀，如图 2-1-10 所示。

④ 铣键槽时，为了保证槽的尺寸精度，一般用两刃键槽铣刀，如图 2-1-11 所示。

⑤ 孔加工时，可采用钻头、镗刀等孔加工类刀具，如图 2-1-12 所示。

（2）铣刀结构选择

铣刀一般由刀片、定位元件、夹紧元件和刀体组成。由于刀片在刀体上有多种定位与夹紧方式，刀片定位元件的结构又有不同类型，因此铣刀的结构形式有多种，分类方法也较多。主要可根据刀片排列方式选用。刀片排列方式可分为平装结构和立装结构两大类。

① 平装结构（刀片径向排列）　平装结构铣刀（如图 2-1-13 所示）的刀体结构工艺性好，容易加工，并可采用无孔刀片（刀片价格较低，可重磨）。由于需要夹紧元件，刀片的一部分被覆盖，容屑空间较小，且在切削力方向上的硬质合金截面较小，故平装结构的铣刀一般用于轻型和中量型的铣削加工。

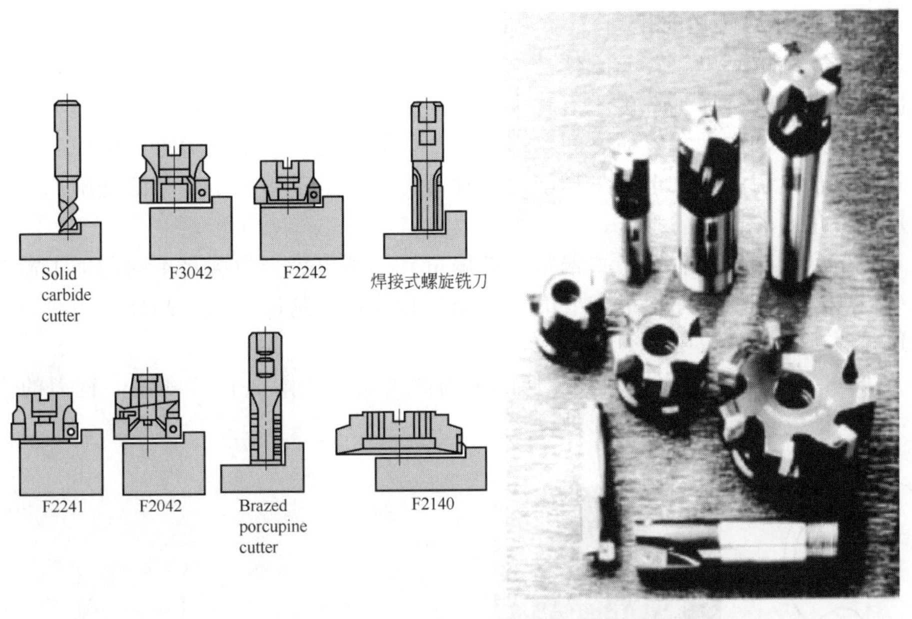


图 2-1-10　加工台阶面铣刀

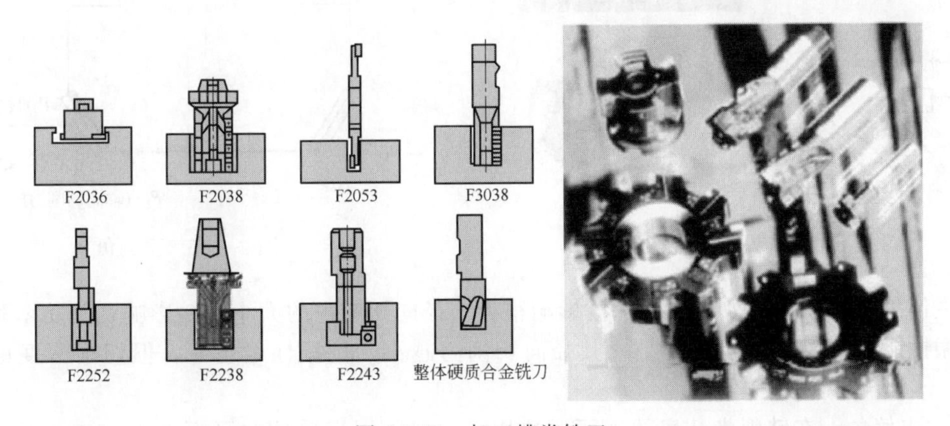


图 2-1-11　加工槽类铣刀

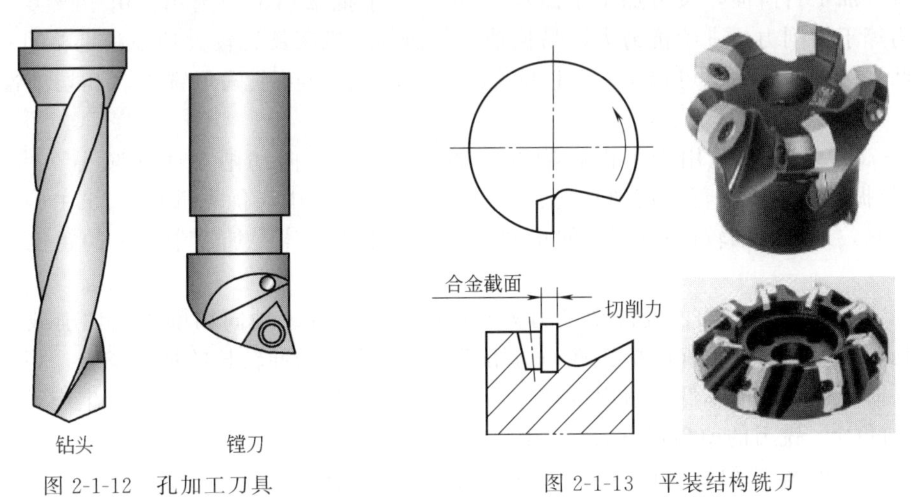


图 2-1-12　孔加工刀具

图 2-1-13　平装结构铣刀

② 立装结构（刀片切向排列） 立装结构铣刀（如图 2-1-14 所示）的刀片只用一个螺钉固定在刀槽上，结构简单，转位方便。虽然刀具零件较少，但刀体的加工难度较大，一般需用五坐标加工中心进行加工。由于刀片采用切削力夹紧，夹紧力随切削力的增大而增大，因此可省去夹紧元件，增大了容屑空间。由于刀片切向安装，在切削力方向的硬质合金截面较大，因而可进行大切深、大走刀量切削，这种铣刀适用于重型和中量型的铣削加工。

（3）铣刀角度的选择

铣刀的角度有前角、后角、主偏角、副偏角、刃倾角等。为满足不同的加工需要，有多种角度组合形式。各种角度中最主要的是主偏角和前角（制造厂的产品样本中对刀具的主偏角和前角一般都有明确说明）。

① 主偏角 $\kappa_r$ 主偏角为切削刃与切削平面的夹角，如图 2-1-15 所示。铣刀的主偏角有 90°、88°、75°、70°、60°、45°等几种。

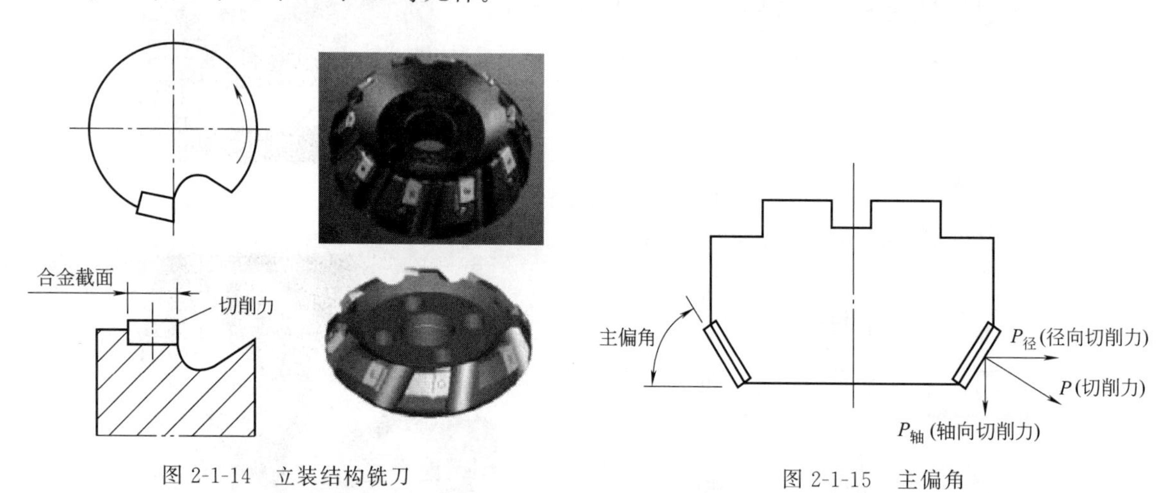


图 2-1-14 立装结构铣刀

图 2-1-15 主偏角

主偏角对径向切削力和切削深度影响很大。径向切削力的大小直接影响切削功率和刀具的抗振性能。铣刀的主偏角越小，其径向切削力越小，抗振性也越好，但切削深度也随之减小。

90°主偏角，在铣削带凸肩的平面时选用，一般不用于单纯的平面加工。该类刀具通用性好（既可加工台阶面，又可加工平面），在单件、小批量加工中选用。由于该类刀具的径向切削力等于切削力，进给抗力大，易振动，因而要求机床具有较大功率和足够的刚性。在加工带凸肩的平面时，也可选用 88°主偏角的铣刀，较之 90°主偏角铣刀，其切削性能有一定改善。

60°～75°主偏角，适用于平面铣削的粗加工。由于径向切削力明显减小（特别是 60°时），其抗振性有较大改善，切削平稳、轻快，在平面加工中应优先选用。75°主偏角铣刀为通用型刀具，适用范围较广；60°主偏角铣刀主要用于镗铣床、加工中心上的粗铣和半精铣加工。

45°主偏角，此类铣刀的径向切削力大幅度减小，约等于轴向切削力，切削载荷分布在较长的切削刃上，具有很好的抗振性，适用于镗铣床主轴悬伸较长的加工场合。用该类刀具加工平面时，刀片破损率低，耐用度高；在加工铸铁件时，工件边缘不易产生崩刃。

② 前角 $\gamma$ 铣刀的前角可分解为径向前角 $\gamma_f$ [图 2-1-16（a）]和轴向前角 $\gamma_p$ [图 2-1-16（b）]，径向前角 $\gamma_f$ 主要影响切削功率；轴向前角 $\gamma_p$ 则影响切屑的形成和轴向力的方向，当 $\gamma_p$ 为正值时切屑即飞离加工面。径向前角 $\gamma_f$ 和轴向前角 $\gamma_p$ 正负的判别见图 2-1-16。常用

的前角组合形式如下。

a. 双负前角。双负前角的铣刀通常均采用方形（或长方形）无后角的刀片，刀具切削刃多（一般为 8 个），且强度高、抗冲击性好，适用于铸钢、铸铁的粗加工。由于切屑收缩比大，需要较大的切削力，因此要求机床具有较大功率和较高刚性。由于轴向前角为负值，切屑不能自动流出，当切削韧性材料时易出现积屑瘤和刀具振动。

凡能采用双负前角刀具加工时建议优先选用双负前角铣刀，以便充分利用和节省刀片。当采用双正前角铣刀产生崩刃（即冲击载荷大）时，在机床允许的条件下亦应优先选用双负前角铣刀。

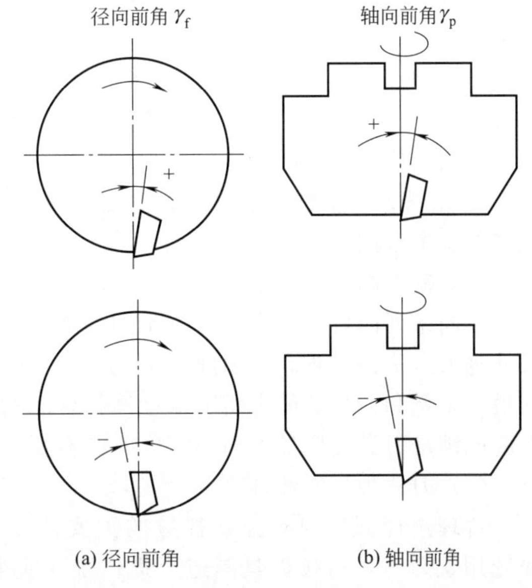


图 2-1-16　前角

b. 双正前角。双正前角铣刀采用带有后角的刀片，这种铣刀楔角小，具有锋利的切削刃。由于切屑收缩比小，所耗切削功率较小，切屑呈螺旋状排出，不易形成积屑瘤。这种铣刀最宜用于软材料和不锈钢、耐热钢等材料的切削加工。对于刚性差（如主轴悬伸较长的镗铣床）、功率小的机床和加工焊接结构件时，也应优先选用双正前角铣刀。

c. 正负前角（轴向正前角、径向负前角）。这种铣刀综合了双正前角和双负前角铣刀的优点，轴向正前角有利于切屑的形成和排出；径向负前角可提高刀刃强度，改善抗冲击性能。此种铣刀切削平稳、排屑顺利、金属切除率高，适用于大余量铣削加工。WALTER 公司的切向布齿重切削铣刀 F2265 就是采用轴向正前角、径向负前角结构的铣刀。

（4）铣刀齿数（齿距）的选择

铣刀齿数多，可提高生产效率，但受容屑空间、刀齿强度、机床功率及刚性等的限制，不同直径的铣刀的齿数均有相应规定。为满足不同用户的需要，同一直径的铣刀一般有粗齿、中齿、密齿三种类型。

① 粗齿铣刀　适用于普通机床的大余量粗加工和软材料或切削宽度较大的铣削加工；当机床功率较小时，为使切削稳定，也常选用粗齿铣刀。

② 中齿铣刀　系通用系列，使用范围广泛，具有较高的金属切除率和切削稳定性。

③ 密齿铣刀　主要用于铸铁、铝合金和有色金属的大进给速度切削加工。在专业化生产（如流水线加工）中，为充分利用设备功率和满足生产节奏要求，也常选用密齿铣刀（此时多为专用非标铣刀）。

为防止工艺系统出现共振，使切削平稳，还有一种不等分齿距铣刀。如 WALTER 公司的 NOVEX 系列铣刀均采用了不等分齿距技术。在铸钢、铸铁件的大余量粗加工中建议优先选用不等分齿距的铣刀。

（5）铣刀直径的选择

铣刀直径的选用视产品及生产批量的不同差异较大，刀具直径的选用主要取决于设备的规格和工件的加工尺寸。

① 平面铣刀　选择平面铣刀直径时主要需考虑刀具所需功率应在机床功率范围之内，

也可将机床主轴直径作为选取的依据。平面铣刀直径可按 $D=1.5d$（$d$ 为主轴直径）选取。在批量生产时，也可按工件切削宽度的 1.6 倍选择刀具直径。

② 立铣刀　立铣刀直径的选择主要应考虑工件加工尺寸的要求，并保证刀具所需功率在机床额定功率范围以内。如系小直径立铣刀，则应主要考虑机床的最高转速能否达到刀具的最低切削速度（60m/min）。

③ 槽铣刀　槽铣刀的直径和宽度应根据加工工件尺寸选择，并保证其切削功率在机床允许的功率范围之内。

（6）铣刀的最大切削深度

不同系列的可转位面铣刀有不同的最大切削深度。最大切削深度越大的刀具所用刀片的尺寸越大，价格也越高，因此从节约费用、降低成本的角度考虑，选择刀具时一般应按加工的最大余量和刀具的最大切削深度选择合适的规格。当然，还需要考虑机床的额定功率和刚性应能满足刀具使用最大切削深度时的需要。

（7）刀片牌号的选择

合理选择刀片硬质合金牌号的主要依据是被加工材料的性能和硬质合金的性能。一般选用铣刀时，可按刀具制造厂提供加工的材料及加工条件，来配备相应牌号的硬质合金刀片。

由于各厂生产的同类用途硬质合金的成分及性能各不相同，硬质合金牌号的表示方法也不同，为方便用户，国际标准化组织规定，切削加工用硬质合金按其排屑类型和被加工材料分为三大类：P 类、M 类和 K 类。根据被加工材料及适用的加工条件，每大类中又分为若干组，用两位阿拉伯数字表示，每类中数字越大，其耐磨性越低、韧性越高。

P 类合金（包括金属陶瓷）用于加工产生长切屑的金属材料，如钢、铸钢、可锻铸铁、不锈钢、耐热钢等。其中，组号越大，则可选用越大的进给量和切削深度，而切削速度则应越小。

M 类合金用于加工产生长切屑和短切屑的黑色金属或有色金属，如钢、铸钢、奥氏体不锈钢、耐热钢、可锻铸铁、合金铸铁等。其中，组号越大，则可选用越大的进给量和切削深度，而切削速度则应越小。

K 类合金用于加工产生短切屑的黑色金属、有色金属及非金属材料，如铸铁、铝合金、铜合金、塑料、硬胶木等。其中，组号越大，则可选用越大的进给量和切削深度，而切削速度则应越小。

上述三类牌号的选择原则如表 2-1-6 所示。

**表 2-1-6　P、M、K 类合金切削用量的选择**

| P 类 | P01 | P05 | P10 | P15 | P20 | P25 | P30 | P40 | P50 |
|---|---|---|---|---|---|---|---|---|---|
| M 类 | M10 | M20 | M30 | M40 | | | | | |
| K 类 | K01 | K10 | K20 | K30 | K40 | | | | |
| 进给量 | → | | | | | | | | |
| 背吃刀量 | → | | | | | | | | |
| 切削速度 | ← | | | | | | | | |

各厂生产的硬质合金虽然有各自编制的牌号，但都有对应国际标准的分类号，选用十分方便。

### 4. 数控铣床的加工工艺范围

铣削加工是机械加工中最常用的加工方法之一，它主要包括平面铣削和轮廓铣削，也可以对零件进行钻、扩、铰、镗、锪加工及螺纹加工等。数控铣削主要适合于下列几类零件的加工。

① 平面类零件，见图 2-1-17。

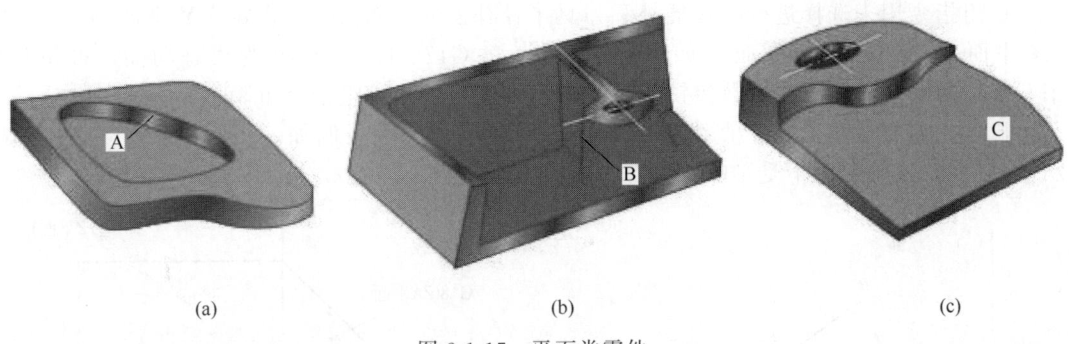


图 2-1-17　平面类零件

② 直纹曲面类零件，见图 2-1-18。

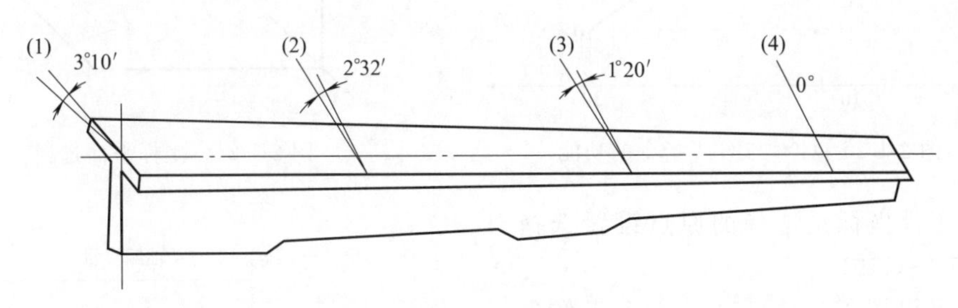


图 2-1-18　直纹曲面类零件

③ 立体曲面类零件，见图 2-1-19。

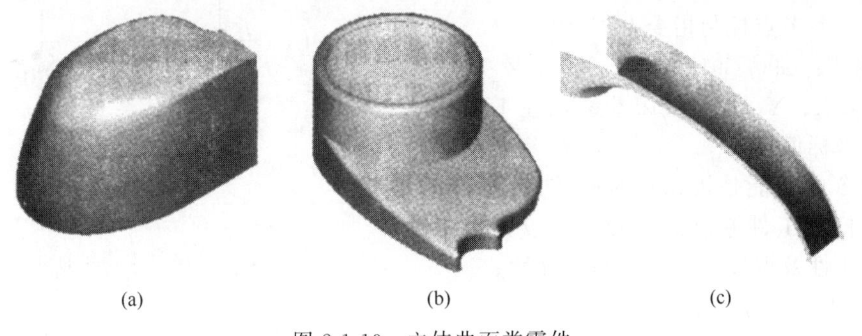


图 2-1-19　立体曲面类零件

## 四、工件坐标系设定

（1）绝对值指令和相对值指令

G90 为绝对值编程指令，表示程序段中给出的刀具运动坐标尺寸为绝对坐标值，即给出的坐标值相对于坐标原点。

G91 为相对值编程指令，表示程序段中给出的刀具运动坐标尺寸为增量坐标值。如图 2-1-20 所示，若刀具从 $A$ 点沿直线运动到 $B$ 点，则：

用绝对值方式编程时，程序段为 G90 G01 X10.0 Y20.0；

用增量值方式编程时，程序段为 G91 G01 X-20.0 Y15.0；

G90、G91 为模态功能，可相互注销，G90 为缺省值。

（2）坐标平面指定指令（G17，G18，G19）

该组指令用来选择进行圆弧插补和刀具半径补偿的平面。G17 指定 $XY$ 平面，G18 指定 $ZX$ 平面，G19 指定 $YZ$ 平面，如图 2-1-21 所示。G17、G18、G19 为模态功能，可相互注销，G17 为缺省值。故立式数控铣床（含数控加工中心）该组指令可隐含不写。

此外，需要注意的是：直线移动指令与平面选择无关。例如，当执行指令：G17 G01 Z10.0；时，$Z$ 轴移动不受影响。

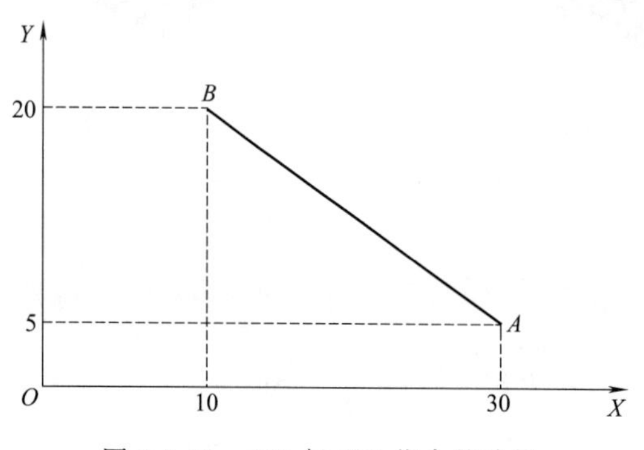


图 2-1-20　G90 与 G91 指令的功能

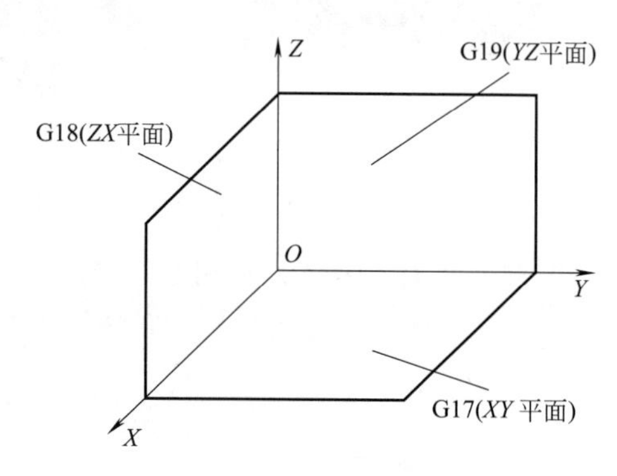


图 2-1-21　坐标平面选择

（3）工件坐标系选择的原点设置选择指令（G54～G59）

工件可设置 G54～G59 共六个工作坐标系原点，如图 2-1-22。工作原点数据值可通过对刀操作后，预先输入机床的偏置寄存器中，编程时不体现。

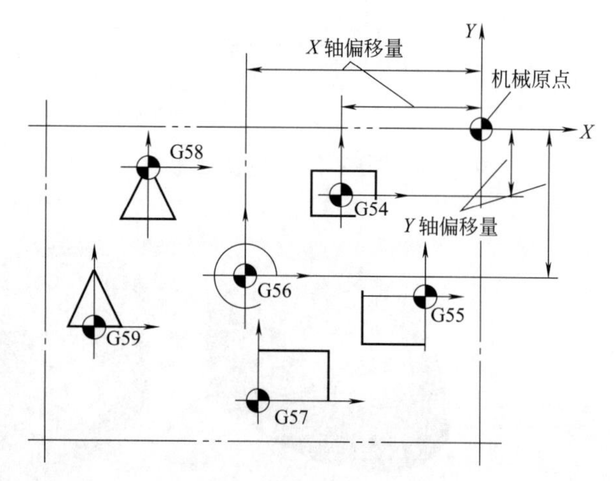


图 2-1-22　设定工作坐标系

（4）回参考点控制指令

① 格式：G90/G91 G27 X __ Y __ Z __；

其中，X、Y、Z 为机床参考点在工件坐标系的坐标值。

该指令的功能是机床长时间连续运转后，用来检查工件原点的正确性，以提高加工的可靠性及保证工件尺寸的正确性。

② 自动返回到参考点 G28

格式：G90/G91 G28 X __ Y __ Z __；

其中，X、Y、Z 为指令的终点位置。

该指令的终点称之为“中间点”，而非参考点。在 G90 时为终点在工件坐标系中的坐标；在 G91 时为终点相对于起点的位移量。由该指令指定的轴能够自动地定位到参考点上。

例：图 2-1-23 中 G28 的使用如下：

G91 G28 X100. Y150. ；

G90 G28 X300. Y250. ;

③ 自动从参考点返回 G29

格式：G90/G91 G29 X _ Y _ Z _

其中，X、Y、Z 为指令的定位终点。

在 G90 时为终点在工件坐标系中的坐标；在 G91 时为终点相对于中间点的位移量。由此功能可使刀具从参考点经由一个中间点而定位于指定点。通常该指令紧跟在一个 G28 指令之后。用 G29 的程序段的动作，可使所有被指令的轴以快速进给经由以前用 G28 指令定义的中间点，然后再到达指定点。

G29 指令仅在其被规定的程序段中有效。

例：图 2-1-24 中 G29 的使用如下：

```
M06 T02;
……
G90 G28 Z50.0;
M06 T03;
G29 X35. Y30. Z5.;
```

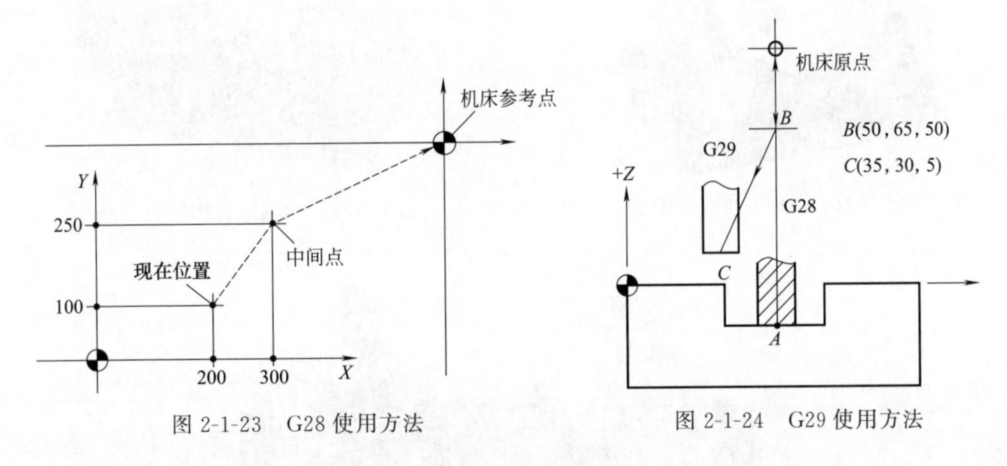


图 2-1-23　G28 使用方法　　图 2-1-24　G29 使用方法

## 五、常用夹具种类

常用的夹具包括通用夹具、组合夹具、专用夹具和可调整夹具等。

### 1. 通用夹具

通用夹具一般为可装夹各种零件的机床附件或通用装夹工具，如各种虎钳、分度头、花盘和三爪卡盘等通用夹具。如图 2-1-25 所示为单轴分度盘，一次安装工件，可从四面进行加工。如图 2-1-26 所示为单轴数控回转台（座），可以一次安装工件，而从四面进行加工，又可形成五轴联动加工，如加工圆柱凸轮的空间成形面和平面凸轮等。如图 2-1-27 所示为两轴分度盘，可用于加工在表面上成不同角度布置的孔等，可做五个方向和倾斜方向的加工。

如图 2-1-28 所示为两轴数控回转台（座），既可做五个方向的加工，又可做五轴联动加工。如图 2-1-29 所示为带托盘交换的两轴可倾斜数控转台，此转台的托盘尺寸为 1000mm×1000mm，最大负载为 6000kg，允许转动惯量 $J$ 为 2000kg·m$^2$，其 A 轴倾斜角范围为 −115°（CCW）～5°（CW），最大转速为 6r/min，同轴度为 0.01mm，轴向跳动为 0.02mm，定位精度为 ±10″。$B$ 轴（回转轴）台面尺寸为 1000mm×1000mm，最大转速为 6r/min，同轴度为 0.01mm，轴向跳动为 0.015mm，定位精度为 ±7″。

图 2-1-25　单轴分度盘

图 2-1-26　单轴数控回转台（座）

图 2-1-27　两轴分度盘

图 2-1-28　两轴数控回转台（座）

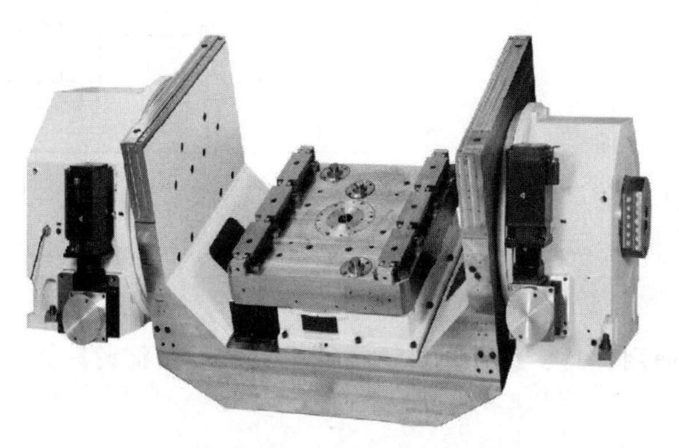

图 2-1-29　带托盘交换的两轴可倾斜数控转台

**2. 组合夹具**

组合夹具由一套结构已经标准化、尺寸已经规格化的通用元件和组合元件构成，可以按工件的加工需要组成各种功用的夹具。组合夹具可以分为孔系组合夹具和槽系组合夹具，如图 2-1-30 所示为槽系组合夹具。组合夹具一般可以满足标准化、系列化、通用化要求，具有组合性、可调性、模拟性、柔性、应急性和经济性的优点，使用寿命长，能适应产品加工中的周期短、成本低等要求，比较适合加工中心应用。

但是，组合夹具是由各种通用标准元件组合而成的，各元件间相互配合的环节较多，夹具精度、刚性比不上专用夹具，尤其是元件连接的接合面刚度，对加工精度影响较大。通常，采用组合夹具时其加工尺寸精度只能达到 IT9 和 IT8 级。此外，组合夹具总体显得笨

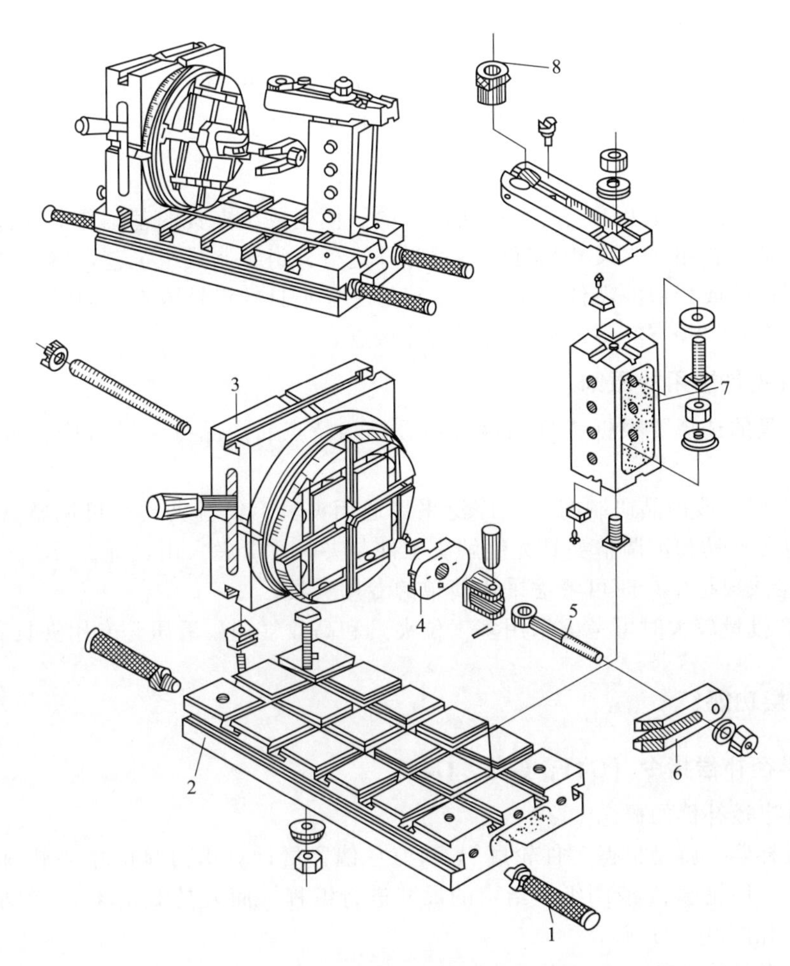


图 2-1-30　槽系组合夹具

1—其他件；2—基础件；3—合件；4—定位件；
5—紧固件；6—压紧件；7—支撑件；8—导向件

重，还有排屑不便等不足。

**3. 专用夹具**

专用夹具是特别为某一项或类似的几项加工专门设计制造的夹具，结构合理，刚性强，装夹稳定可靠，操作方便，安装精度高，装夹速度快。选用这种夹具，一批工件加工后尺寸比较稳定，互换性也较好，可大大提高生产率。但是，专用夹具只能为一种零件的加工所专用，和产品品种不断变型更新的形势不相适应，特别是专用夹具的设计和制造周期长，花费的劳动量较大，加工简单零件经济性差。一般对于工厂的主导产品，批量较大，精度要求较高的关键性零件，在加工中心上加工时，可以选用专用夹具。专用夹具中的夹紧机构一般采用气动或液压夹紧机构，能减轻工人劳动强度和提高生产率。

**4. 可调整夹具**

可调整夹具是组合夹具和专用夹具的结合。可调整夹具能有效地克服以上两种夹具的不足，既能满足加工精度要求，又有一定的柔性。可调整夹具与组合夹具主要不同之处是它具有一系列整体刚性好的夹具体，在夹具体上设置了具有定位、夹紧等多功能的 T 形槽及台阶式光孔、螺孔，配制有多种夹压、定位元件。

**5. 多工位夹具**

多工位夹具可以同时装夹多个工件，减少换刀次数，也便于一边加工，一边装卸工件，有利于缩短辅助时间，提高生产率，较适宜于中批量生产。

**6. 成组夹具**

成组夹具是随成组加工工艺的发展而出现的。使用成组夹具的基础是对零件的分类。通过工艺分析，把形状相似、尺寸相近的各种零件进行分组编制成组工艺，然后把定位、夹紧和加工方法相同的或相似的零件集中起来，统筹考虑夹具的设计方案。对结构外形相似的零件，采用成组夹具，具有经济、夹压精度高等特点。

**7. 数控铣夹具选用的原则**

数控铣夹具的选择要根据零件精度等级、结构特点、产品批量及机床精度等情况综合考虑。

① 在单件生产或产品研制时，应广泛采用通用夹具、组合夹具和可调整夹具，只有在通用夹具、组合夹具和可调整夹具无法解决工件装夹时才考虑采用其他夹具。

② 小批量或成批生产时可考虑采用简单的专用夹具。

③ 在生产批量较大时可考虑采用多工位夹具和高效气动、液压等专用夹具。

## 六、刀具补偿功能

**1. 刀具半径补偿指令（G41，G42，G40）**

（1）刀具半径补偿的概念

刀具半径补偿，就是根据工件轮廓 $A$ 和刀具偏置量计算出刀具的中心轨迹 $B$，这样编程者就以根据工件轮廓 $A$ 或图纸上给定的尺寸进行编程，而刀具沿轮廓 $B$ 运动，加工出所需的轮廓 $A$，如图 2-1-31 所示。

（2）刀具半径补偿的工作过程（见图 2-1-32）

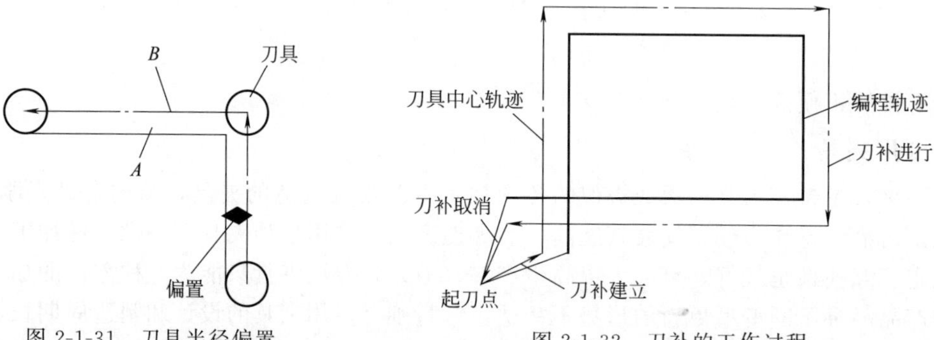


图 2-1-31　刀具半径偏置

图 2-1-32　刀补的工作过程

① 刀补的建立　刀具从起点接近工件时，刀具中心从与编程轨迹重合过渡到与编程轨迹偏移一个偏置量的过程。该过程不应进行零件加工。

② 刀补的执行　刀具的中心轨迹与编程轨迹始终偏移一个刀具偏置量的距离。

③ 刀补的取消　刀具撤离工件，使刀具中心轨迹的终点与编程轨迹的终点重合。它是刀补建立的逆过程，同样该过程不应进行零件的加工。

格式　G41 G00/G01 X _ Y _ D _；

　　　G42 G00/G01 X _ Y _ D _；

G40 G00/G01 X __ Y __ Z __；

G41 为刀具半径左补偿，沿刀具运动方向向前看，刀具位于零件左侧，如图 2-1-33（a）所示。

G42 为刀具半径右补偿，沿刀具运动方向向前看，刀具位于零件右侧，如图 2-1-33（b）所示。

G40 为撤销刀具补偿指令。

D 为控制系统存放刀具半径补偿量寄存器单元的代码（称为刀补号）。

G41、G42、G40 都是模态代码，可相互注销，G40 为缺省值。

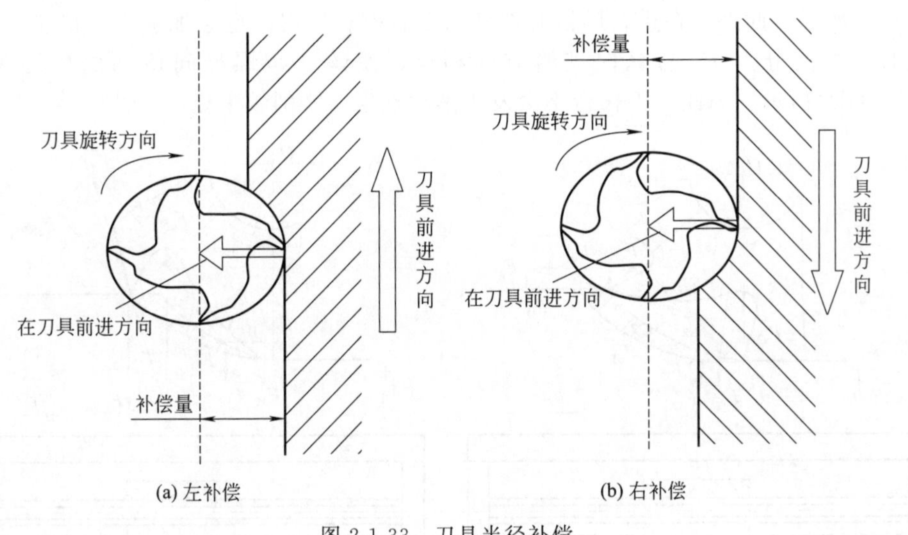


图 2-1-33 刀具半径补偿

注意事项：

① 刀具半径补偿平面的切换必须在补偿取消方式下进行；

② 刀具半径补偿值，由操作者输入到刀具补偿寄存器中；

③ 刀具半径补偿的建立与取消，只能用 G00 或 G01 指令，而不能是 G02 或 G03 指令。所谓刀具半径补偿建立，就是刀具从无半径补偿运动到所希望的刀具半径补偿起点的过程，而刀具半径补偿取消则恰好与此相反。

**2. 逆铣、顺铣**

（1）逆铣与顺铣的概念

铣刀的旋转方向和工件的进给方向相反时称为逆铣，相同时称为顺铣。

（2）逆铣与顺铣的特点

如图 2-1-34（a）所示，逆铣时，刀具从已加工表面切入，切削厚度从零逐渐增大。铣刀刃口有一钝圆半径 $r_\beta$，当 $r_\beta$ 大于瞬时切削厚度时，实际切削前角为负值，刀齿在加工表面上挤压、滑行，不能切削，使这段表面产生严重冷硬层。下一个刀齿切入时，又在冷硬层表面挤压、滑行，刀齿容易磨损，同时使工件表面粗糙度值增大。并且，刀齿在已加工表面处切入工件时，由于切屑变形大，切屑作用在刀具上的力使刀具实际切深加大，可能会产生“挖刀”式的多切，造成后续加工余量不足，这种现象称为“挖刀”现象。“挖刀”现象对大型复杂零件毛坯危害极大，严重时可能造成零件报废。同时刀齿切离工件时垂直方向的分力 $F_{v1}$ 的方向使工件脱离工作台，需较大的夹紧力。逆铣的优点是刀齿从已加工表面切入，不会造成直接从过硬的毛坯面切入而打刀的问题。如

图 2-1-34（b）所示，顺铣时，刀具从待加工表面切入，刀齿的切削厚度从最大开始逐渐降为零，避免了挤压、滑行现象的产生，切屑分离时切削力很小，加工表面处光滑且不会产生“挖刀”，同时垂直方向的分力 $F_{v2}$ 始终压向工作台，减小了工件的上下振动，因而提高了铣刀耐用度和工件加工表面质量。

铣床工作台的纵向进给运动一般是依靠工作台下面的丝杠和螺母来实现的，螺母固定不动，丝杠一边转动，一边带动工作台移动。如果在丝杠与螺母传动副中存在着间隙情况下采用顺铣，当纵向分力 $F_1$ 逐渐增大至超过工作台摩擦力时，工作台带动丝杠向左窜动，丝杠与螺母传动副右侧面出现间隙，造成工作台颤动和进给不均匀，严重时会使铣刀崩刃，如图 2-1-34（c）所示。此外，在进行顺铣时遇到加工表面有硬皮，也会加速刀齿磨损甚至打刀。在逆铣时，纵向分力 $F_1$ 与纵向进给方向相反，使丝杠与螺母间传动面始终紧贴，如图 2-1-34（d）所示。因此，工作台不会发生窜动现象，铣削较平稳。

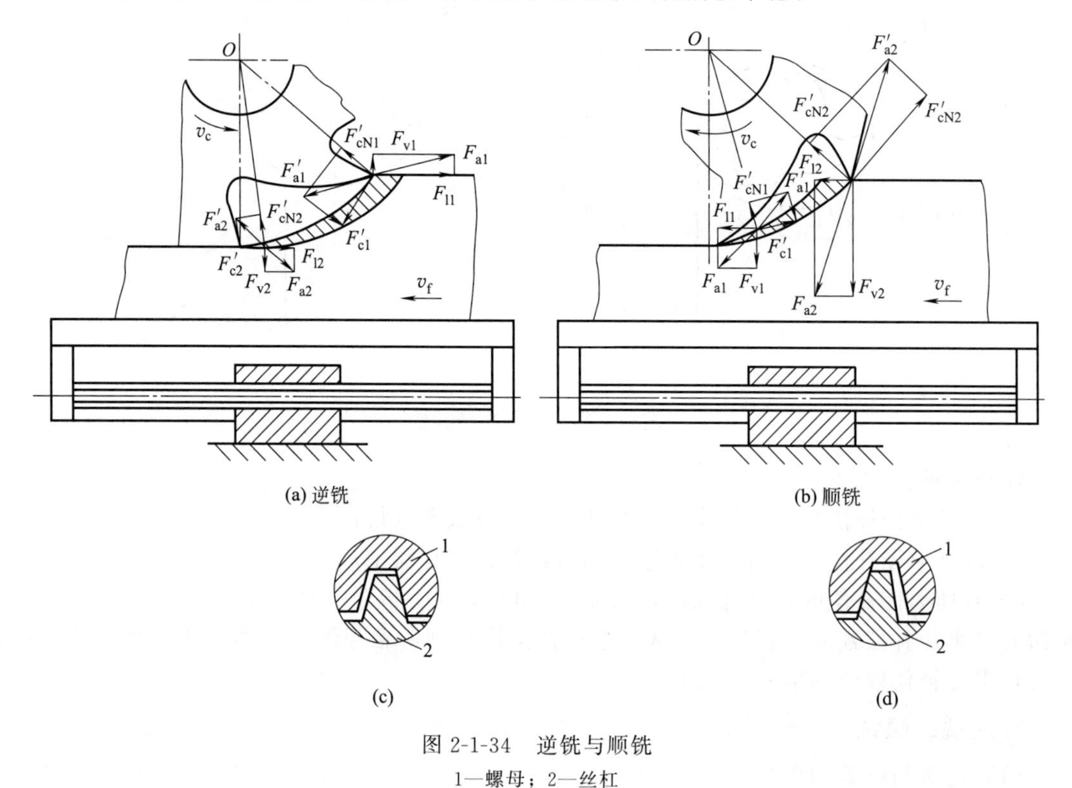


图 2-1-34　逆铣与顺铣

1—螺母；2—丝杠

（3）逆铣、顺铣的确定

根据上面分析，当工件表面有硬皮，机床的进给机构有间隙时，应选用逆铣，按照逆铣方式安排进给路线。逆铣时，刀齿从已加工表面切入，不会崩刃，机床进给机构的间隙不会引起振动和爬行，这正符合粗铣的要求，因此，粗铣时应尽量采用逆铣。当工件表面无硬皮，机床进给机构无间隙时，应选用顺铣，按照顺铣方式安排进给路线。采用顺铣加工后，零件已加工表面质量好，刀齿磨损小，这正符合精铣的要求，因此，精铣时，尤其是零件材料为铝镁合金、钛合金或耐热合金时，应尽量采用顺铣。

但是，这里要强调，数控铣削加工时使用的数控设备基本都采用滚珠丝杠螺母副传动，进给机构一般无间隙，这时，如果加工的毛坯硬度不高，尺寸大，形状复杂，成本大，即使粗加工，也应采用顺铣，这样有利于对减少刀具的磨损，避免粗加工时逆铣可能产生“挖刀”式多切而造成的余量不足。

在主轴正向旋转，刀具为右旋铣刀时，顺铣正好符合左刀补（G41），逆铣正好符合右刀补（G42），因此，一般情况下，精铣用G41建立刀具半径补偿，粗铣用G42建立刀具半径补偿。

**3. 刀具长度补偿指令（G43，G44，G49）**

刀具长度补偿使用格式如下：

```
G43 G00/G01 Z __ H __;
G44 Z __ H __;
G49 Z __;
```

G43为刀长正补，即 $Z$ 坐标实际移动的坐标值为将 $Z$ 坐标尺寸字与刀具长度补偿值相加所得的量，如图2-1-35（a）所示。执行G43时：Z实际值=Z指令值+(H××)。

G44为刀长负补，即 $Z$ 坐标实际移动的坐标值为将 $Z$ 坐标尺寸字与刀具长度补偿值相减所得的量，如图2-1-35（b）所示。执行G44时：Z实际值=Z指令值−(H××)。

H为控制系统存放刀具长度补偿量寄存器单元的代码。

G43，G44，G49都是模态代码，可相互注销，G49为缺省值。

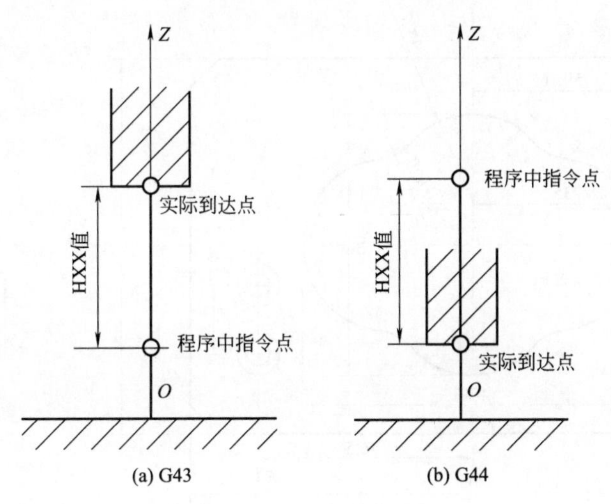


图2-1-35　刀具长度偏置

**习题**

1. 数控铣床的分类有哪些？
2. 数控铣床的主要功能是什么？
3. 常见的数控铣刀有哪几种？
4. 刀片牌号如何选择？
5. 简述数控铣床的加工工艺范围。
6. 绝对值指令和相对值指令的区别是什么？
7. 常用夹具种类有哪些？
8. 刀具半径补偿指令包括什么？
9. 逆铣、顺铣的区别是什么？
10. 刀具长度补偿指令包括哪些？

# 任务二 底板的加工

## 一、图样与技术要求

如图 2-2-1、图 2-2-2、表 2-2-1、表 2-2-2 所示。

图 2-2-1 底板实物

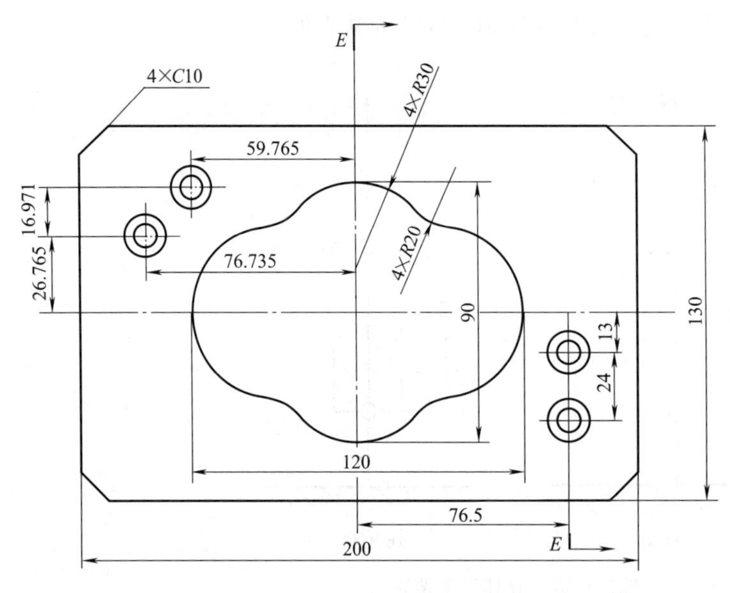


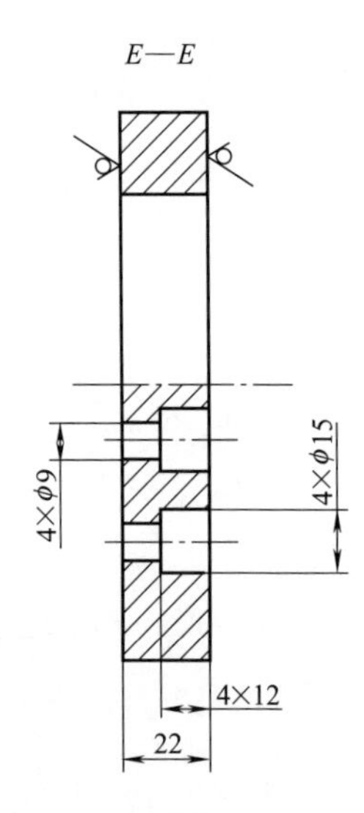


图 2-2-2 底板

**表 2-2-1 训练用具清单**

| 序号 | 类别 | 名称 | 规格 | 数量 | 备注 |
| --- | --- | --- | --- | --- | --- |
| 1 | 材料 | 6061 | 210mm×140mm×30mm | 1 | |
| 2 | 刀具 | 高速钢立铣刀 | $\phi$12mm、$\phi$6mm | 各1支 | |
| | | 麻花钻 | $\phi$9mm | 1支 | |
| | | 中心钻 | A3 | 1支 | |
| 3 | 夹具 | 精密平口虎钳 | 0～300mm | 1套 | |
| 4 | 工具 | 铣夹头 | 0～13 | 1个 | |
| | | 弹簧夹套 | $\phi$12mm、$\phi$6mm | 各1个 | 与刀具配套 |
| | | 平行垫铁 | | 1副 | |
| | | 油石 | | 1支 | |

表 2-2-2 零件检测项目及评分表（配分 100 分） 得分________

| 序号 | 考核项目 | 考核内容及精度要求 | 配 分 | 评分标准 | 实测结果 | 得分 |
|---|---|---|---|---|---|---|
| 1 | 轮廓尺寸 | $200_{-0.05}^{0}$ | 7 | 超差全扣 | | |
| 2 | | $130_{-0.05}^{0}$ | 7 | 超差全扣 | | |
| 3 | | 22±0.02 | 7 | 超差全扣 | | |
| 4 | | $120_{0}^{+0.05}$ | 7 | 超差全扣 | | |
| 5 | | $90_{0}^{+0.05}$ | 6 | 超差全扣 | | |
| 6 | | 76.5±0.02 | 6 | 超差全扣 | | |
| 7 | | $\phi$15(4 处) | 10 | 超差全扣 | | |
| 8 | | $\phi$9(4 处) | 10 | 超差全扣 | | |
| 9 | | 12±0.05(4 处) | 7 | 超差全扣 | | |
| 10 | | 59.7±0.02 | 7 | 超差全扣 | | |
| 11 | | 76.7±0.02 | 6 | 超差全扣 | | |
| 12 | 其他 | 表面粗糙度 3.2μm | 5 | 超差面扣分 | | |
| 13 | | 棱边倒钝 | 5 | 超差全扣 | | |
| 14 | | 图形完整度 | 5 | 不完整全扣 | | |
| 15 | | 文明生产 | 5 | 违规操作全扣 | | |

## 二、图纸分析

通过识图，该零件主要外形尺寸为 210mm×140mm×30mm，中间圆弧内轮廓；底部四个沉头孔，沉头 $\phi$15 深 12mm，通孔 $\phi$9mm 深 10mm。

零件形状较简单，最高公差要求是 0.05mm。

### 1. 刀具选择

刀具材料的选择及合理应用是十分重要的，目前切削加工中常用的刀具材料主要有高速钢和硬质合金等材料。本例工件材料为硬铝，刀具选择刃口锋利、直线度好、精度高的高速钢整体立铣刀。

根据图纸，考虑零件的结构，选用切削加工刀具，见表 2-2-3。

表 2-2-3 刀具卡片

| 刀具名称 | 刀具规格 | 材料 | 数量 | 刀具用途 | 备注 |
|---|---|---|---|---|---|
| 立铣刀 | $\phi$12mm | 高速钢 | 1 | 平面加工，轮廓粗、精加工 | |
| 立铣刀 | $\phi$6mm | 高速钢 | 1 | 沉头部分粗、精加工 | |
| 中心钻 | $\phi$3mm | 高速钢 | 1 | 钻中心孔 | |
| 麻花钻 | $\phi$9mm | 高速钢 | 1 | 钻孔 | |

### 2. 切削参数选择

根据加工对象的材质，刀具的材质和规格，查找刀具切削速度、每齿进给量，确定选用刀具的转速、进给速度，参考切削参数见表 2-2-4。

表 2-2-4 切削参数卡片

| 刀具 | 切削速度 $v$(m/min) | 每刃进给量 $f$/(mm/刃) | 主轴转速 $S$/(r/min) | 进给速度 $F$/(mm/min) | 备注 |
|---|---|---|---|---|---|
| $\phi$12m 立铣刀 | 50 | 0.04 | 1300 | 200 | 粗加工 |
| | 80 | 0.03 | 2100 | 240 | 精加工 |
| $\phi$6m 立铣刀 | 50 | 0.04 | 2000 | 200 | 粗加工 |
| | 80 | 0.03 | 2500 | 240 | 精加工 |
| $\phi$3mm 中心钻 | 30 | 0.03 | 3200 | 200 | |
| $\phi$9mm 钻头 | 25 | 0.05 | 1400 | 140 | |

### 3. 切削深度 $a_p$

切削深度在粗加工时主要受机床和刀具刚度的限制，一般情况下，径向切削量较大时切削深度取（0.6～0.8）$D_刀$，否则切削深度可较大一些。

该零件轮廓加工量不大，每个轮廓加工深度按图纸标注尺寸加工即可，不需分层加工。

### 4. 工艺规程安排

零件各个轮廓加工工艺安排如表 2-2-5。

表 2-2-5 零件工序卡片

| 单位 | | 产品名称及型号 | 零件名称 | 零件图号 |
|---|---|---|---|---|
| | | | 简单零件 | |
| 工序 | 程序编号 | 夹具名称 | 使用设备 | 工件材料 |
| 1 | O0001 | 精密平口钳 | VMC850 | LY12 |
| 工步 | 工步内容 | 刀号 | 刀具及切削用量 | 备注 |
| 1 | 铣上表面 | T01 | $\phi$12mm 立铣刀<br>$n$=2100r/min<br>$F$=240mm/min<br>$a_p$=0.3mm | 按精加工方式铣削 |
| 2 | 加工原点设定在工件上表面中心 | T01 | | 采用试切法 |
| 3 | 粗加工 210×130 为主要尺寸的外轮廓，留余量 0.2mm | T01 | $\phi$12mm 立铣刀<br>$n$=1300r/min<br>$F$=200mm/min<br>$a_p$=6.8mm | |
| 4 | 粗加工圆弧内轮廓，留余量 0.2mm | T01 | $\phi$12mm 立铣刀<br>$n$=1300r/min<br>$F$=200mm/min<br>$a_p$=14.8mm | |
| 5 | 精加工 210×130 为主要尺寸的外轮廓，至规定尺寸 | T01 | $\phi$12mm 立铣刀<br>$n$=1300r/min<br>$F$=200mm/min<br>$a_p$=4.8mm | |
| 6 | 精加工圆弧内轮廓，至规定尺寸 | T01 | $\phi$12mm 立铣刀<br>$n$=2100r/min<br>$F$=240mm/min<br>$a_p$=7mm | |

续表

| 工步 | 工步内容 | 刀号 | 刀具及切削用量 | 备注 |
| --- | --- | --- | --- | --- |
| 7 | 钻四个 $\phi$9mm 孔的定位中心孔 | T02 | $\phi$3mm 中心钻<br>$n$=3200r/min<br>$F$=200mm/min<br>$a_p$=3mm | |
| 8 | 钻三个 $\phi$9mm 通孔 | T03 | $\phi$5.8mm 钻头<br>$n$=1400r/min<br>$F$=140mm/min<br>$a_p$=2mm | |
| 9 | 粗加工沉头孔部分，<br>留余量 0.2mm | T04 | $\phi$6mm 立铣刀<br>$n$=2000r/min<br>$F$=200mm/min<br>$a_p$=14.8mm | |
| 10 | 精加工沉头孔部分，至规定尺寸 | T04 | $\phi$6mm 立铣刀<br>$n$=2000r/min<br>$F$=200mm/min<br>$a_p$=14.8mm | |

**5. 加工程序**

```
O0001(主程序)
G69  G40;
G28  G91  Z0;
G54  G90  G00  X0  Y0;
Z100;
M03  S1500;
X0  Y0;
Z5;
G01  Z-5  F100;
G01  G42  D01  X-60  Y0;
G02  X-33.63  Y29.77  R30;
G03  X-27.12  Y31.74  R20;
G02  X27.12  Y31.74  R30;
G03  X33.63  Y29.77  R20;
G02  X33.63  Y-29.77  R20;
G03  X27.12  Y-31.74  R20;
G02  X27.12  Y-31.74  R30;
G03  X-33.63  Y-29.77  R20;
G02  X-60  Y0  R30;
G01  G40  X0 Y0;
M98  P0002  L3;
G01  Z-22  F100;
G01  G42  D01  X-60  Y0;
G02  X-33.63  Y29.77  R30;
G03  X-27.12  Y31.74  R20;
G02  X27.12  Y31.74  R30;
G03  X33.63  Y29.77  R20;
G02  X33.63  Y-29.77  R20;
G03  X27.12  Y-31.74  R20;
G02  X27.12  Y-31.74  R30;
G03  X-33.63  Y-29.77  R20;
```

```
G02  X-60  Y0  R30;
G01  G40  X0  Y0;
G00  Z100;
M05;
M30;
O0002(子程序)
Z5;
G01  Z-5   F100;
G01  G42  D01  X-60  Y0;
G02  X-33.63  Y29.77  R30;
G03  X-27.12  Y31.74  R20;
G02  X27.12  Y31.74  R30;
G03  X33.63  Y29.77  R20;
G02  X33.63  Y-29.77  R20;
G03  X27.12  Y-31.74  R20;
G02  X27.12  Y-31.74  R30;
G03  X-33.63  Y-29.77  R20;
G02  X-60  Y0  R30;
G01  G40  X0  Y0;
Z5;
M99;
```

## 三、数控铣削加工工艺分析

数控铣床加工的程序是数控铣床的指令性文件。数控铣床受控于程序指令，加工的全过程都是按程序指令自动进行的。因此，数控铣床加工程序与普通铣床工艺规程有较大差别，涉及的内容也较广。数控铣床加工程序不仅要包括零件的工艺过程，而且还要包括切削用量、走刀路线、刀具尺寸以及铣床的运动过程。因此，要求编程人员对数控铣床的性能、特点、运动方式、刀具系统、切削规范以及工件的装夹方法都要非常熟悉。工艺方案的好坏不仅会影响铣床效率的发挥，而且将直接影响到零件的加工质量。

### 1. 数控铣削加工工艺的主要内容

（1）数控铣削加工工艺

① 选择适合在数控铣床上加工的零件，确定工序内容。

② 分析被加工零件的图纸，明确加工内容及技术要求。

③ 确定零件的加工方案，制订数控加工工艺路线。

④ 加工工序的设计。如选取零件的定位基准、夹具方案的确定、工步划分、刀具选择和确定切削用量等。

⑤ 数控加工程序的调整。如选取对刀点和换刀点、确定刀具补偿及确定加工路线等。

（2）选择并确定数控铣削的加工部位及内容

以下几方面适宜采用数控铣削加工。

① 由直线、圆弧、非圆曲线及列表曲线构成的内外轮廓。

② 空间曲线或曲面。

③ 形状虽然简单，但尺寸繁多，检测困难的部位。

④ 用普通机床加工时难以观察、控制及检测的内腔、箱体内部等。

⑤ 有严格位置尺寸要求的孔或平面。

⑥ 能够在一次装夹中顺带加工出来的简单表面或形状。

⑦ 采用数控铣削加工能有效提高生产率，减轻劳动强度的一般加工内容。

### 2. 数控铣削加工零件的工艺性分析

（1）零件图及其结构工艺性分析

① 分析零件的形状、结构及尺寸的特点，确定零件上是否有妨碍刀具运动的部位，是否有会产生加工干涉或加工不到的区域，零件的最大形状尺寸是否超过机床的最大行程，零件的刚性随着加工的进行是否有太大的变化等。

② 检查零件的加工要求，如尺寸加工精度、形位公差及表面粗糙度在现有的加工条件下是否可以得到保证，是否还有更经济的加工方法或方案。

③ 在零件上是否存在对刀具形状及尺寸有限制的部位和尺寸要求，如过渡圆角、倒角、槽宽等，这些尺寸是否过于凌乱，是否可以统一。尽量使用最少的刀具进行加工，减少刀具规格、换刀及对刀次数和时间，以缩短总的加工时间。

④ 对于零件加工中使用的工艺基准应当着重考虑，它不仅决定了各个加工工序的前后顺序，还将对各个工序加工后各个加工表面之间的位置精度产生直接的影响。应分析零件上是否有可以利用的工艺基准，对于一般加工精度要求，可以利用零件上现有的一些基准面或基准孔，或者专门在零件上加工出工艺基准。当零件的加工精度要求很高时，必须采用先进的统一基准定位装夹系统才能保证加工要求。

⑤ 分析零件材料的种类、牌号及热处理要求，了解零件材料的切削加工性能，才能合理选择刀具材料和切削参数。

⑥ 构成零件轮廓的几何元素（点、线、面）的条件（如相切、相交、垂直和平行等），是数控编程的重要依据。因此，在分析零件图样时，务必要分析几何元素的给定条件是否充分，发现问题及时与设计人员协商解决。

制订工艺方案会影响到零件的加工质量，改进零件结构提高工艺性能有效地解决一些工艺问题，表 2-2-6 为几种典型提高工艺性方法。

**表 2-2-6　改进零件结构提高工艺性**

| 提高工艺性方法 | 结构 | | 结果 |
|---|---|---|---|
| | 改进前 | 改进后 | |
| 铣加工 | | | |
| 改进内壁形状 | 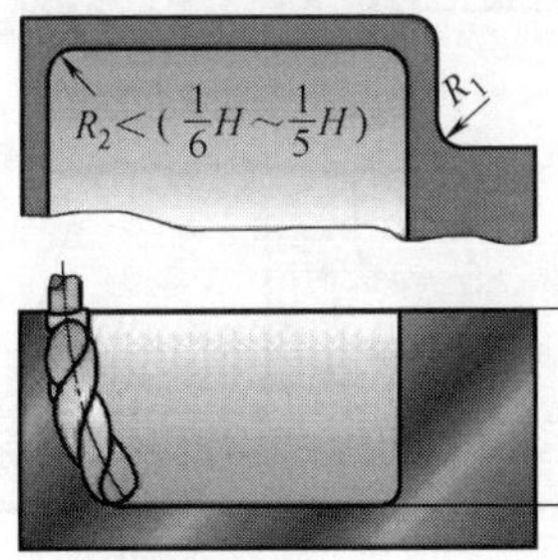 $R_2<(\frac{1}{6}H\sim\frac{1}{5}H)$ $R_1$ $H$ | 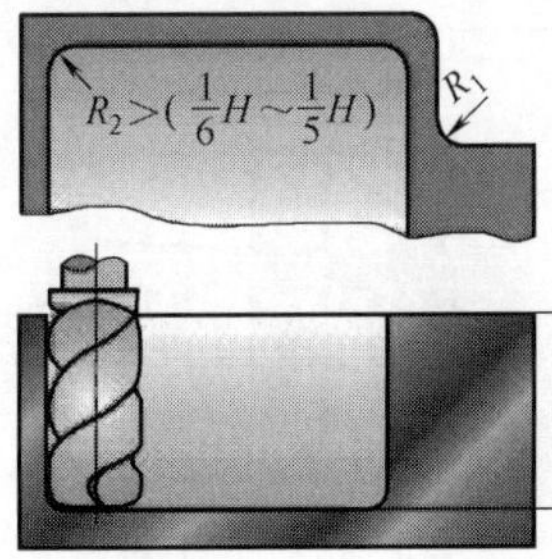 $R_2>(\frac{1}{6}H\sim\frac{1}{5}H)$ $R_1$ $H$ | 可采用较高刚性刀具 |
| 统一圆弧尺寸 | 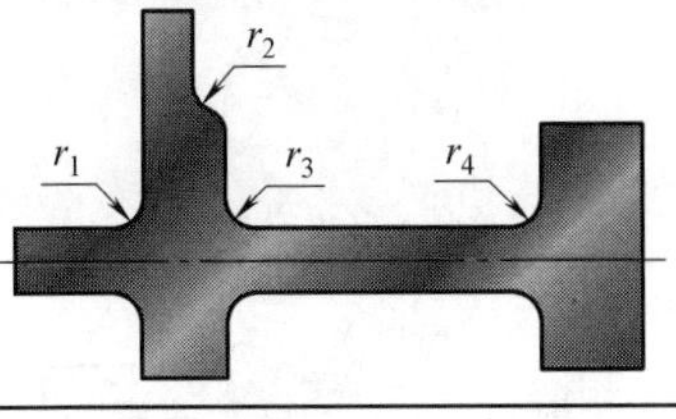 $r_1$ $r_2$ $r_3$ $r_4$ | 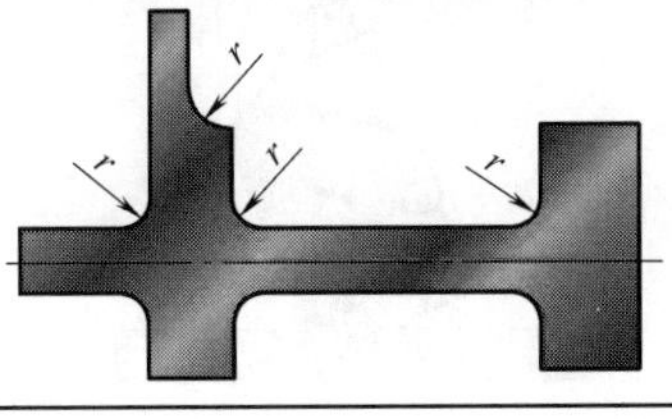 $r$ $r$ $r$ $r$ | 减少刀具数和更换刀具次数，减少辅助时间 |

续表

| 提高工艺性方法 | 结构 | | 结果 |
| --- | --- | --- | --- |
| | 改进前 | 改进后 | |
| 铣加工 | | | |
| 选择合适的圆弧半径 $R$ 和 $r$ | 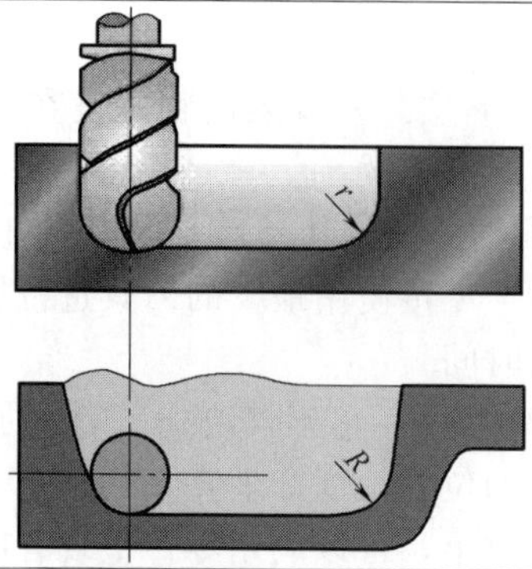 | 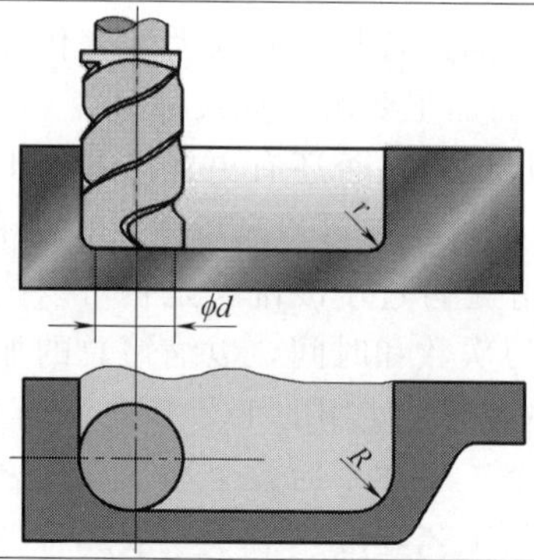 | 提高生产效率 |
| 用两面对称结构 | 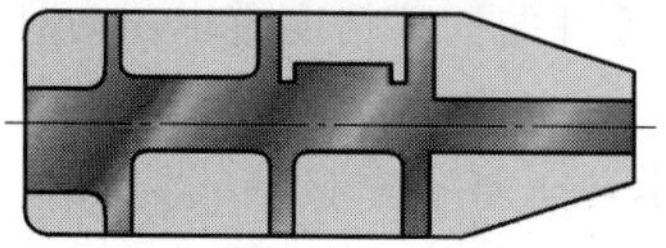 | 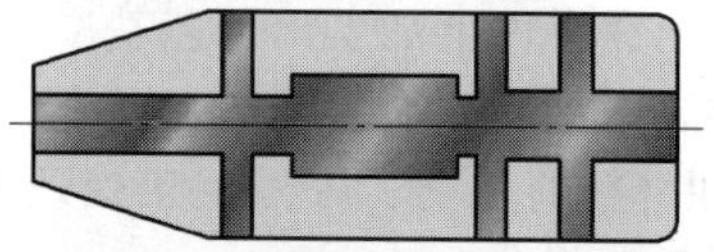 | 减少编程时间，简化编程 |
| 合理改进凸台分布 | 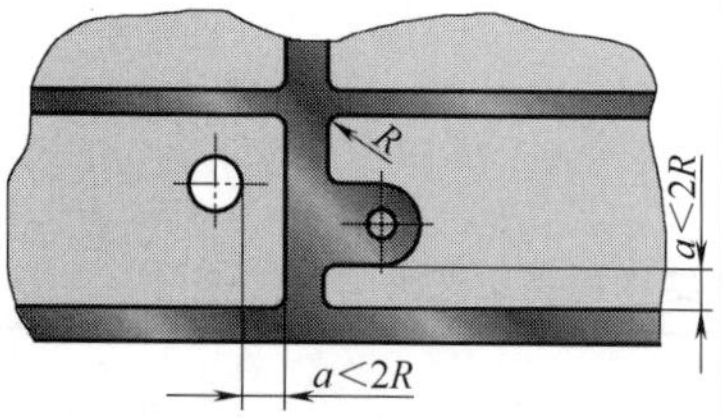 | 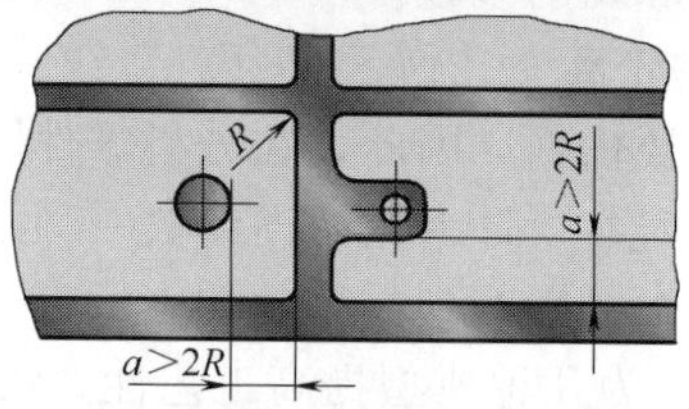 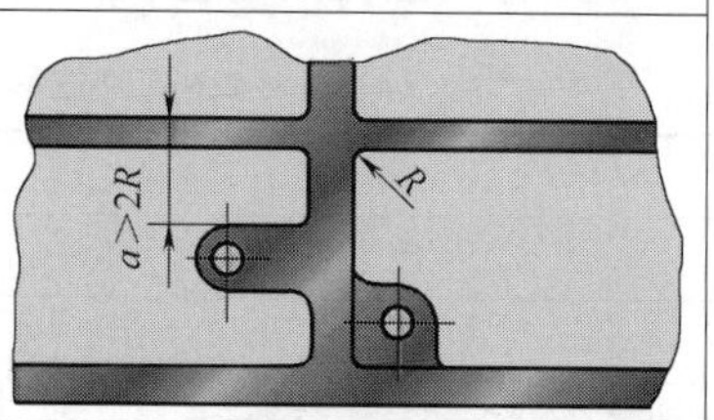 | 减少加工劳动量 |
| 改进结构形状 | 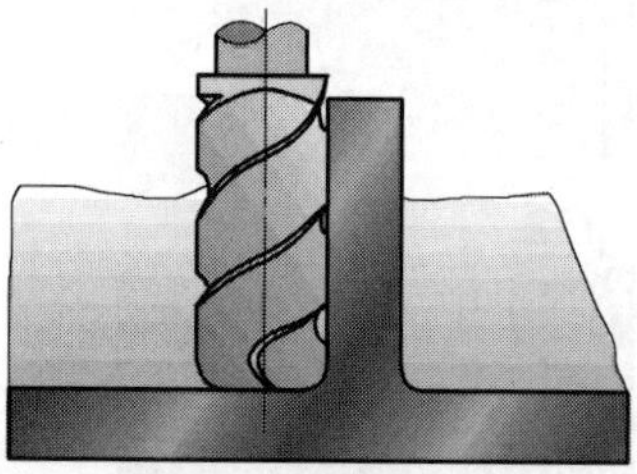 | 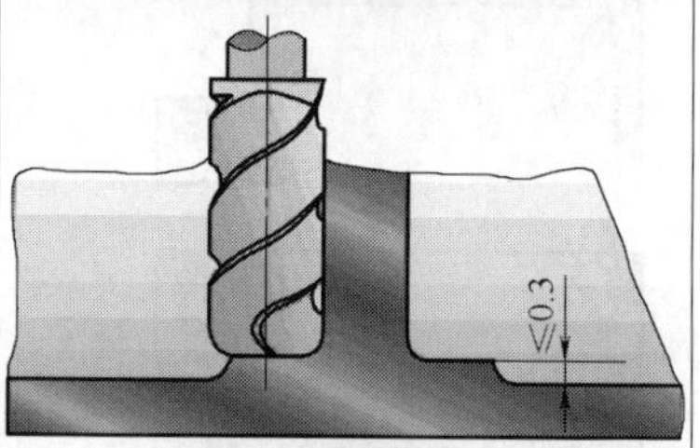 | 减少加工劳动量 |

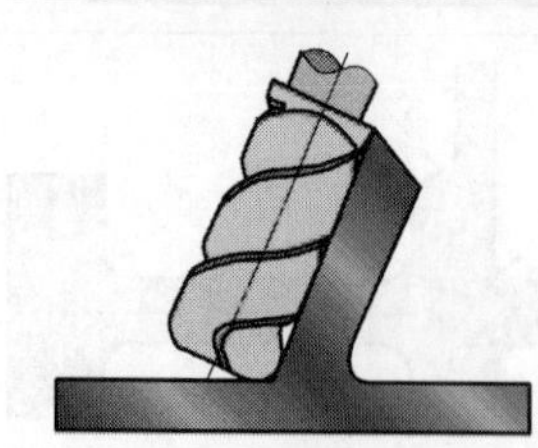

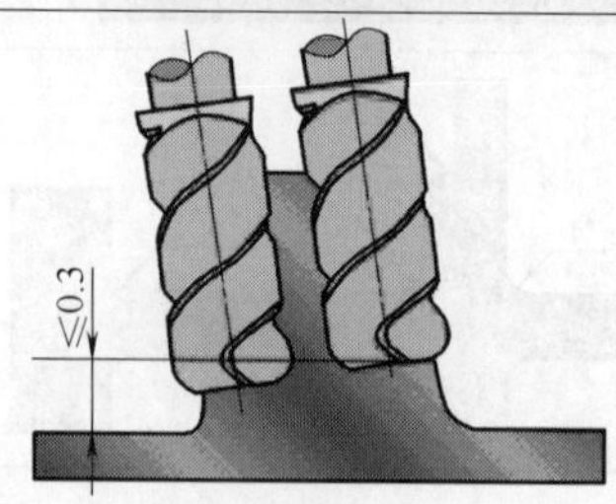

续表

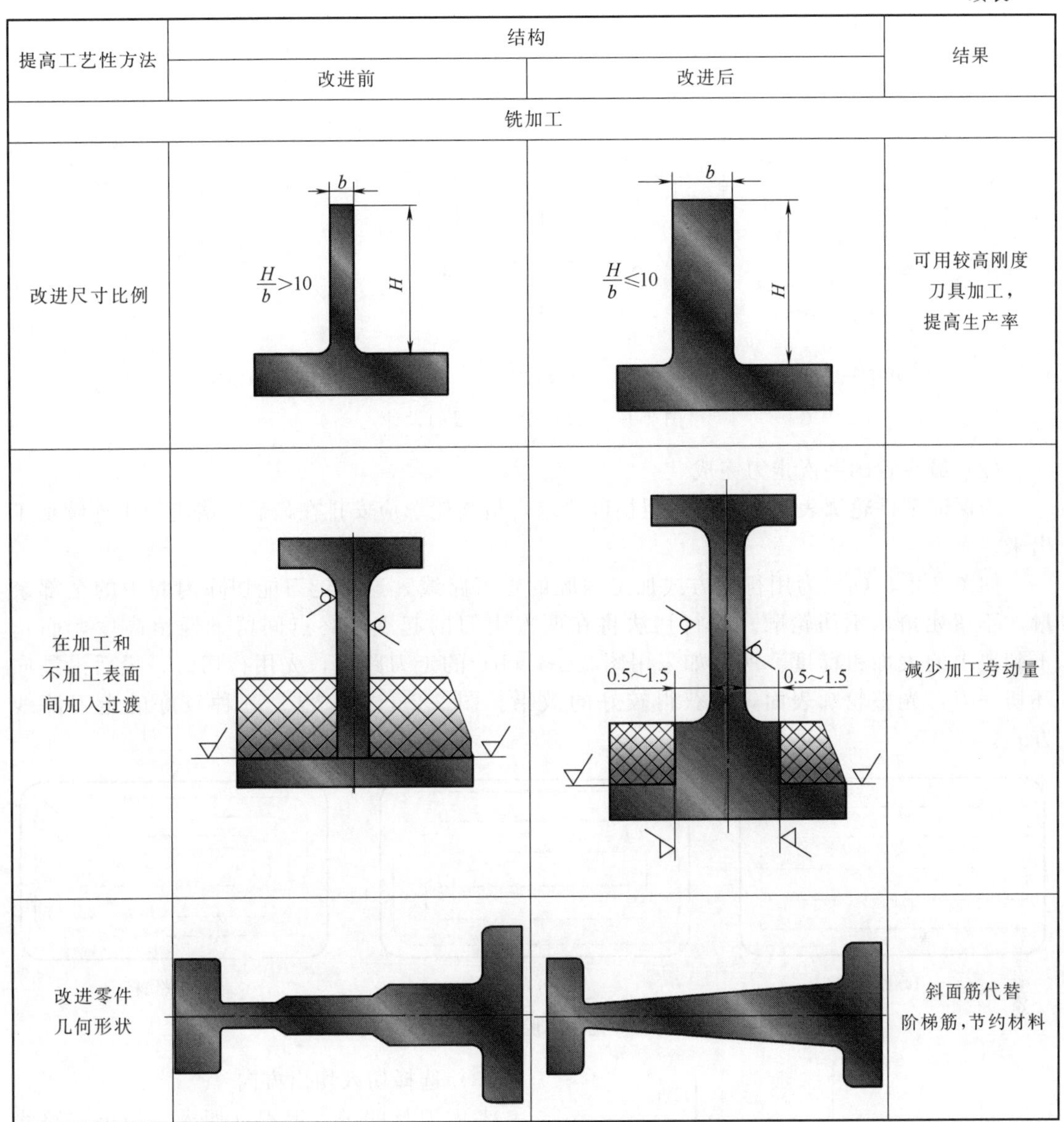

| 提高工艺性方法 | 结构 | | 结果 |
|---|---|---|---|
| | 改进前 | 改进后 | |
| 铣加工 | | | |
| 改进尺寸比例 | $\frac{H}{b}>10$ | $\frac{H}{b}\leqslant10$ | 可用较高刚度刀具加工，提高生产率 |
| 在加工和不加工表面间加入过渡 | | 0.5~1.5　0.5~1.5 | 减少加工劳动量 |
| 改进零件几何形状 | | | 斜面筋代替阶梯筋，节约材料 |

（2）零件毛坯的工艺性分析

① 毛坯应有充分、稳定的加工余量。

② 分析毛坯的装夹适应性。

③ 分析毛坯的余量大小及均匀性。

**3. 确定走刀路线和安排加工顺序**

走刀路线就是刀具在整个加工工序中的运动轨迹，它不但包括了工步的内容，也反映出工步顺序。走刀路线是编写程序的依据之一。确定走刀路线时应注意以下几点。

（1）寻求最短加工路线

如加工图 2-2-3（a）所示零件上的孔系。图 2-2-3（b）的走刀路线为先加工完外圈孔后，再加工内圈孔。若改用图 2-2-3（c）的走刀路线，减少空刀时间，则可节省定位时间近一倍，提高了加工效率。

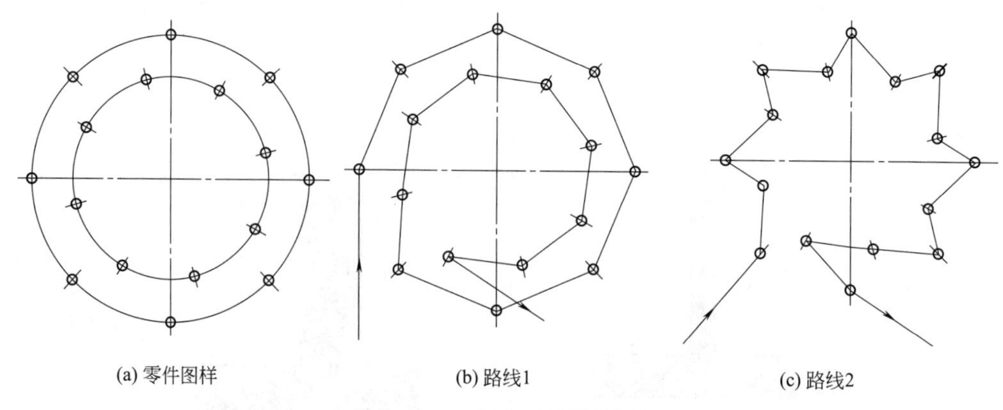


图 2-2-3 最短走刀路线的设计

（2）最终轮廓一次走刀完成

为保证工件轮廓表面加工后的粗糙度要求，最终轮廓应安排在最后一次走刀中连续加工出来。

如图 2-2-4（a）为用行切方式加工内腔的走刀路线，这种走刀能切除内腔中的全部余量，不留死角，不伤轮廓。但行切法将在两次走刀的起点和终点间留下残留高度，而达不到要求的表面粗糙度。所以如采用图 2-2-4（b）的走刀路线，先用行切法，最后沿周向环切一刀，光整轮廓表面，能获得较好的效果。图 2-2-4（c）也是一种较好的走刀路线方式。

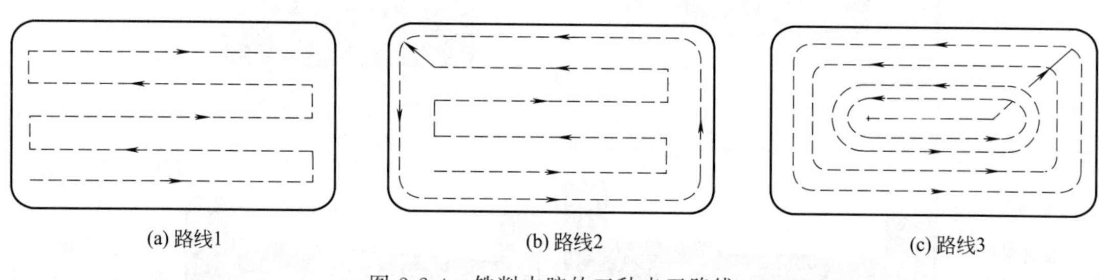


图 2-2-4 铣削内腔的三种走刀路线

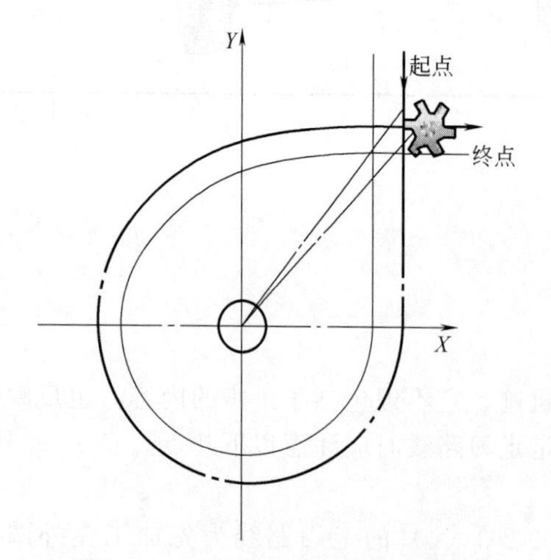


图 2-2-5 刀具切入和切出时的外延

（3）选择切入切出方向

考虑刀具的进、退刀（切入、切出）路线时，刀具的切出或切入点应在沿零件轮廓的切线上，以保证工件轮廓光滑；应避免在工件轮廓面上垂直上、下刀而划伤工件表面；尽量减少在轮廓加工切削过程中的暂停（切削力突然变化造成弹性变形），以免留下刀痕，如图 2-2-5 所示。

（4）选择使工件在加工后变形小的路线

对横截面积小的细长零件或薄板零件应采用分几次走刀加工到最后尺寸或对称去除余量法安排走刀路线。安排工步时，应先安排对工件刚性破坏较小的工步。

## 四、数控铣基本指令

下面以FANUC 0i系统为例，来介绍数控铣床中G指令字（见表2-2-7）和常用的M指令（见表2-2-8）。

表 2-2-7 FANUC 0i 中常用 G 指令字及含义

| G代码 | 组别 | 功 能 | 备 注 |
|---|---|---|---|
| G00※ | 01 | 快速点定位 | |
| G01※ | | 直线插补 | |
| G02※ | | 顺时针圆弧插补 | G02 X Y I J 或 G02 X Y R 中，X、Y 表示终点坐标，I、J 表示圆心相对于起点在 X、Y 方向的距离，R 表示圆半径 |
| G03※ | | 逆时针圆弧插补 | |
| G04 | 00 | 暂停(延时) | G04 P 中 P 表示程序停留时间 |
| G17※ | 02 | XY 平面选择 | |
| G18 | | ZX 平面选择 | |
| G19 | | ZY 平面选择 | |
| G20 | 06 | 英制输入 | |
| G21 | | 公制输入 | |
| G40※ | 07 | 取消刀具半径补偿 | |
| G41※ | | 刀具半径左补偿 | 刀补必须在直线段进行 |
| G42※ | | 刀具半径右补偿 | |
| G43 | 08 | 刀具长度正补偿 | |
| G44 | | 刀具长度负补偿 | |
| G49※ | | 取消刀具长度补偿 | |
| G50 | 11 | | |
| G51 | | | G51 X Y Z I J K 中，I、J、K 表示 X、Y、Z 轴向缩放系数 |
| G50.1 | | 取消坐标系镜像 | |
| G51.1 | | 镜像 | G51.1 X 以平行于 X 轴的直线为对称轴<br>G51.1 Y 以平行于 Y 轴的直线为对称轴<br>G51.1 Z 以(X,Y)为对称点 |
| G53 | 00 | 设置为机床坐标系模式 | |
| G54～G59※ | 14 | 工件坐标系 | |
| G65 | 12 | 子程序调用 | G65 P L 中 P 表示子程序号，L 表示调用次数 |
| G68 | | 坐标系旋转 | G68 X Y R 中 X、Y 表示基准点，R 表示旋转角度 |
| G69 | | 取消坐标系旋转 | |
| G70 | | 圆周均布点钻削循环 | I J L 中 I 表示圆弧半径，J 表示起点到圆心的直线与 X 轴的夹角，L 表示圆上共均布的点数 |
| G71 | | 周均布点钻削循环 | I J K L 中 I 表示圆弧半径，J 表示起点到圆心的直线与 X 轴的夹角，L 表示圆上共均布的点数，K 表示每等分夹角 |

续表

| G 代码 | 组别 | 功 能 | 备 注 |
|---|---|---|---|
| G72 | | 线均布点钻削循环 | I J L 中 I 表示等分距离,J 表示直线与 $X$ 轴夹角,L 表示等分点 |
| G80※ | 09 | 取消固定钻削循环 | |
| G81※ | | 普通钻削循环 | G81 X Y Z R F L 中 X、Y 表示加工点 $X$、$Y$ 坐标,Z 表示钻孔深度,R 表示参考平面位置,F 表示切削速度,L 表示反复钻削次数 |
| G82※ | | 钻削循环(孔底有停留) | G82 X Y Z R F L P 中 P 表示停留时间 |
| G83※ | | 钻削循环(间隙进给) | G83 X Y Z R F L P Q I J K 中,Q 表示每次下降高度 ,L 表示第一次切削深度,J 表示每一次切削后切削量的减少值,K 表示最少切削量 |
| G84 | | 攻螺纹循环 | G84 X Y Z R F L P |
| G85 | | 精钻削循环 | G85 X Y Z R F L P |
| G86 | | 镗孔循环 | G86 X Y Z R F L P |
| G87 | | 反向镗孔循环 | G87 X Y Z R F L P |
| G88 | | 反向攻螺纹循环 | G88 X Y Z R F L P |
| G90※ | 03 | 绝对值编程 | |
| G91※ | | 相对值编程 | |
| G92 | 00 | 坐标系设定 | |
| G94 | 05 | 每分钟进给 | |
| G95 | | 每转进给 | |
| G98 | 05 | 钻削循环返回到初始点 | |
| G99 | 10 | 环返回到 $R$ 点 | |

注;带※号的为常用指令。

**表 2-2-8 FANUC 0i 中常用的 M 指令及含义**

| M 指令 | 功能 | 备注 |
|---|---|---|
| M00 | 程序停止 | 按循环启动按钮,可以再启动 |
| M01 | 选择停止 | 程序是否停止取决于机床操作面板上的跳步开关 |
| M02 | 程序结束 | 程序结束后不返回到程序开头的位置 |
| M03※ | 主轴顺时针转 | 从主轴尾端向主轴前端看时,为顺时针 |
| M04※ | 主轴逆时针转 | 从主轴尾端向主轴前端看时,为逆时针 |
| M05※ | 主轴停止 | |
| M06 | 刀具交换 | |
| M08 | 切削液开 | |
| M09 | 切削液关 | |
| M13 | 主轴顺时针转切削液开 | |
| M14 | 主轴逆时针转切削液开 | |
| M30※ | 程序结束 | 程序结束后,自动返回到程序开头的位置 |
| M98 | 子程序调用 | M98 P L,P 表示程序地址,L 表示调用次数 |
| M99※ | 子程序返回 | |

在程序编制中辅助M指令在需要时在程序段中加入就可以了，编程人员仅需牢记其功能及含义，使用比较简单。而准备功能G指令还需按照规定的指令格式进行编程，相对比较复杂，下面将介绍一下数控铣床中常用的准备功能G指令的定义和使用方法。

（1）G00快速定位指令

G00快速定位指令的指令格式为：

```
G00 X__Y__Z__;
```

其中，X、Y、Z是快速定位至终点的坐标值，在G90编程方式下，终点为相对于工件坐标系原点的坐标，在G91编程方式下，终点为相对于起点的位移量；G00为模态功能指令，可由G01、G02或G03功能指令注销。

注意：在执行G00指令时，由于各轴以各自速度移动，联动直线轴的合成轨迹不一定是直线。

如图2-2-6所示，使用G00编程，要求刀具从$A$点快速定位到$B$点。

绝对坐标编程为G90 G00 X90 Y45.0；

增量坐标编程为G91 G00 X70 Y30.0；

为避免刀具与工件发生碰撞，常见的做法是将$Z$轴移动到安全高度，再执行G00指令，其实际路径如图2-2-6所示。

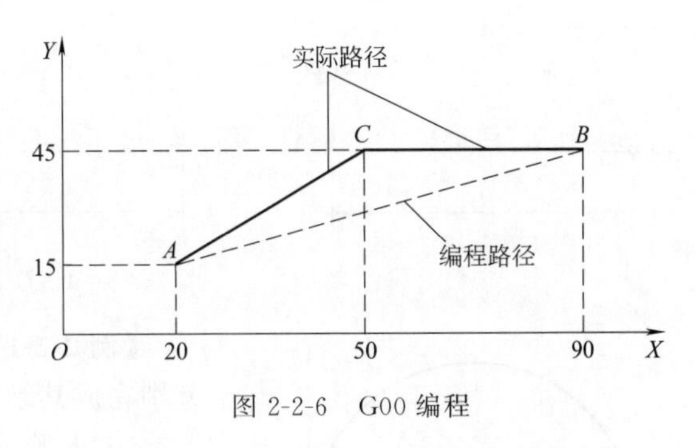


图2-2-6　G00编程

（2）G01直线插补指令

G01直线插补指令的指令格式为

```
G01 X__Y__Z__F__;
```

其中，X、Y、Z是直线插补进给终点，在G90编程方式下，终点为相对于工件坐标系原点的坐标，在G91编程方式下，终点为相对于起点的位移量；F为合成进给速度，在没有新的F指令以前一直有效，不必在每个程序段中都写入F指令。

G01是模态代码指令，可由G01、G02或G03功能指令注销。G01指令刀具以联动的方式，按F规定的合成进给速度，从当前位置按线性路线（联动直线轴的合成轨迹为直线）移动到程序段指令的终点。

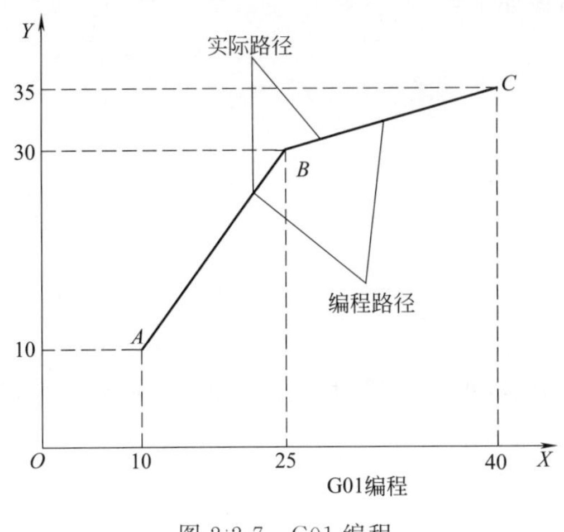


图2-2-7　G01编程

如图2-2-7所示，使用G01编程，要求刀具从$A$点经$B$点线性进给到$C$点（此时进给路线是从$A$—$B$—$C$的折线）。

绝对坐标编程为

```
G90 G01 X25.0 Y30.0 F100;
X40.0 Y35.0;
```

增量坐标编程为

```
G91 G01 X15.0 Y20.0 F100;
X15.0 Y5.0;
```

（3）G02/G03圆弧插补指令

G02/G03圆弧插补指令的指令格式为

$$\begin{Bmatrix} G17 \\ G18 \\ G19 \end{Bmatrix} \begin{Bmatrix} G02 \\ G03 \end{Bmatrix} \begin{Bmatrix} X\_\_ Y\_\_ \\ X\_\_ Z\_\_ \\ Y\_\_ Z\_\_ \end{Bmatrix} \begin{Bmatrix} I\_\_ J\_\_ \\ I\_\_ K\_\_ \\ J\_\_ K\_\_ \\ R\_\_ \end{Bmatrix} F\_\_$$

其中，G02 为顺时针圆弧插补；G03 为逆时针圆弧插补；X、Y、Z 是圆弧终点坐标值；I、J、K 分别表示 $X(U)$、$Y(V)$、$Z(W)$ 轴圆心的坐标减去圆弧起点的坐标，如图 2-2-8 所示。圆心位置亦可用圆弧半径 R 表示（当圆弧的圆心角≤180°时，R 为正值；圆弧的圆心角＞180°时，R 为负值；圆弧的圆心角＝360°时，不能用 R 编程，只能用 I、J、K 编程）。

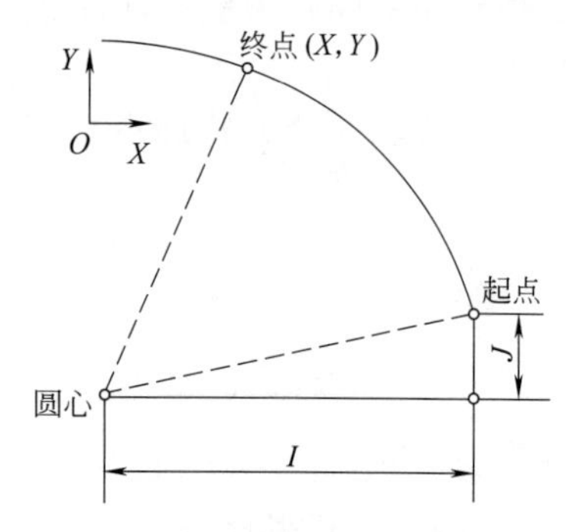


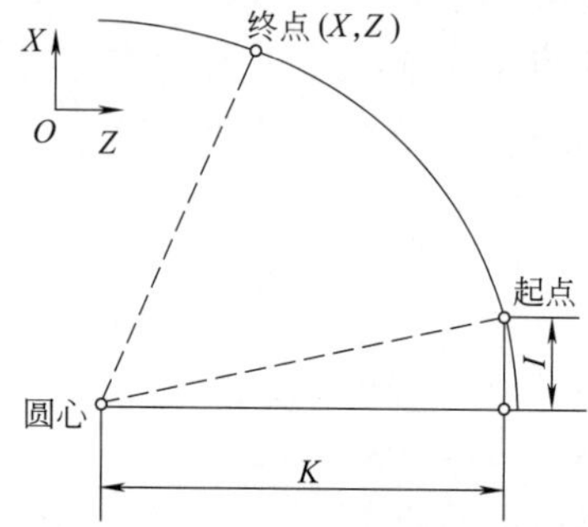


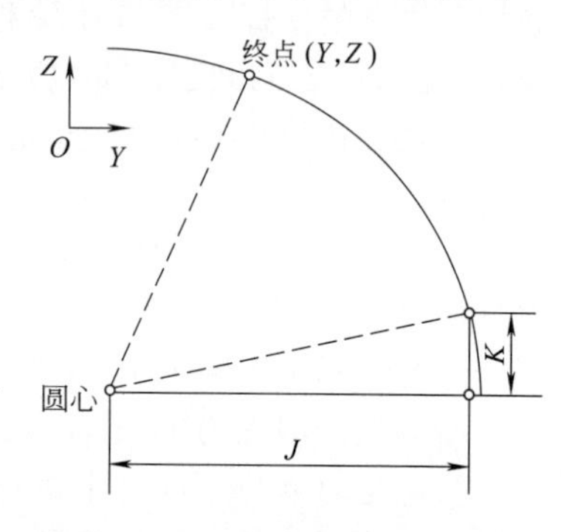


图 2-2-8 I、J、K 表示方法

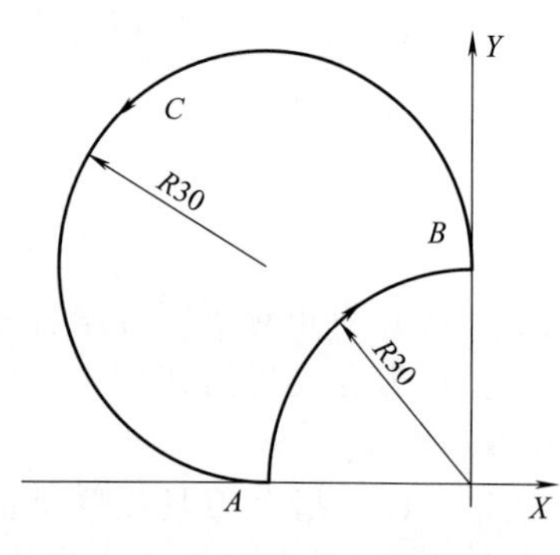


图 2-2-9 劣弧 AB、优弧 BCA

**【例 2-2-1】** 如图 2-2-9 所示为劣弧 $AB$、优弧 $BCA$，分别完成其绝对编程与增量编程。

① 劣弧 $AB$ 的绝对坐标编程与增量坐标编程如下：

绝对坐标编程为

```
G90 G02 X0 Y30.0 R30.0 F80;
或  G90 G02 X0 Y30.0 I30.0 F80;
```

增量坐标编程为

```
G91 G02 X30.0 Y30.0 R30.0 F80;
或  G91 G02 X30.0 Y30.0 I30.0 F80;
```

② 优弧 $BCA$ 的绝对坐标编程与增量坐标编程如下：

绝对坐标编程为

```
G90 G03 X-30.0 Y0 R-30.0 F80;
或  G90 G03 X-30.0 Y0 J-30.0 F80;
```

增量坐标编程为

```
G91 G03 X-30.0 Y-30.0 R-30.0 F80;
或  G91 G03 X-30.0 Y-30 J-30.0 F80;
```

**【例 2-2-2】** 如图 2-2-10 所示整圆加工，分别完成如下的绝对坐标编程与增量坐标编程。

① 从 $A$ 点顺时针一周的绝对坐标编程与增量坐标编程如下：

绝对坐标编程为

```
G90 G02 X30.0 Y0 I- 30.0 F80;
```

增量坐标编程为

```
G91 G02 X0 Y0 I- 30.0 F80;
```

② 从 B 点逆时针一周的绝对编程与增量坐标编程如下：

绝对坐标编程为

```
G90 G03 X0 Y- 30.0 J30 F80;
```

增量坐标编程为

```
G91 G03 X0 Y0 J30.0 F80;
```

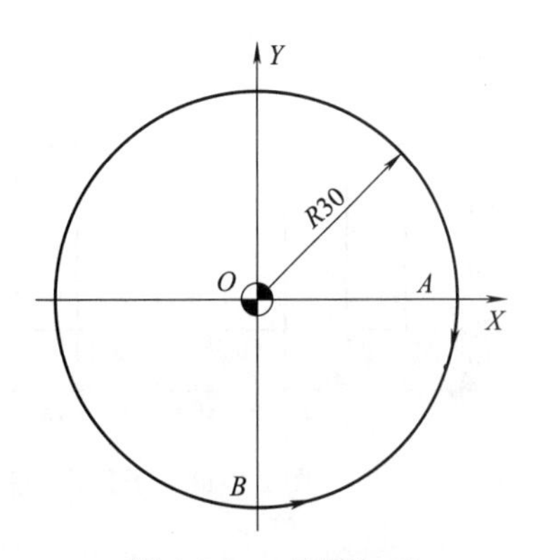

图 2-2-10　整圆加工

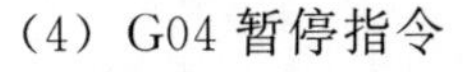
(4) G04 暂停指令

G04 暂停指令的指令格式为

```
G04 P __;
```

其后面跟整数值，单位为 ms。

```
或 G04 X __;
```

其后面带小数点的数，单位为 s。

该指令可使刀具短时间无进给地光整加工，用于打盲孔、车槽等加工。

G04 为非模态指令，仅在其被规定的程序段中有效。

## 五、简化编程指令

(1) 调用子程序

M98 用来调用子程序，M99 表示子程序结束，执行 M99 使控制返回到主程序。

子程序的格式如下：

```
OXXXX  ;子程序号
……
……  ;  子程序体
……
M99  ;  子程序结束,返回主程序
```

在子程序开头，必须规定子程序号，以作为调用入口地址。在子程序的结尾用 M99，以控制执行完该子程序后返回主程序。

调用子程序的格式如下：

```
M98 P __ L __
```

其中，P 为被调用的子程序号；L 为重复调用次数。

**【例 2-2-3】** 如图 2-2-11 所示，Z 起始高度 100mm，切削深度 20mm，轮廓外侧切削。试编制加工程序。

程序：O6001；

```
N1;(主程序)
```

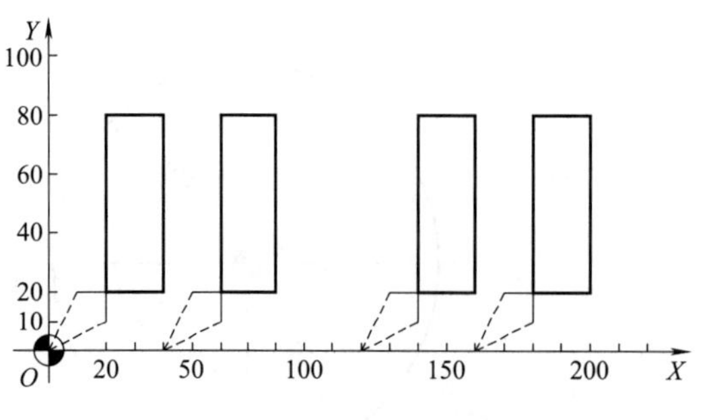

图 2-2-11　子程序编程

```
G90 G54 G00 X0 Y0 S500 M03;
G00 Z100.0;
M98 P100 L2;
G90 X120.0;
M98 P100 L2;
G90 G00 X0 Y0 M05;
M30;
O0100;(子程序)
G91 G00 Z-95.0;
G41 X20.0 Y10.0 D01;
G01 Z-25.0 F50;
Y70.0;
X20.0;
Y-60.0;
X-30.0;
G00 G40 X-10.0 Y-20.0;
Z120.0;
X40.0;
M99;
```

（2）镜像功能指令（G51.1，G50.1）

编程格式：G51.1 X __ Y __ Z __；

G50.1 X __ Y __ Z __；

其中，G51.1 为建立镜像指令；G50.1 为取消镜像指令；X，Y，Z 为镜像位置（X0 表示 Y 轴对称；Y0 表示 X 轴对称；X0 Y0 表示原点对称）。G51.1、G50.1 为模态指令，可相互注销。

**【例 2-2-4】** 运用镜像功能编制程序（见图 2-2-12）。

```
O6002 主程序
N01 G92 X0 Y0 Z10;
N02 G91 G17 M03;
N03 M98 P100;加工①
N04 G51.1 X0;以 Y 轴镜像
N05 M98 P100;加工②
N06 G50.1 X0;取消 Y 轴镜像
N07 G51.1 X0 Y0;以位置点为(0,0)
N08 M98 P100;加工③
N09 G50.1 X0 Y0;取消点(0,0)镜像
N10 G51.1 Y0;以 X 轴镜像
N11 M98 P100;加工④
N12 G50.1 Y0;取消 X 轴镜像
N13 M05;
N14 M30;
% 100 子程序
N01 G01 Z-5 F50;
N02 G00 G41 X20 Y10 D01;
```

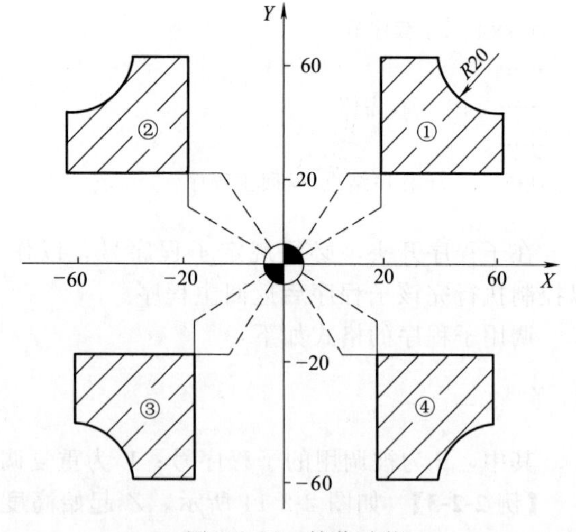

图 2-2-12　镜像功能

```
N03 G01 Y60;
N04 X40;
N05 G03 X60 Y40 R20;
N06 Y20;
N07 X10;
N08 G00 X0 Y0;
N09 Z10;
N10 M99;
```

（3）缩放功能指令（G50，G51）

编程格式：G51 X __ Y __ Z __ P __；

其中，G51 为建立缩放指令；G50 为取消缩放指令；X、Y、Z 为缩放中心的坐标值；P 为缩放倍数，其范围为 0.001～999.999，0.001＜$P$＜1 为缩小，1＜$P$＜999.999 为放大。

在有刀具补偿的情况下，先进行缩放，然后才进行刀具半径补偿、刀具长度补偿。

G51、G50 为模态指令，可相互注销。

（4）旋转变换指令（G68，G69）

编程格式：G68 X __ Y __ R __；

其中，G68 为建立旋转指令，G69 为取消旋转指令；X、Y、Z 为旋转中心的坐标值；R 为旋转角度，单位是度（°），0≤$R$≤360°，顺时针为负，逆时针为正。

在有刀具补偿的情况下，先旋转后刀补（刀具半径补偿、长度补偿），在有缩放功能的情况下，先缩放后旋转。G68，G69 为模态指令，可相互注销，G69 为缺省值。

**【例 2-2-5】** 编制如图 2-2-13 所示的旋转变换功能程序。

```
O6003                    主程序
N10 G90 G17 M03;
N20 M98 P100;            加工①
N30 G68 X0 Y0 P45;  旋转 45°
N40 M98 P100;            加工②
N50 G69;                 取消旋转
N60 G68 X0 Y0 P90;  旋转 90°
N70 M98 P100;            加工③
N80 G69 M05 M30;    取消旋转
% 100 子程序(①的加工程序)
N100 G90 G01 X20 Y0 F100;
N110 G02 X30 Y0 I5;
N120 G03 X40 Y0 I5;
N130 X20 Y0 I10;
N140 G00 X0 Y0;
N150 M99;
```

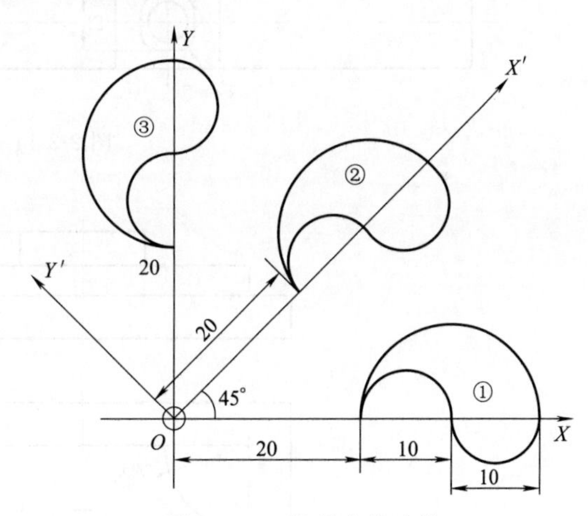


图 2-2-13　旋转变换功能

**习题**

1. 数控铣削加工零件的工艺性分析包括哪些？
2. 零件毛坯的工艺性分析包括哪些内容？
3. 走刀路线的加工顺序如何确定？
4. G00 和 G01 指令的区别是什么？

5. 如何判断顺时针、逆时针圆弧插补指令？
6. 简化编程指令包括哪些？
7. 调用子程序的格式是什么？
8. 解释旋转变换指令中的指令含义。
9. 半圆块加工如图 2-2-14 所示，试编写数控加工程序。
10. 双层轮廓件加工如图 2-2-15 所示，试编写数控加工程序。

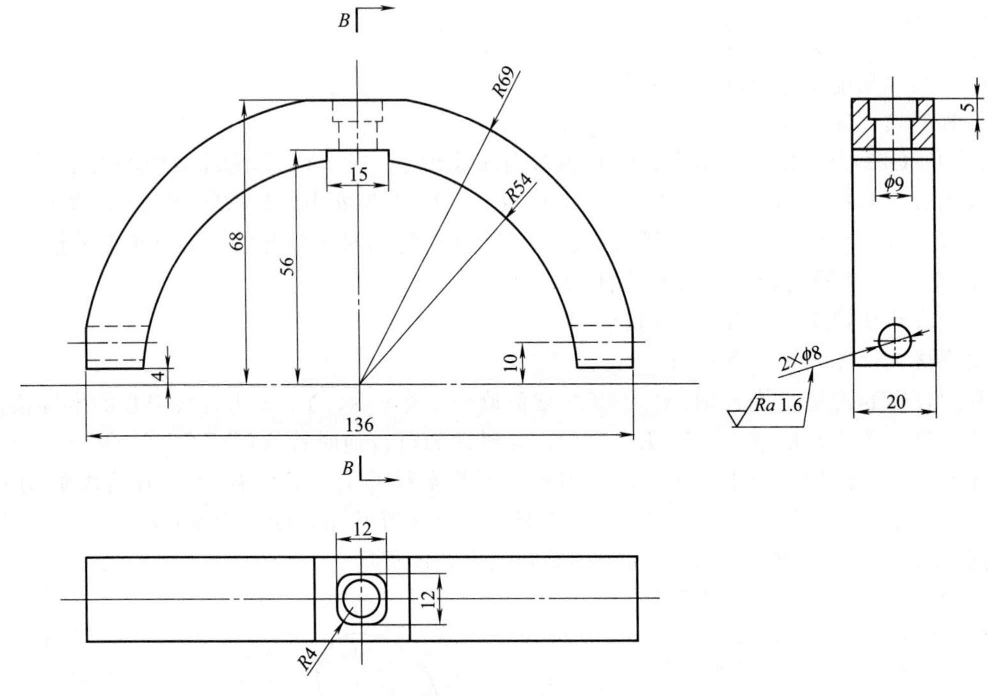


图 2-2-14　半圆块

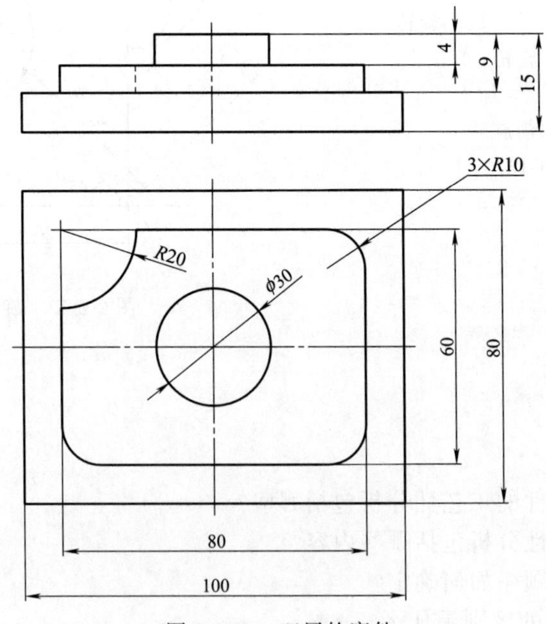


图 2-2-15　双层轮廓件

# 任务三 轮廓及孔零件的加工

## 一、图样与技术要求

如图 2-3-1 所示轮廓与孔零件，材料为 45 钢，规格为 $\phi$60mm 的圆柱棒料，正火处理，硬度 200HB。

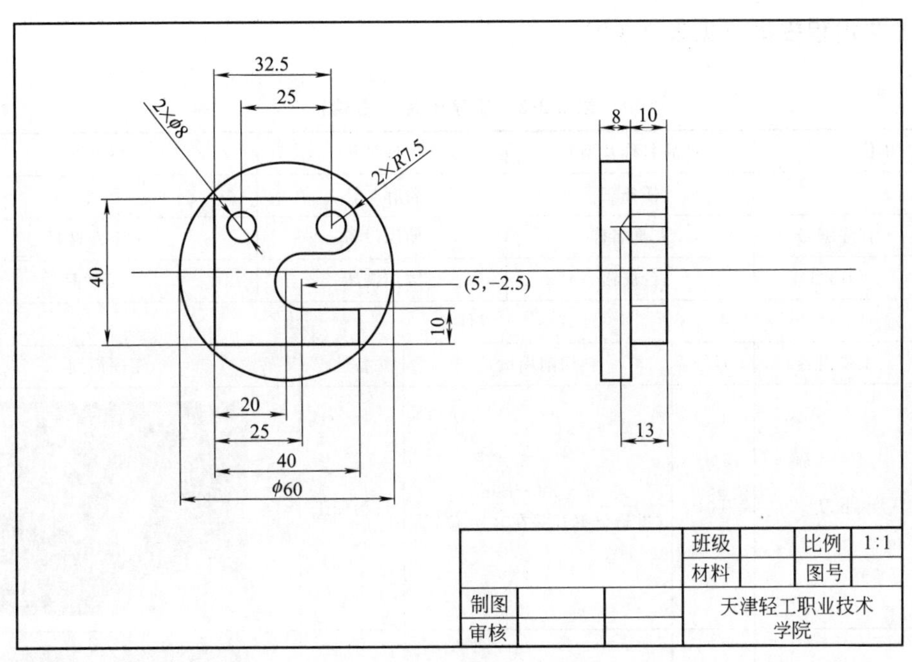


图 2-3-1 轮廓及孔零件

## 二、图纸分析

教学策略：分组讨论、小组汇报、教师总结。

**1. 零件图分析**

零件的外轮廓是一个 $\phi$60mm 的圆柱面，由外轮廓及 2×$\phi$8 的孔组成，属于轮廓和孔类零件。零件是根据训练、学习的前后顺序而设计的。

**2. 工艺分析**

① 结构分析：零件为简单外轮廓和孔类零件。

② 精度分析：零件的重点尺寸在两个圆弧尺寸 2×$R$7.5，其中有一点坐标（5，−2.5），在计算点的坐标时需要注意，外轮廓深度为 10mm，孔的深度为 13mm，加工时需要注意。

③ 定位及装夹分析：本零件采用台虎钳进行装夹。工件装夹时的夹紧力要适中，既要防止工件的变形和夹伤，又要防止工件在加工时的松动。装夹时需要在零件下方垫垫铁，用木榔头敲击并垫实。

④ 加工工艺分析：经过以上分析，本任务零件加工时总体安排顺序是，先加工零件的外轮廓，后打孔加工。

### 3. 主要刀具选择（见表 2-3-1）

表 2-3-1　刀具卡片

| 刀具名称 | 刀具规格 | 材料 | 数量 | 刀具半径 |
|---|---|---|---|---|
| 立铣刀 | $\phi$12mm | YT15 | 1 | 6mm |
| 钻头 | $\phi$8mm | YT15 | 1 | 4mm |

### 4. 工艺规程安排（见表 2-3-2）

表 2-3-2　工序卡片（右端）

| 单位 | | 产品名称及型号 | 零件名称 | | 零件图号 |
|---|---|---|---|---|---|
| | | 任务三 | 轮廓与孔 | | 图 2-3-1 |
| 工序 | 程序编号 | 夹具名称 | 使用设备 | | 工件材料 |
| 001 | O1301 | 台虎钳 | 数控铣床 | | 45 钢 |
| 件 1 | | | | | |
| 工步 | 工步内容 | 刀号 | 切削用量 | 备注 | 工序简图 |
| 1 | 下刀 | T01 | $n=500\text{r/min}$<br>（进给倍率开关在×10 位置） | 自动加工 |  |
| 2 | 沿外轮廓切削 | T01 | $n=500\text{r/min}$<br>$f=50\text{mm/min}$<br>$a_p=5.0\text{mm}$ | 自动加工 |  |
| 3 | 加工轮廓尺寸和圆弧尺寸 | T01 | $n=500\text{r/min}$<br>$f=50\text{mm/min}$<br>$a_p=5.0\text{mm}$ | 自动加工 |  |
| 4 | 左上方孔加工 | T02 | $n=500\text{r/min}$<br>$f=50\text{mm/min}$<br>$a_p=13.0\text{mm}$ | 自动加工 | 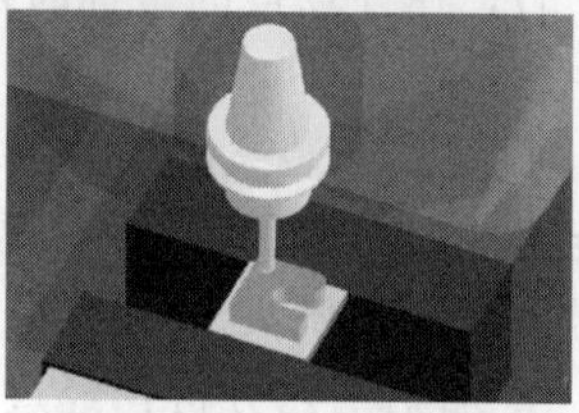 |

续表

| 件 1 | | | | | |
|---|---|---|---|---|---|
| 工步 | 工步内容 | 刀号 | 切削用量 | 备注 | 工序简图 |
| 5 | 右上方孔加工 | T02 | $n=500$r/min<br>$f=50$mm/min<br>$a_p=13.0$mm | 自动加工 | |
| 6 | 完成整个零件加工 | T02 | $n=500$r/min<br>$f=50$mm/min<br>$a_p=13.0$mm | 自动加工 | |

## 5. 加工程序

```
O1301;
G91 G28 Z0;
T1 M6;
G54 G90 G00 X0 Y-40 M03 S500;
G43 Z100.0 H1;
Z10.0;
G01 Z-5.0 F50;
M98 P100 D1;
M98 P100 D2;
G01 Z-10.0 F50;
G00 Z100.0;
M05;
G91 G28 Z0;
T2 M6;
G90 G54 G00 X0 Y0 M03 S500;
G43 Z100.0 H2;
G01 X-12.5 Y12.5;
G98 G83 Z-13 R4 Q3 F50;
X12.5 Y12.5;
G80 G00 Z100;
M05;
M30;
O100;
G41 G01 X20 Y-40 F120;
G03 X0 Y-20 R20;
G01 X-20 Y-20;
Y20;
X12.5;
```

```
G02 X12.5 Y5 R7.5;
G01 X5 Y5;
G03 X5 Y-10 R7.5;
G01 X20 Y-10;
Y-20;
X0;
G03 X-20 Y-40 R20;
G40 G01 X0 Y40 F120;
M99;
```

## 三、数控铣削相关工艺问题

**1. 程序起始点、返回点和切入点（进刀点）、切出点（退刀点）的概念**

① 程序起始点。程序起始点是指程序开始时，刀尖（设刀尖为刀位点）的初始停留点。

② 程序返回点。程序返回点是指一把刀执行程序完毕后，刀尖返回后的停留点。一般为换刀点。

③ 切入点（进刀点）。切入点是指在曲面的初始切削位置上，刀具与曲面的接触点（也称切触点）。

④ 切出点（退刀点）。切出点是指曲面切削完毕后，刀具与曲面的接触点。

**2. 程序起始点、返回点和切入点、切出点的确定原则**

① 起始点、返回点确定原则。在同一个程序中起始点和返回点最好相同，如果一个零件的加工需要几个程序来完成，那么这几个程序的起始点和返回点也最好相同，以免引起加工操作繁琐。起始点和返回点的坐标值最好设 $X$ 和 $Y$ 值均为零，这样能使操作方便。起始点和返回点的 $Z$ 坐标应定义在使刀尖高出被加工零件的最高点 50～100mm 左右的某一位置上，即起始平面、退刀平面所在的位置。这主要是因为数控加工的安全性，防止碰刀，同时也考虑了数控加工的效率，使非切削时间控制在一定的范围内。

② 切入点选择的原则。在进刀或切削曲面的过程中，应使刀具不受损坏。一般来说，对粗加工而言，应选择曲面内的最高角点作为曲面的切入点（初始切削点）。因为该点的切削余量较小（特别对余量不均匀的方型等型材和锻件毛坯），进刀时不易损坏刀具。对精加工而言，选择曲面内某个曲率比较平缓的角点作为曲面的切入点。因为在该点处，余量已均匀，刀具所受的弯矩较小，不易折断刀具。总之，要避免将铣刀当钻头使用，否则会因受力大、排屑不便而使刀具受损。

③ 切出点选择的原则。切出点选择主要考虑曲面能连续完整的加工及曲面与曲面间的非切削加工时间尽可能短，换刀方便，以提高机床的有效工作时间。被加工曲面为开放型曲面，有曲面的两个角点可作为切出点，按上述原则选择其中一个角点做切出点。若被加工曲面为封闭型曲面，则只有曲面的一个角点为切出点，自动编程时系统一般自动确定。

**3. 起始平面、返回平面、进刀平面、退刀平面和安全平面的概念**

① 起始平面。起始平面是程序开始时刀尖的初始位置所在的 $Z$ 平面，定义在被加工零件的最高点之上 50～100mm 的某一位置上，一般高于安全平面。其对应的高度称为起始高度。在此平面上刀具以 G00 速度行进。

② 返回平面。返回平面是指程序结束时，刀尖所在的 $Z$ 平面，它也定义在高出被加工表面最高点 50～100mm 的某个位置上，一般与起始平面重合。因此，刀具处于返回平面上

时是安全的。其对应的高度称为返回高度。刀具在此平面上也以 G00 速度行进。

③ 进刀平面。刀具以高速（G00）下刀，至要切到材料时，变成以进刀速度下刀，以免撞刀，此速度转折点的位置即为进刀平面，其高度为进刀高度，也称为接近高度，即其转折速度称为进刀速度或接近速度。此高度一般在加工面和安全平面之间，离加工面 5～10mm（刀尖到加工面间的距离），加工面为毛坯面时取大值，加工面为已加工面时取小值。若无此项设置，则数控铣床会发生撞刀事故。

④ 退刀平面。零件（零件区域）加工结束后，刀具以切削进给速度离开工件表面一段距离（5～10mm）后转为以高速（G00）返回安全平面，此转折位置即为退刀平面，其高度为退刀高度。

⑤ 安全平面。安全平面是指当一个曲面切削完毕后，刀具沿刀轴方向返回运动一段距离后，刀尖所在的 $Z$ 平面。它一般被定义在高出被加工零件最高点 10～50mm 的某个位置上，刀具处于安全平面时是安全的，在此平面上也以 G00 速度行进。这样设定安全平面既能防止刀具碰伤工件，又能使非切削加工时间控制在一定的范围内。其对应的高度称为安全高度。刀具在一个位置加工完成后，退回至安全高度，然后沿安全高度移动到下一个位置再下刀进行另一个表面的加工。

**4. 切削方向和切削（走刀）方式的确定**

切削方向是指在切削加工时刀具的运动方向。切削（走刀）方式是指生成刀具运动轨迹时，刀具运动轨迹的分布方式。这两个概念在数控铣削工艺分析时是非常重要的，选择是否合理会直接影响零件的加工精度和生产成本。其选择原则为：根据被加工零件表面的几何形状，在保证加工精度的前提下，使切削加工时间尽可能短。下面分别讨论二维线框的轮廓加工和三维曲面的区域加工中的切削方式、切削方向的选择方法。

（1）二维线框轮廓加工中的切削方向选择

在制订零件轮廓的粗加工工艺时，考虑到零件表面的加工余量大，应采用逆铣方法，以便减少机床的振动。而在制订零件轮廓的精加工工艺时，考虑到精加工的目的是保证零件的加工精度和表面粗糙度，应采用顺铣方法。同时应注意防止刀具直接切入工件表面，留下驻刀痕迹，影响被加工表面的粗糙度，应沿零件轮廓的切线方向切入和切出。

（2）三维曲面区域加工中的切削方向、走刀方式的选择

在三维曲面区域加工的刀具运动轨迹生成技术中，可采用如下三种走刀方式。

① 往复型走刀方式。该走刀方式的特点是：在切削加工过程中顺铣、逆铣交替进行，表面质量较差但加工效率高，如图 2-3-2 所示。

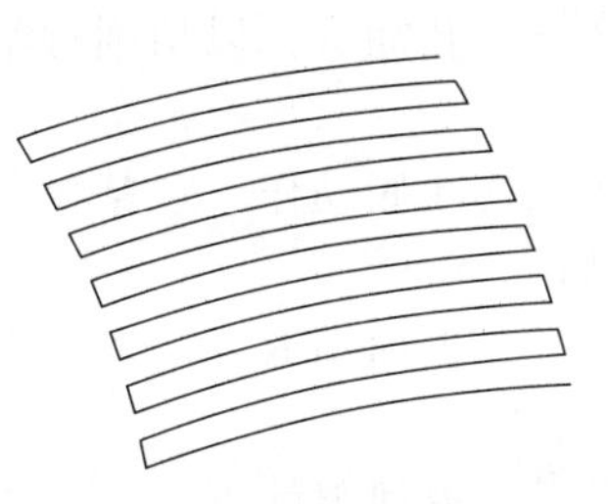

图 2-3-2　往复型走刀轨迹

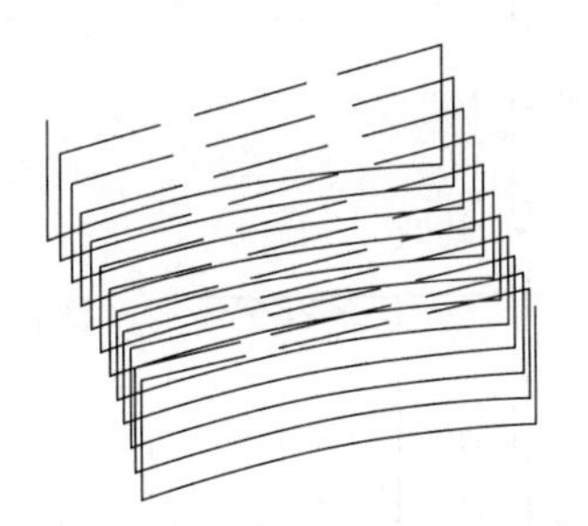

图 2-3-3　单方向走刀轨迹

② 单方向走刀方式。在切削加工过程中能保证顺铣或逆铣的一致性，编程员可根据实际加工要求选择顺铣或逆铣一种走刀方式。由于该走刀方式在完成一条切削轨迹后，附加了一条非切削运动轨迹，因此，延长了机床的加工时间，如图 2-3-3 所示。往复型走刀方式和

单方向走刀方式又称行切走刀方式。

③ 环切走刀方式。该刀具运动轨迹是一组被加工曲面的等参数封闭曲线，如图 2-3-4 所示。环切轨迹又分为等距环切、依外形环切和螺旋环切等，可以从外向内环切，也可以从内向外环切。环切走刀方式既可保证顺铣或逆铣的一致性，又无非切削运动轨迹，加工效率高，且加工轨迹均匀，因此，它是生成封闭环状曲面刀具运动轨迹的主要方法。

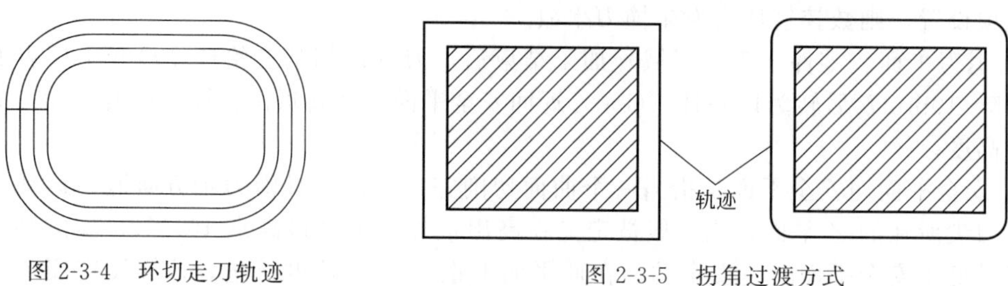


图 2-3-4　环切走刀轨迹

图 2-3-5　拐角过渡方式

外轮廓拐角过渡是指在切削过程中遇到外轮廓拐角时的过渡方式，一般为尖角和圆弧两种过渡方式，如图 2-3-5 所示。

尖角过渡方式是指刀具从轮廓的一边到另一边的过程中，以直线的方式过渡。此方式刀具易产生“超程”现象且工件边角比较尖锐。

圆弧过渡方式是指刀具从轮廓的一边到另一边的过程中，以圆弧的方式过渡。此方式刀具不易产生“超程”现象但工件边角比较圆滑。

## 四、孔加工指令

数控加工中，某些加工动作循环已经典型化。例如，钻孔、镗孔的动作是孔位平面定位、快速进给、工作进给、快速退回等，这样一系列典型的加工动作已经预先编好程序，存储在内存中，可用包含 G 代码的一个程序段调用，从而简化编程工作。这种包含了典型动作循环的 G 代码称为循环指令 。

常用的固定循环指令能完成的工作有钻孔、攻螺纹和镗孔等。这些循环通常包括下列 6 个基本操作动作（见图 2-3-6）。

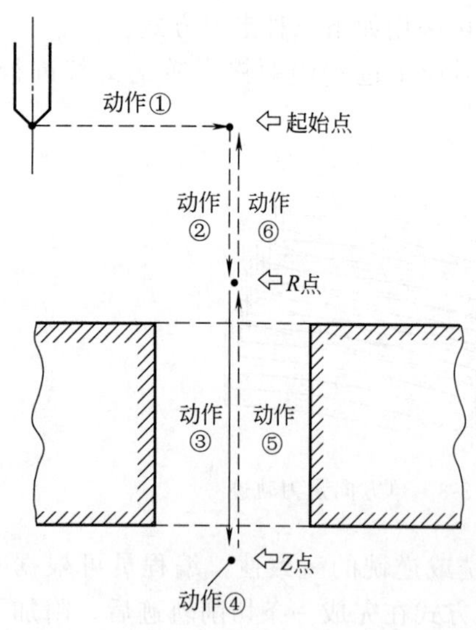


图 2-3-6　孔加工 6 个动作

动作①——$X$ 轴和 $Y$ 轴定位：使刀具快速定位到孔加工的位置。

动作②——快进到 $R$ 点：刀具自起始点快速进给到 $R$ 点。

动作③——孔加工：以切削进给的方式执行孔加工的动作。

动作④——孔底动作：包括暂停、主轴准停、刀具移动等动作。

动作⑤——返回到 $R$ 点：继续加工其他孔时，安全移动刀具。

动作⑥——返回起始点：孔加工完成后一般应返回起始点。

固定循环的程序格式包括数据形式、返回点平面、孔加工方式、孔位置数据、孔加工数据和循环次数。数据形式（G90 或 G91）在程序开始时就已指定，因此，在固定循环程序格式中可不注出。固定循环的程序格式如下：

$\begin{Bmatrix} G90 \\ G91 \end{Bmatrix}$ $\begin{Bmatrix} G98 \\ G99 \end{Bmatrix}$ G□□ X__ Y__ Z__ R__ Q__ P__ F__ L__;

式中 G——孔加工固定循环（G73～G89），见表2-3-3；

X，Y——孔在 *XY* 平面的坐标位置（绝对值或增量值）；

Z——孔底的 *Z* 坐标值（绝对值或增量值）；

R——*R* 点的 *Z* 坐标值（绝对值或增量值）；

Q——每次进给深度（G73、G83），或刀具位移量（G76、G87）；

P——暂停时间，ms；

F——切削进给的进给量，mm/min；

L——固定循环的重复次数，只循环一次时 L 可不指定。

注意：

① G73～G89 是模态指令。G01～G03 取消。

② 固定循环中的参数（Z、R、Q、P、F）是模态的。

③ 在使用固定循环指令前要使主轴启动。

④ 固定循环指令不能和后指令 M 代码同时出现在同一程序段。

⑤ 在固定循环中，刀具半径尺寸补偿无效，刀具长度补偿有效。

⑥ 当用 G80 取消固定循环后，那些在固定循环之前的插补模态恢复。

**表 2-3-3　常用孔加工指令**

| G 代码 | 钻孔操作（−Z 方向） | 在孔底位置的操作 | 退刀操作（+Z 方向） | 用途 |
| --- | --- | --- | --- | --- |
| G73 | 间歇进给 | — | 快速进给 | 高速深孔钻循环 |
| G74 | 切削进给 | 暂停→主轴正转 | 切削进给 | 反攻螺纹 |
| G76 | 切削进给 | 主轴准确停止 | 快速进给 | 精镗 |
| G80 | — | — | — | 取消固定循环 |
| G81 | 切削进给 | — | 快速进给 | 钻孔、锪孔 |
| G82 | 切削进给 | 暂停 | 快速进给 | 钻孔、阶梯镗孔 |
| G83 | 切歇进给 | — | 快速进给 | 深孔钻循环 |
| G84 | 切削进给 | 暂停→主轴反转 | 切削进给 | 攻螺纹 |
| G85 | 切削进给 | — | 切削进给 | 镗削 |
| G86 | 切削进给 | 主轴停止 | 快速进给 | 镗削 |
| G87 | 切削进给 | 主轴正转 | 快速进给 | 背削 |
| G88 | 切削进给 | 暂停→主轴停止 | 手动 | 镗削 |
| G89 | 切削进给 | 暂停 | 切削进给 | 镗削 |

（1）G81：钻孔循环（定点钻）（见图2-3-7）

```
G98(G99)G81 X__Y__Z__R__F__K__
```

（2）G82：带停顿的钻孔循环（见图2-3-8）

```
G98(G99)G82 X__Y__Z__R__P__F__K__
```

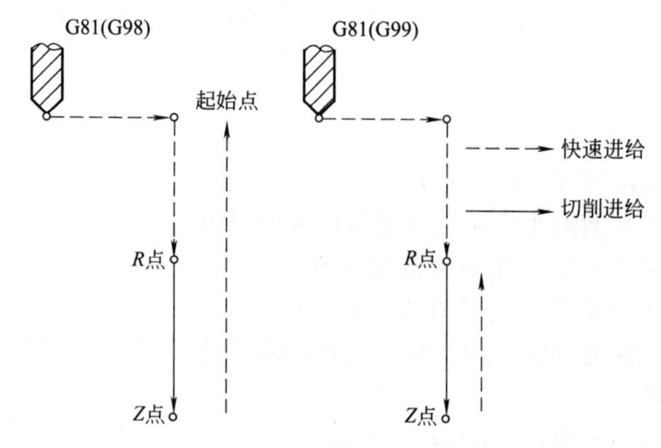


图 2-3-7　钻孔循环 G81 指令动作

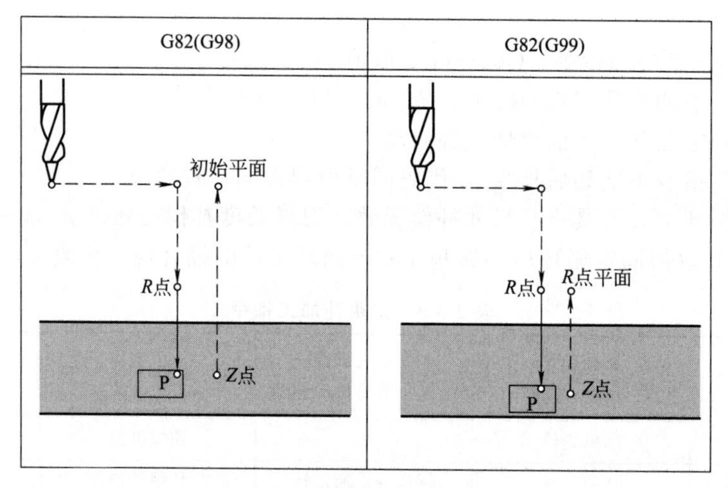


图 2-3-8　钻孔循环 G82 指令动作

功能：此指令主要用于加工沉孔、盲孔，以提高孔深精度。该指令除了要在孔底暂停外，其他动作与 G81 相同。

（3）G83：深孔加工循环（见图 2-3-9）

```
G98(G99)G83X__Y__Z__R__Q__F__K__
```

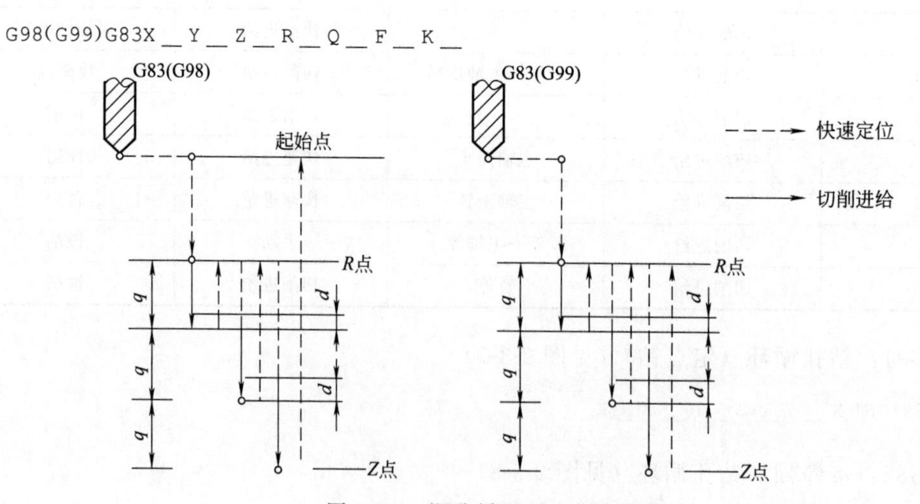


图 2-3-9　深孔循环 G83 指令动作

功能：该固定循环用于 $Z$ 轴的间歇进给，每向下钻一次孔后，快速退到参照 $R$ 点，然后快进到距已加工孔底上方为 $K$ 的位置，再工进钻孔。使深孔加工时更利于排屑、冷却。

（4）G73：高速深孔加工循环

```
G98(G99)G73 X__Y__Z__R__Q__F__K__
```

功能：该固定循环用于 $Z$ 轴的间歇进给，使深孔加工时容易排屑，减少退刀量，可以进行高效率的加工。

（5）G74：攻左旋螺纹循环指令（见图 2-3-10）

```
G74  X__Y__  Z__  R__  F__  ;
```

式中，F 为攻螺纹的进给速度，mm/min，其计算公式为螺纹导程 $P$(mm)×主轴转速 $n$(r/min)。

G74 指令用于切削左旋螺纹孔。主轴反转进刀，正转退刀，正好与 G84 指令中的主轴转向相反，其他运动均与 G84 指令相同。

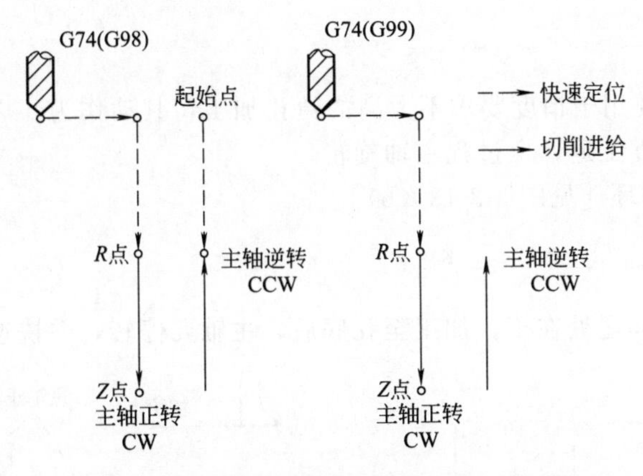


图 2-3-10　攻左旋螺纹循环 G74 指令动作

（6）G76：精镗循环（见图 2-3-11）

```
G98(G99)G76 X__Y__Z__Q__F__K__;
```

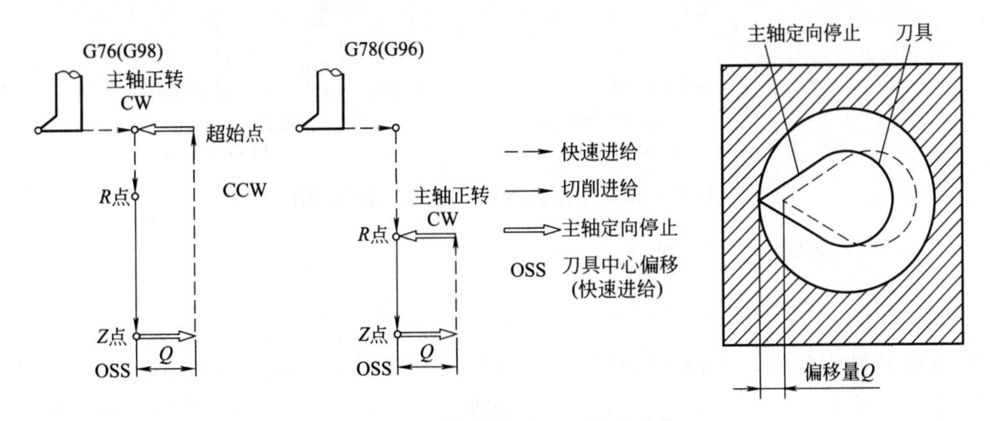


图 2-3-11　精镗循环 G76 指令动作

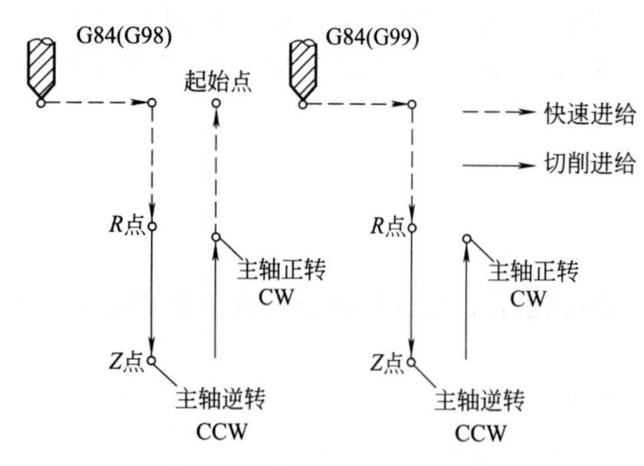


图 2-3-12 攻右旋螺纹循环 G84 指令动作

G76 指令用于精镗孔加工。镗削至孔底时，主轴停止在定向位置上，即准停，再使刀尖偏移离开加工表面，然后再退刀。这样可以高精度、高效率地完成孔加工而不损伤工件已加工表面。

程序格式中，Q 表示刀尖的偏移量，一般为正数，移动方向由机床参数设定。

（7）G84：攻右旋螺纹循环指令（见图 2-3-12）

```
G98(G99)G84 X__Y__Z__R__P__F__K__
```

G84 指令用于切削右旋螺纹孔。向下切削时主轴正转，孔底动作是变正转为反转，再退出。F 表示导程，在 G84 切削螺纹期间速率修正无效，移动将不会中途停顿，直到循环结束。

（8）G85：粗镗循环［见图 2-3-13（a）］

```
G98(G99)G85 X__Y__Z__R__F__K__
```

功能：该指令主要用于精度要求不太高的镗孔加工，其动作为：以 F 的速度工进镗孔、孔底延时，以 F 的速度工退，全过程主轴旋转。

（9）G86：镗孔循环［见图 2-3-13（b）］

```
G98(G99)G86 X__Y__Z__R__F__K__
```

G86 与 G85 的不同之处在于，加工至孔底后，主轴要停转，并快速退刀。

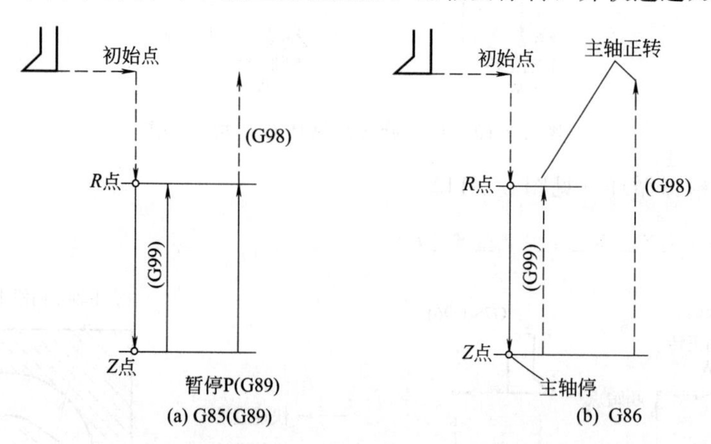


图 2-3-13 镗孔循环 G85 和 G86 指令动作

（10）G87 反镗循环指令（见图 2-3-14）

```
G87  X__Y__Z__R__Q__F__K__;
```

*X* 轴和 *Y* 轴定位后，主轴定向停止，刀具以与刀尖相反的方向按 *Q* 值给定的偏移量偏移并快速定位到孔底（*R* 点），在这里刀具按原偏移量（*Q* 值）返回，然后主轴正转，沿 *Z* 轴向上加工到 *Z* 点，在这个位置主轴再次定向停止后，刀具再次按原偏移量反向移动，然后主轴向孔的上方快速移动到达初始平面，并按原偏移量返回后主轴正转，继续执行下一个

程序段。采用这种循环方式时，只能让刀具返回到初始平面而不能返回到 $R$ 点平面，因为 $R$ 点平面低于 $Z$ 点平面。本指令的参数设定与 G76 通用。

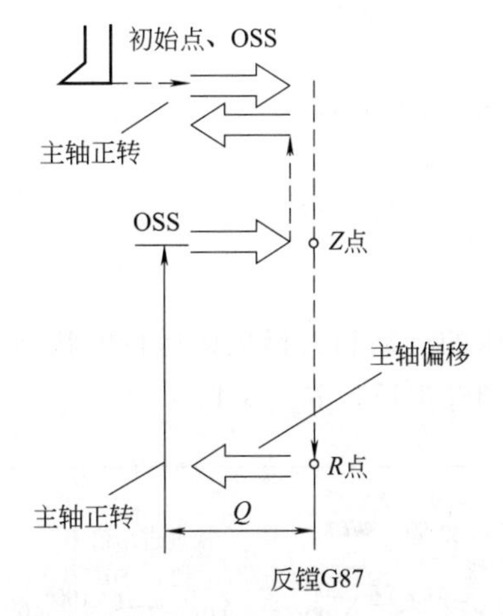


图 2-3-14　反镗孔循环 G87 指令动作

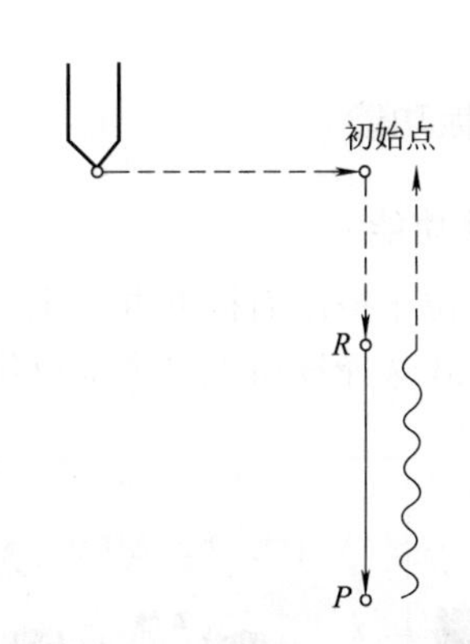


图 2-3-15　镗孔循环 G88 指令动作

（11）G88：镗孔循环（手镗）（见图 2-3-15）

```
G98(G99)G88 X__Y__Z__R__P__F__K__
```

刀具到达孔底后延时，主轴停止且系统进入进给保持状态，在此情况下可以执行手动操作，但为了安全起见应当先把刀具从孔中退出，为了再启动加工，手动操作后应再转换到纸带方式或存储器方式，按循环启动按钮，刀具快速返回到 $R$ 点（G99）或初始点（G98），然后主轴正转。

（12）G89：锪镗循环、镗阶梯孔循环

```
G98(G99)G89 X__Y__Z__R__P__F__K__
```

此指令与 G86 指令相同，但在孔底有暂停（孔底延时、停主轴）。

（13）G80

当用 G80 取消孔加工固定循环后，固定循环指令中的孔加工数据也被取消。那些在固定循环之前的插补模态恢复。

例对图 2-3-16 所示的 5 × $\phi$8mm、深为 50mm 的孔进行加工。显然，这属于深孔加工。利用 G73 进行深孔钻加工，运用 G80 指令，程序为：

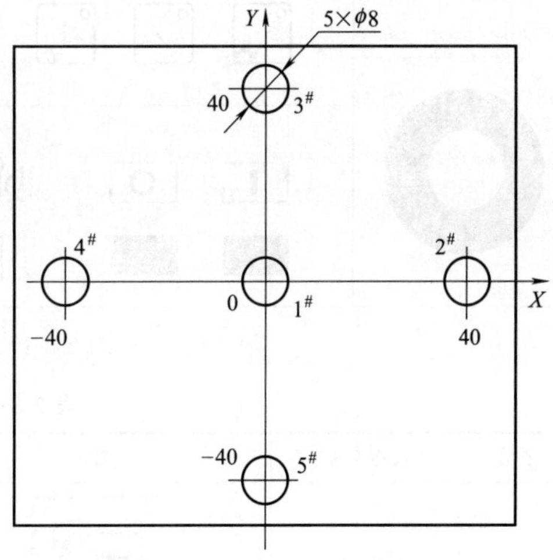


图 2-3-16　G80 指令的应用

```
O40
N10 G56 G90 G1 Z60 F2000;          选择 2 号加工坐标系，到 Z 向起始点
N20 M03 S600;                       主轴启动
N30 G98 G73 X0 Y0 Z-50 R30 Q5 F50;  选择高速深孔钻方式加工 1 号孔
N40 G73 X40 Y0 Z-50 R30 Q5 F50;     选择高速深孔钻方式加工 2 号孔
N50 G73 X0 Y40 Z-50 R30 Q5 F50;     选择高速深孔钻方式加工 3 号孔
N60 G73 X-40 Y0 Z-50 R30 Q5 F50;    选择高速深孔钻方式加工 4 号孔
```

```
N70 G73 X0 Y-40 Z-50 R30 Q5 F50;      选择高速深孔钻方式加工 5 号孔
N80 G01 Z60 F2000;                    返回 Z 向起始点
N90 G80;                              取消孔加工固定循环
N100 M05;                             主轴停
N100 M30;                             程序结束并返回起点
```

## 五、控制面板功能

### 1. 功能键功能

功能键区位于控制面板下方，主要控制机床的运动和选择机床运行的状态，由数控程序运行开关、方式选择按钮等多个部分组成，见图 2-3-17、表 2-3-4。

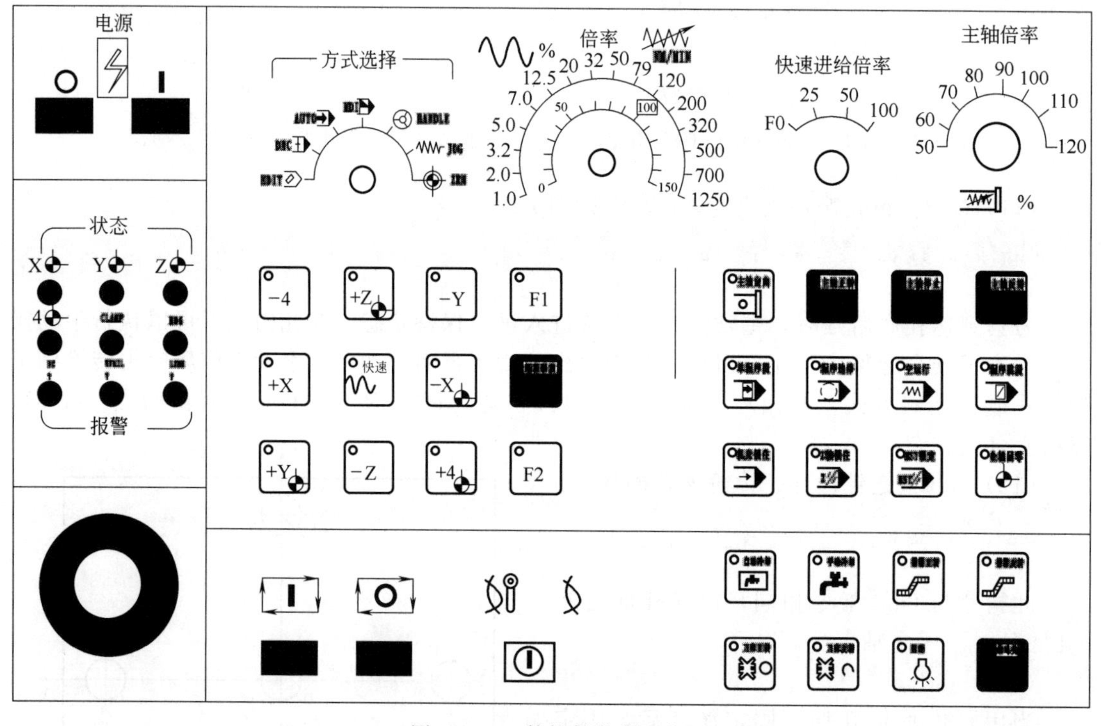


图 2-3-17　数控铣控制面板

**表 2-3-4　功能区说明**

| 序号 | 按键名称 | 图标 | 功能说明 |
|---|---|---|---|
| 1 | 自动加工 | | 按下此键，可以按【循环启动】键，完成程序的自动运行 |
| 2 | 编辑模式 | | 按下此键，可以进行数控程序的输入与编辑 |

续表

| 序号 | 按键名称 | 图标 | 功能说明 |
|---|---|---|---|
| 3 | MDI 模式 |  | 按下此键，MDI 手动数据输入，可操作系统面板并设置必要的参数 |
| 4 | DNC 传输 |  | 按下此键，DNC 远程控制 |
| 5 | 单段运行 |  | 按下此键，进入单段运行方式，每次执行一行数控指令 |
| 6 | 跳步功能 |  | 按下此键，在单行前加上“/”符号，表示此行跳步，即不执行此行的动作 |
| 7 | M01 选择性停止 |  | 按下此键，表示为 M01 选择性停止按钮，若此键开启，表示 M01 指令起作用。若此键关闭，表示 M01 指令不起作用 |
| 8 | 回零 |  | 处于此位置，操作【+X】、【+Y】、【+Z】等相应键，可以使机床返回参考点 |
| 9 | 手动进给 |  | 按下此键，可以进行手动连续进给或步进进给 |
| 10 | 手轮 |  | 按下此键，可以通过操作手轮，在 $X$、$Y$、$Z$ 三个方向进行精确的移动。对刀时常用 |
| 11 | 进给保持 |  | 自动运转时刀具减速并停止进给。再按【循环启动】键，机床继续进给 |
| 12 | 循环启动 | 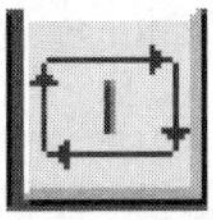 | 按下此键，自动运转启动并执行程序。在自动运转中自动运转指示灯亮 |
| 13 | 回原点显示 |  X 原点灯 Y 原点灯 Z 原点灯 | 各轴回原点显示为灯亮 |

续表

| 序号 | 按键名称 | 图标 | 功能说明 |
| --- | --- | --- | --- |
| 14 | 手动轴选择 | X Y Z | 手动模式时，按下相应的X/Y/Z按键，然后按+/-键，手动控制机床沿相应的坐标轴运动 |
| 15 | 多轴联动 | 4 5 6 | 为4、5、6轴预留手动轴按钮 |
| 16 | 快速进给 | + ∿ - | 按下此键，配合当前轴的正负方向，刀具快速进给 |
| 17 | 主轴正/停止/反转 |  | 按下对应键，主轴分别正/停止/反转 |
| 18 | 急停 | EMERGENCY STOP | 当遇到突发事件或紧急情况时按下此键，机床则处于紧急停止状态，排除故障后，需要朝按钮上箭头的方向旋转才能使紧急按钮键复位。紧急停止状态的复位还需按MDI面板上的"RESET"键 |
| 19 | 进给速率修调 | 0 1 2 4 6 8 10 15 20 30 40 50 60 70 80 90 95 100 105 110 120 (%) | 选择自动运行和手动运行时进给速度的倍率。可以调节数控程序运行中的进给速度，调节的范围在0～120% |
| 20 | 主轴转速修调 | 50 60 70 80 90 100 110 120 (%) | 选择自动运行和手动运行时主轴转速的倍率。可以调节数控程序运行中的进给速度，调节的范围在50%～120% |
| 21 | 启动 | 启动 | 按下此键，接通CNC的电源 |
| 22 | 停止 | 停止 | 按下此键，断开CNC的电源 |

续表

| 序号 | 按键名称 | 图标 | 功能说明 |
| --- | --- | --- | --- |
| 23 | 超程解锁 | 超程/释放 | 按下此键，在超程的情况下可解锁 |
| 24 | 手轮 |  | 单步进给量倍率控制键，与手轮配合使用，表示工作台面的每一步进给量，×1 为 0.001mm，×10 为 0.01mm，×100 为 0.1mm。手轮顺时针旋转为机床正方向移动，手轮逆时针旋转为机床负方向移动。手轮刻度盘上的每一格表示移动倍率的一个单位，移动的每个单位分别有 0.1mm、0.01mm、0.001mm 三个挡位和速度变化的三个键配合使用 |

**2. 输入区功能**

地址输入键区，见图 2-3-18。

地址输入区域是用于输入数据到输入域中。该系统能够自动判别选取的是数字还是字母。举例说明，如直接输入上面的地址键 GR，将在 CRT 屏幕上显示字母 G。如果先按下 SHIFT 键后，再按下 GR 键，则将在 CRT 的屏幕上显示字母 R，用同样的方法也可以切换数字和字母，如 F 和 L，9 与 D 等。

图 2-3-18　地址输入键区

(1) 地址输入键区图标及功能（见表 2-3-5）

**表 2-3-5　地址输入键区图标及功能**

| 序号 | 图标 | 功能说明 |
| --- | --- | --- |
| 1 | POS | 该按键显示机床现有的位置，可以显示坐标位置屏幕，按下该按键可以显示位置屏幕，位置显示有三种方式：相对坐标、绝对坐标和综合坐标。其中综合坐标又包括了机械坐标、相对坐标、绝对坐标和剩余进给四项内容 |
| 2 | PROG | 该按键必须要配合在 EDIT 的方式下进行编辑和修改程序，它是程序显示与编辑的按键，可以显示存储器里的程序 |
| 3 | OFFSET SETTING | 用于坐标系设定、显示补偿值和宏程序变量。按第一次进入坐标系设定页面，按第二次进入刀具补偿参数页面 |

续表

| 序号 | 图标 | 功能说明 |
| --- | --- | --- |
| 4 | SYSTEM | 系统参数设定页面键，按下此按键可以显示系统参数屏幕及自诊断数据的显示等 |
| 5 | MESSAGE | 信息页面键，按下此键可以显示屏幕中的信息，如"报警"信息等 |
| 6 | CUSTOM GRAPH | 图形显示页面键，通过该键可以显示用户加工完成的图形 |

（2）编辑键图标及功能（见表 2-3-6）

**表 2-3-6　编辑键图标及功能**

| 序号 | 图标 | 功能说明 |
| --- | --- | --- |
| 1 | SHIFT | 换挡键，在键盘中按键一般具备两种功能，按下"SHIFT"键可以在两个功能之间相互切换 |
| 2 | ALTER | 替代键，此键可用于输入域内的数据替代光标所在位置处的数据 |
| 3 | CAN | 取消键，取消已输入到缓冲器中的最后一个字符或数据。如当前输入 G02 X30 Z500 后按下此键，最后的数字 0 就删除了，变成 G02 X30 Z50 |
| 4 | INSERT | 插入键，此键是将把输入区域的数据插入到当前光标位置之后 |
| 5 | INPUT | 输入键，把输入域内的数据输入参数的页面或者输入一个外部的数控程序 |
| 6 | DELETE | 删除键，此键可以删除光标处所在的数据，也可删除一个程序或者删除全部程序 |

（3）翻页键图标及功能（见表 2-3-7）

表 2-3-7 翻页键图标及功能

| 序号 | 图标 | 功能说明 |
|---|---|---|
| 1 | ↑PAGE | 此键用于屏幕显示时页面向前翻页 |
| 2 | PAGE↓ | 此键用于屏幕显示时页面向后翻页 |
| 3 | ↑ | 此键用于光标向上移动 |
| 4 | ↓ | 此键用于光标向下移动 |
| 5 | ← | 此键用于光标向左移动 |
| 6 | → | 此键用于光标向右移动 |

(4) 其他图表及功能(见表 2-3-8)

表 2-3-8 其他图表及功能

| 序号 | 图标 | 功能说明 |
|---|---|---|
| 1 | HELP | 系统帮助键，操作者在对系统控制面板不熟悉或不清楚时可按下此键，能获得帮助 |
| 2 | RESET | 复位键，此键具备以下功能：①当机床自动运行时按下此键，机床的所有操作都将停止；②能取消机床报警；③可以使机床复位，光标回到整段程序的句首；④在 MDI 方式下清除编辑的程序 |
| 3 | EOB E | 结束程序段键，结束一行程序段的输入，在编程中其中一行的最后出现“;”，则进行换行 |

(5) 加工步骤流程(见表 2-3-9)

表 2-3-9 加工步骤流程

| 序号 | 流程步骤 | 流程内容 |
| --- | --- | --- |
| 1 | 步骤一 | 电源总开关开启，机床正常送电，控制面板中的数控系统正常送电 |
| 2 | 步骤二 | 选择所用机床及操作系统，将 $X$ 轴、$Y$ 轴、$Z$ 轴分别返回到参考点 |
| 3 | 步骤三 | 输入，编辑、检查加工程序，保证程序正确及输入无误 |
| 4 | 步骤四 | 选择工件、装夹工件，选择各种刀具，用试切法测量各个刀具的补偿值并输入刀具补偿表中，注意正负号及小数点 |
| 5 | 步骤五 | 调出当前要加工的程序，进行自动加工，按下循环启动键开始加工，利用单段程序功能边检查边加工，还要配合适当的进给倍率来调节速度，减少因程序或对刀错误而引发的事故 |
| 6 | 步骤六 | 首件加工完成后测量各加工部位的尺寸，修改各刀具补偿值来加工第二个工件，完成后确定无误即可成批量地生产工件 |

## 习题

1. 切入点、切出点的确定原则是什么？
2. 切削（走刀）方式有哪几种？
3. 孔加工指令的动作有哪些？
4. G81 和 G82 指令的区别是什么？
5. 深孔加工有哪些指令？各表示什么？
6. 控制面板中急停的功能是什么？
7. 控制面板中“ALTER”按键表示什么？如何使用？
8. 简述数控铣床加工步骤。
9. 在如图 2-3-19 所示的孔类零件上，钻削 5 个 $\phi10$ 的孔。试选用合适的刀具，并编写加工程序。

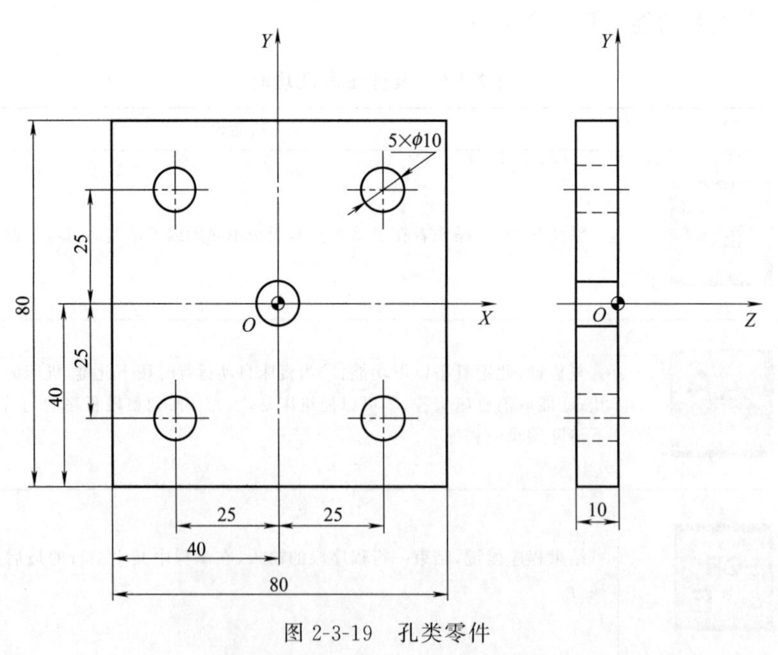


图 2-3-19 孔类零件

10. 在数控铣床上完成盖板零件（图 2-3-20）的加工，已知毛坯为 120mm×80mm×

30mm，材料为 45 钢。

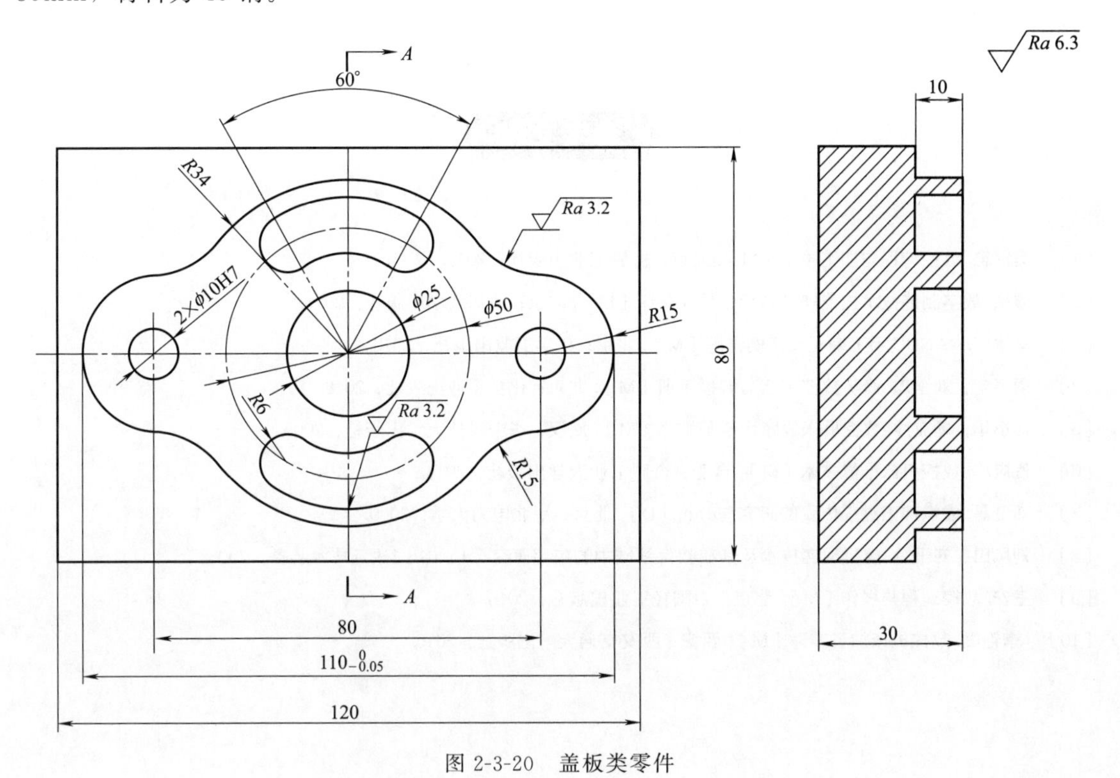


图 2-3-20　盖板类零件

# 参考文献

[1] 吴宜忠. 数控编程与加工技术 [M]. 北京：机械工业出版社，2018.

[2] 谭斌. 数字制造技术技能实训教程：数控车床 [M]. 北京：清华大学出版社，2015.

[3] 李体仁. FANUC 手工编程及实例详解 [M]. 北京：化学工业出版社，2018.

[4] 周湛学，刘玉忠. 简明数控工艺与编程手册 [M]. 北京：化学工业出版社，2018.

[5] 郑小年，杨克冲. 数控机床故障诊断与维修 [M]. 武汉：华中科技大学出版社，2005.

[6] 贺应和. 数控机床维修技术 [M]. 合肥：合肥工业大学出版社，2014.

[7] 刘金磊. 虚拟加工过程仿真的研究与分析 [D]. 北京：华北电力大学，2009.

[8] 刘凤田，刘玉兰. 虚拟现实技术及其在教育领域中的应用研究 [J]. 中国教育技术装备，2005.

[9] 王安. 数控编程与操作 [M]. 北京：中国轻工业出版社，2015.

[10] 李艳霞. 数控机床加工零件 [M]. 西安：西安交通大学出版社，2016.

Bilingual Textbooks of Vocational Education

职业教育双语教材

# NC Machining Technology

# 数控加工技术

Edited by Hui Zhao An Wang

赵 慧 王 安 主编

Subeditor Wanju Liu

刘万菊 副主编

Co-edited by MOHAMED AHMED ALI BAIOUMY MOHAMED

MOHAMED AHMED ALI BAIOUMY MOHAMED 参编

Reviewed by Yunmei Li

李云梅 主审

Chemical Industry Press

化学工业出版社

· Beijing ·

· 北 京 ·

# Foreword

In order to expand the vocational education cooperation with the countries along the routes and implement the requirements of program "outputting excellent vocational education results abroad and sharing with the world" which is launched and carried out by Tianjin, as a form of education which is most closely connected with manufacturing, vocational education is playing a decisive role. The research group compiles the teaching material NC Machining Technology in order to cooperate with the theory and practical teaching of "Luban Workshop" in India, carry out exchange and cooperation, improve the international influence of Chinese vocational education, innovate international cooperation mode of vocational colleges and output the excellent resources of vocational education in China.

CNC machine tool is a kind of cold processing equipment to process components by executing the programme code and achieving the coordination of various components through computer with the digital operation as the center. It is the core technology of intelligent manufacturing. It is mainly used in IT, automotive, light industry, medical, aviation and other industries and is a kind of machining equipment with high speed, high precision and high reliability.

Taking Luban Workshop as the background, the CNC machining equipment of Luban Workshop as the carrier and the training of senior skilled talents as the goal, the book emphasizes practical education links, tries to achieve profound and simple learning and be easy to teach and highlights the characteristics of higher vocational education while paying attention to basic theoretical education. It is suitable for engineering students in higher vocational colleges and is also suitable for Cairo Advanced Maintenance Technology School, Egypt.

CNC technology is composed of two parts, programming and operation of CNC lathe and programming and operation of CNC milling machine.

There are two projects in this book. The Project Ⅰ was prepared by Zhao Hui and the Project Ⅱ was prepared by Wang An. Professor Liu Wanju gave a careful guidance on the preparation of the book. The Egyptian teacher, MOHAMED AHMED ALI BAIOUMY MOHAMED, rationally adjusted the content by contacting the actual situation of Egypt on the content of the textbook. Teachers of Tianjin Light Industry Vocational Technical College Wang Juan, Wang Danyang, Guo Shijie, Li Xiaotong participated in the translation work of this book.

Due to the editor's limited knowledge and limited time, there are many omissions and mistakes in the book, so readers are urged to criticize and correct.

**Editor**

**Oct, 2019**

# Contents

# Project Ⅰ

## CNC Lathe Machining

**【Project Description】**

CNC lathe is a kind of mechanical-electrical integrated automatic machine tool with high-precision and high-efficiency, and is also the most widely used CNC machine tools, accounting for about 25% of the total number of CNC machine tools. CNC lathe is mainly applied for machining rotational parts including shaft sleeves and plates, through the operation of NC machining program, which can automatically complete the internal and external cylindrical surface, conical surface, forming surface, thread and end-face and other cutting processes and can carry out operations such as slot turning, drilling, expanding hole, reaming and so on. The turning center can complete the more machining processes in one time so as to improve machining precision and production efficiency, especially suitable for machining rotational parts with complex shape. FANUC 0i system is a kind of CNC system which is widely used in CNC lathe currently, whose programming method and instruction formats are quite typical. The teaching in this project is developed based on the CNC lathe (equipped with FANUC 0iFD system) of Shandong Chenbang CNC Equipment Co., Ltd.

**【Capability Goals】**

1. Understand the composition of CNC lathes.
2. Understand the machining characteristics of NC turning.
3. Master common instructions of G code and M code.
4. Master the establishment principle of coordinate system.
5. Be proficient in the usage of instructions for CNC lathe G00, G01, G02, G03, etc..
6. Be proficient in the usage of cyclic instructions such as G71, G70, etc..
7. Master thread turning instructions for G32, G92, etc..

## Task Ⅰ  Master Safe and Civilized Production of CNC Lathe

### Ⅰ. Key Points of Civilized Production

Civilized production is a quite important content in enterprise management, which directly affects product quality, the using effect and service life of the equipment, tools, jigs and measurement tools, and also influences the operating skill of the workers. The students in the technical school are the backup workers for the factory, who should pay attention to the cultivation of the good habit of civilized production when studying this course at the beginning. Therefore, the operator must do the followings in the whole process of operation.

① After entering the NC practice workshop, operator should obey the arrangement and comply with the commands. Do not to start or operate the CNC system without authorization.

② Before operating, check carefully whether the parts of the machine are in good conditionor not, and whether the position of the transmission handle and gear shifting lever (mainly refers to the economical CNC lathe) is correct. The plugs and sockets of CNC system and electrical accessories should also be carefully checked to ensure reliable connection.

③ For a modified dual-purpose (automatic control and manual operation) CNC machine tool, it is necessary to check whether its transmission mechanism interferes with each other before use in order to ensure that the equipment is not damaged and can run normally.

④ The main body of the machine tool shall be used and maintained in accordance with the relevant requirements of the general machine tool.

⑤ When the CNC system is not in use, it shall be covered with a cloth to prevent entering the dust, and shall, under the guidance of professional personnel, regularly carry out internal dust removal or careful cleaning.

⑥ Before operating the CNC system, check whether the cooling fan on both sides is running normally to ensure good heat dissipation effect.

⑦ When operating a CNC system, the strength for operating the buttons and switches shall not be too severe, and no wrench or other tool shall be allowed to operate it.

⑧ In the absence of thread turning, the spindle pulse generator shall be disconnected from the spindle in order to prolong its service life.

⑨ When the auto-rotating tool-rest is not in place, it shall not be forced to position the tool rest abnormally by external force so as to prevent damage to the internal structure of the tool-rest.

⑩ Although NC machining is done automatically not belonging to the unmanned processing, it still requires constant observation by the operator, and does not permit the casual departure from the production position.

⑪ When going off work, in addition to being shut down according to regulations, operator should also do the shift work carefully and keep a well written record (such as the processing procedure and the execution of the program).

## Ⅱ. Safe Operating Techniques

The CNC machine have the characteristics of high precision, high efficiency and high adaptability, and its operation efficiency, failure rate of the accessories and service life depend on the correct use and maintenance of the users to a great extent. Good working environment, good users and maintainers will greatly extend trouble-free working hours, improve productivity, reduce the wear and tear of mechanical components and avoid unnecessary mistakes, thus reducing the burden on maintenance personnel. When operating, it is necessary to improve the self-awareness for implementing discipline, strictly abide by the various rules and regulations relying mainly on safety technical requirements, and earnestly achieve the followings:

① After entering the training workshop for CNC turning, operator shall be subject to the arrangement and shall not start or operate the CNC system of lathe without authorization.

② Dress and wear labor protection supplies according to regulations.

③ No high heels, slippers, gloves or scarves are allowed in operation.

④ Before starting the machine, check carefully whether all parts of the lathe are in

good condition, whether the position of the transmission handle and gear shifting lever is correct, and carefully lubricate and maintain the machine tool in accordance with the requirements.

⑤ When operating the CNC panel, the strength for operating the buttons and switches shall not be too heavy, and no spanner or other tools shall be allowed to operate.

⑥ After the tool setting is finished, a simulated tool changing test should be made to prevent the tool, work-piece and equipment from being damaged during the formal operation.

⑦ In the process of CNC turning, due to the fact that the observation time is more than the operation time, it is necessary for operator to properly choose the observation position, and operator is not allowed to leave the training post at will to ensure the safety.

⑧ When operating NC system panel and NC machine tool, it is forbidden to operate by two people at the same time.

⑨ In automatic operation, the operator should concentrate the thought, with the left hand fingers being placed on the stop button of the program, the eye observing the movement of nose of tool, and the right hand controlling adjust switch in order to control the running speed of machine tool slide, find problems in time and press the stop button of the program, thus ensuring the safety of the cutter and the CNC machine tool and avoiding all kinds of accidents.

⑩ At the end of the training, not only should CNC machine tools be maintained according to the regulations, but also the handover work should be done conscientiously, moreover, the written records should be made when necessary.

⑪ Wear a close protective clothing with cuffs being fastening and hem of the coat being closing before operation. Do not wear gloves. Do not wear or change clothes or wear cloth around the moving machine to prevent being injured from machine. Must wear a hard hat, with braids being put in the hat. Do not wear a skirt and slippers. Wear protective glasses to prevent iron spatters from injuring your eyes.

⑫ Before the lathe is started, labor protection articles must be properly worn in accordance with the requirements of safe operation. All parts and protective devices of the machine must be carefully checked to ensure whether they are in good condition and safe and reliable. Lubricate the machine tool and run it at low speed without load for 2-3 minutes to check whether the machine tool runs normally.

⑬ When loading and unloading the chuck and the large-sized piece, check whether there is obstacle around. Well pad the board to protect the lathe surface, seize it up and keep it be firmly fastened and putted. When the car leans towards heavy objects, it should be balanced by weight. Work-piece and the clamping of the tools should be tightened to prevent the work-piece or tools from flying out of the fixture. The chuck wrench and the socket wrench should be taken off.

⑭ Gloves are strictly prohibited in the operation of machine tools; It is strictly forbidden to touch the rotating part of the machine tool by hand, and transfer objects between lathes in the operation of the lathe. Loading and unloading parts, installing cutter, oiling and cleaning of cuttings should be done when the machine tool is stopped. Use a brush or hook to remove iron scraps instead of hand.

⑮ When the machine is running, it is not allowed to measure the work-piece, and stop the rotating chuck by hand. When using sandpaper, it should be put on file, it is strictly prohibited to use sandpaper with gloves, abrasive sandpaper and file without handle are

not allowed to be used. No forward and reverse electric brakes are used as brake, it should go through the middle brake process.

⑯ The cutting parameters of the machining work-piece should be selected according to the technical requirements so as to avoid accidents caused by the overload of the machine tool.

⑰ In the process of machining and cutting, exit the cutter when stopping the lathe. The cutting for long shaft shall use the center frame to prevent injuries caused by work-piece bending deformation. The bar drawn into the machine head does not exceed the length of vertical shaft of the machine head, with slow processing. When stretching out, pay attention to protection.

⑱ When cutting in high-speed, the protective cover is required, work-piece and tool should be fixed firmly. When iron scraps splash heavily, the baffle should be installed around the machine to isolate it from the operating area.

⑲ When the machine tool is running, the operator can't leave the machine tool. If the machine tool doesn't work properly, stop the machine immediately and ask the repairman to check and repair it. When a sudden power failure occurs, shut down the machine tool immediately and remove the cutter from the workshop.

⑳ When working, the operator must stand sideways in the operating position to prohibit the body from facing the rotating workpiece.

㉑ At the end of the work, cut off the power supply or the main power supply of the machine tool, withdraw the cutter and the work piece from the work position, arrange and place the used tools, jigs and measuring tools, and clean the machine tool.

㉒ Local lighting shall be installed on each machine tool, and the lighting on the machine tool shall use a safe voltage (under 36 V).

## Ⅲ. 6S Occupational Regulations

Based on the management spirit of 6S and the current practical application of the school training workshop, the 6S management norms are formulated as follows.

### 1. 1S——SEIRI

(1) Definition

① Distinguish anything in the work-place as the necessary and the unnecessary;

② Separate the necessary from the unnecessary clearly and strictly;

③ Get rid of unnecessary things as soon as possible.

(2) Purpose

① Make room for flexible use of space.

② Prevent misuse and misdeliver.

③ Form a clean workplace.

### 2. 2S——SEITON

(1) Definition

① Sort out and arrange the necessary items left on the scene in order.

② Clear quantity, make effective identification, and do a good job of registration.

(2) Purpose

① The workplace is absolutely clear.

② Keep a neat working environment.

③ Eliminate time spent on searching for items.

④ Eliminate excessive backlogs.

**3. 3S——SEISO**

(1) Definition

① Clean up the working environment and equipment.

② Keep the workplace clean and bright.

(2) Purpose

① Remove the dirt and keep the workshop clean and bright.

② Stabilize the quality.

③ Reduce industrial injury.

**4. 4S——SEIKETSU**

(1) Definition

Institutionalize and standardize the practices implemented in 3S above.

(2) Purpose

Maintain the results achieved by 3S above.

**5. 5S——SHITSUKE**

(1) Definition

By learning, visiting and other means, improve the manners quality of the staff, enhance the sense of team, and form a good working habit act as required.

(2) Purpose

Promote the quality of teachers, staff members and students to make them be serious about any job.

**6. 6S——SECURITY**

(1) Definition

Attach importance to the safety education of all employees, let them have the concept of safety first all the time and take precautions in the future.

(2) Purpose

Set up a safe production environment, with all work being performed under the premise of safety.

**Exercise**

1. What should be paid attention to during the operation of the machine tool?
2. What points should be done for the safe operation of machine tool?
3. Please elaborate the key points and contents of 6S professional regulations.
4. How to interpret the letters and numbers in CAK6140dj/1000?
5. What are the components of CNC lathe and their corresponding functions?
6. Please give an example to explain what modal instruction is and what non-modal instruction is.
7. What are the components of the program?
8. What are the operation modes on the operation panel?

# Task Ⅱ Basic Operation of CNC Lathe

## Ⅰ. A Preliminary Understanding of CNC Lathe

**1. Mark of CNC lathe**

CAK6140dj/100:

C——Machine category code (lathe Class);

A——Manufacturer code (improved batch);

K——Machine tool structure characteristic code (NC);

6——Machine tool set code (horizontal lathe set);

1——Machine tool series code (horizontal lathe series);

40——Main parameter code (maximum turning diameter is 400mm);

d——System model;

j——Protection form (semi closed);

1000——Main parameter code (maximum workpiece length is 1000mm).

**2. Components and functions of CNC lathes**

① Safety doors: For safety purposes, there is a peephole on the door.

② Chuck: Being used for clamping work-piece.

③ NC operation panel: NC programming, controlling the moving parts and adjusting processing parameters.

④ Tool rest: To Install all kinds of turning cutters.

⑤ Tailstock: To install tools such as centers, drills, etc.

⑥ Lead rail: Mainly act as a guide and a support.

## Ⅱ. Instructions for the Operation of Machine Tools

(1) Controller panel (see Figure 1-2-1)

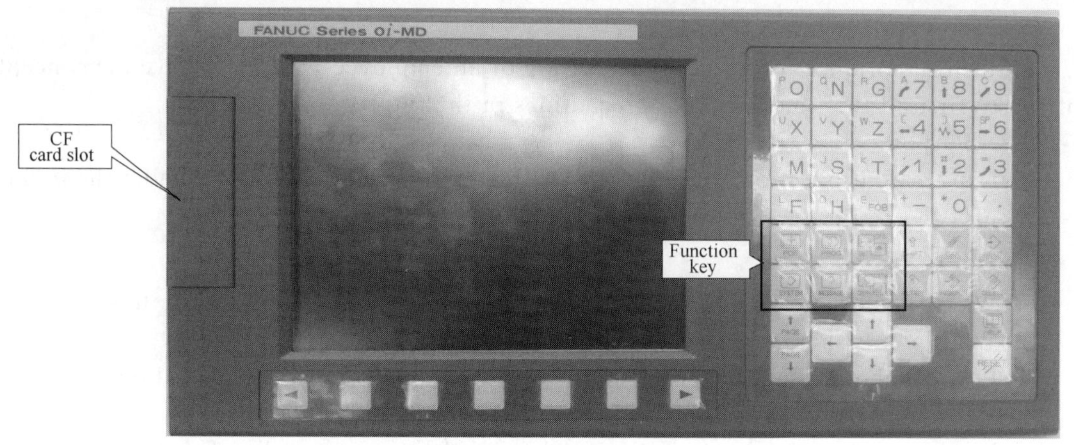


Figure 1-2-1 Controller panel

(2) Operation panel

This machine is equipped with convenient, fast-operating CNC special control panel, panel buttons and other operator schematic diagram as shown Figure 1-2-2.

Figure 1-2-2 Schematic diagram for the operation panel of machine tool

(3) System power on-off switch (see Table 1-2-1)

**Table 1-2-1 System power on-off switch**

| Legend | Description |
| --- | --- |
| | Name: Controller Start Button (Green button)<br>When the button is triggered on the operation panel, the controller power is turned on and the CNC system enters into a starting state |
| | Name: Controller Close button (Red button)<br>When the button is triggered on the operation panel, the controller power is turned off and the CNC system enters into a close state<br>This button is not triggered while the program is running |

(4) Emergency stop (see Table 1-2-2)

**Table 1-2-2 Emergency stop**

| Legend | Description |
| --- | --- |
| EMERGENCY STOP | Name: Emergency Stop Knob (Red knob)<br>The emergency stop knob is red and located in the middle of the left side of the operation panel of this machine tool<br>① In the course of operation or use of this machine tool, in case of emergency, it is necessary to press the button immediately to completely stop the machine tool action and interrupt the current output to the motor, but the machine tool is continuously running<br>② The phenomenon of pushing the emergency stop button<br>a. If the servo shaft is in operation, the running shaft stops moving (if the machine tool is equipped with the fourth shaft which is in operation, the fourth shaft will stop running)<br>b. The spindle in rotation stops rotating |

Continued

| Legend | Description |
| --- | --- |
| 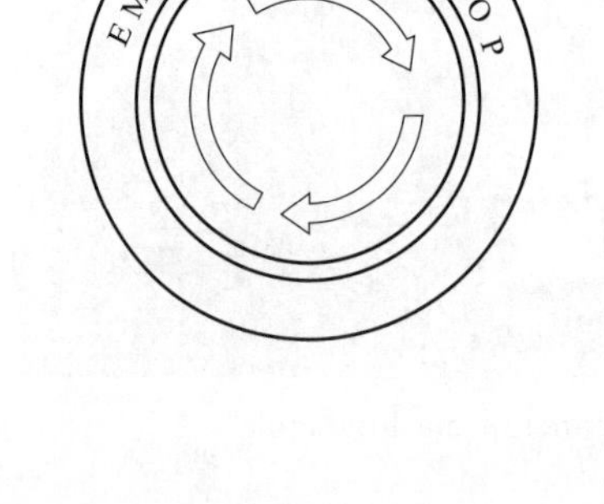  | c. Machine tool displays alarm information screen, with following ALARM information<br>1000 EMG STOP OR OVERTRAVEL<br>d. If press the emergency stop knob when the cutter head is rotating, the cutter head will stop rotating immediately<br>e. If press the emergency stop knob during tool changing process, the tool changing action will stop, and change into the abnormal interruption of tool changing<br>③ Note the following conditions<br>a. This knob shall not be lifted until the emergency has been completely relieved<br>b. All instructions at the time of the stop have been removed from the machine, so the process needs to be re-checked before any related operation are made<br>c. When the automatic tool changing is performed midway, all actions will stop immediately after pressing this knob, so the cutter head may be in an indeterminate position |

(5) Program switch (see Table 1-2-3)

**Table 1-2-3 Program switch**

| Legend | Description |
| --- | --- |
| 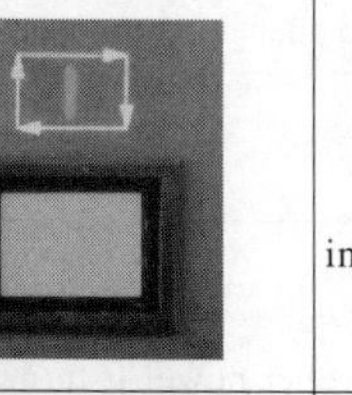 | Name: Program Start Button<br>Select the processing program to be executed under the automatic operation mode (manual input, memory, online), and press the "program start" button to start the program |
|  | Name: Program Pause Button<br>① In the automatic operation mode (manual input, memory, online), press the "program pause" button, and each shaft will immediately decelerate and stop and enter into the operation rest state<br>② After pressing the "program start" button again, the processing program will continue to execute from the single section currently suspended |

(6) The program protection switch function (see Table 1-2-4)

**Table 1-2-4 The program protection switch function**

| Legend | Description |
| --- | --- |
| 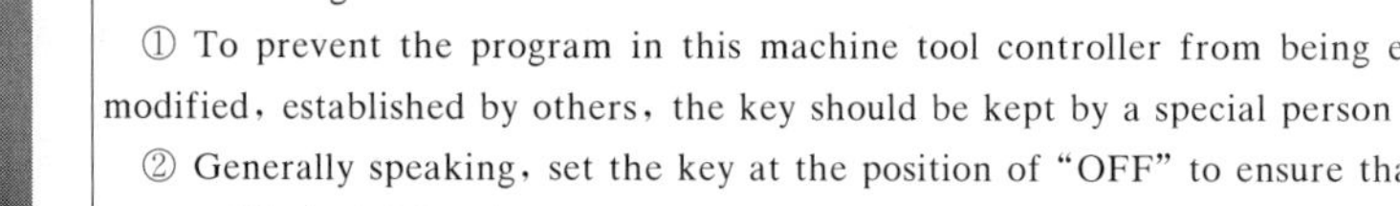 | Name: Program Protection Switch<br>① To prevent the program in this machine tool controller from being edited, canceled, modified, established by others, the key should be kept by a special person<br>② Generally speaking, set the key at the position of "OFF" to ensure that the program is not modified or deleted.<br>③ If want to edit, cancel or modify the program, we should set the key at the position of "ON" |

(7) Operation mode (see Table 1-2-5)

**Table 1-2-5 Operation mode**

| Legend | Description |
|---|---|
| 编辑方式 | Name: Edit Mode<br>The machine tool display reads "edit" in the lower left corner<br>① When the program can be edited, modified, added, or deleted, the program protection switch button is set to "ON" state, which can only be edited in this mode<br>② This mode is only for editing and cannot execute program<br>③ When executing a new editing program, the program must be in automatic mode (manual input, memory) before it can be executed<br>④ The controller is automatically stored when the program is edited, so no storage action is required<br>⑤ In this mode, the machining program, NC parameters, cutter length correction and tool radius correction can be read by individual computer |
| 自动方式 | Name: Auto Mode<br>The machine tool display reads "MEM" in the lower left corner<br>① In this mode, press the "program start" button to execute the currently selected machining program<br>② In this mode, a program in CNC memory can be executed<br>③ The feed rate in this mode can refer to the feed rate adjustment button for instructions<br>④ In this mode, the program ends when it executes the M30 |
| MDI方式 | Name: MDI (Manual Data Input) Mode<br>The machine tool display reads "MDI" in the lower left corner.<br>① In this mode, a single program instruction is entered on the controller MDI panel to be executed<br>② In this mode, after the instruction is executed, parameters can be set to determine whether the programmed program needs to be eliminated<br>③ In this mode, only a portion of the program can be entered |
| 手摇方式 | Name: Hand Wheel Mode<br>The machine tool display reads "HAND" in the lower left corner.<br>① In this mode, each shaft can be moved by a hand-held unit.<br>② In this mode, the axial movements can be controlled by the axial knob and the optional shaft on the hand-held unit so as to control the selected axial movement.<br>③ In this mode, the manual movement speed of each shaft can be determined by the feed-rate knob on the hand-held unit<br>④ The rotation speed of the manual pulse generator should not be greater than 5 laps per second |
| 手动方式 | Name: Hand Motion Mode<br>The machine tool display reads "JOG" in the lower left corner.<br>① In this mode, press each axial button and select the slow feed rate to move each shaft.<br>② In this mode, move the feed rate according to the speed of the slow feed rate. The speed can be adjusted by 0～1000mm/min<br>③ When the axial button is pressed in this mode, it can be moved after specifying axial direction, and the shaft will stop upon releasing the button<br>④ When being used with a fast button, move the feed rate according to the speed of the fast feed rate, press each axial button, move according to the specified shaft, and the shaft will stop upon releasing the button |

Continued

| Legend | Description |
| --- | --- |
|  | Name: Origin Return Pattern Mode<br>The machine tool display reads "REF" in the lower left corner<br>① In manual mode, it is used for the mechanical origin return operation of the servo shaft<br>② After each turn-on, the machine needs to make a origin return action. If the shafts are located near the origin, the shafts need to be manually moved away from the origin before returning to the origin<br>③ In this mode, after selecting the axial button which requires mechanical origin return, the indicator light of the original point continues to flicker until the return action of the origin is completed, then the origin indicator light is in the normal state without flickering<br>④ The mechanical origin return rate of each shaft is multiplied by the zero return speed set by the parameters × the multiple value % of the feed rate switch. |

(8) Auxiliary functions (see Table 1-2-6)

**Table 1-2-6 Auxiliary functions**

| Legend | Description |
| --- | --- |
|  | Name: Single Block<br>This function is only valid in auto-correlation mode<br>① When the light is on inside this button, the single block function button of the program is valid<br>After this function is turned on, the program will be executed by a single block and pause after executing the current single section, then execute the next single block when continuing to press the program start button and so forth<br>② The single block execution function of the program is invalid when the button indicator is not on. Processing program will be executed until the end of the program |
| 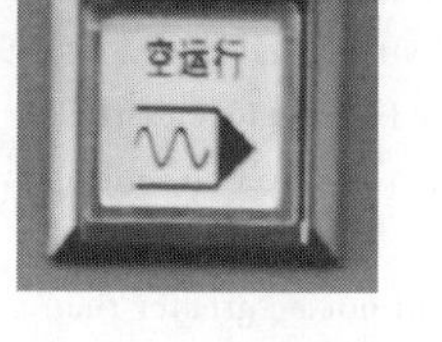 | Name: Dry Running<br>This function is only valid in auto-correlation mode<br>① When the button indicator is on, the $Z$-axis lock function button is effective<br>When this function is turned on, the F value (cutting feed rate) instruction set in the program is invalid and the movement rate of each shaft is displaced at the rate specified by the slow displacement rate<br>② When the function is effective, if the program executes the cyclic program, the slow feed rate or the cutting feed rate can not change the feed rate, and still refers to the F value in the control as a fixed feed rate. |
|  | 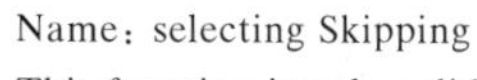Name: selecting Skipping<br>This function is only valid in auto-correlation mode<br>① When the button indicator light is on, the selecting skipping function of the program is valid<br>When this function is turned on, when an "/" (slash) symbol is specified at the beginning of the program fragment, the program fragment will be skipped without being executed<br>② When the light inside the button does not work, the program skipping function is invalid<br>After this function is turned off, that is, when there is a "/" (slash) symbol before a single program section, this program fragment can also be executed normally |

Continued

| Legend | Description |
| --- | --- |
| | Name: Selecting Stop<br>This function is only valid in Auto mode<br>① When the button indicator light is on, the selecting stop function button of the program is valid<br>When this function is turned on, the program will stop in the single block if the M01 instruction is executed. If operator want to continue executing the program, please press the program start button<br>② The program stop function is invalid when the button indicator is not on<br>After this function is turned off, the program will not stop executing even if it has M01 instructions |
| | Name: Mechanical Locking<br>① When this button indicator is on, all shaft mechanical locking function buttons are effective<br>After this function is turned on, move any one shaft in manual mode or auto mode, CNC stops outputting pulses to the actuating motor of shaft (move instruction), but is still doing the instruction assignment, the absolute coordinates and relative coordinates of the corresponding shaft are also updated<br>② The M、S、T、B code will continue to operate and be not limited by mechanical locking<br>③ Undo this function to return to the machine zero again. After returning to zero correctly and complete, do other related operations<br>If relevant operation are performed before returning to zero, it will cause coordinate offset, even appear collider, program running helter-skelter and other abnormal phenomenon, resulting in danger |
| | Name: F1<br>This button is based on the actual configuration of the machine tool<br>It is a preparatory space key which cannot be operated by operator |
| | Name: F2<br>This button is based on the actual configuration of the machine tool<br>It is a preparatory space key which cannot be operated by operator |
| | Name: F3<br>Working light extension<br>Control working light on and off, and be not subject to any operation mode |

(9) Function of spindle (see Table 1-2-7)

**Table 1-2-7 Function of spindle**

| Legend | Description |
|---|---|
| 主轴降速 | Name: Spindle Speed Reduction<br>① This button is located on the operation panel of this machine tool, and used to reduce the S-speed of the spindle programmed. The actual speed = S instruction value given by the program× the ratio of spindle speed reduction<br>② When in use, it cooperates with the keys in the spindle control keys |
| 主轴升速 | Name: Spindle Speed Increase<br>① This button is located on the operation panel of this machine tool and used to increase the S speed of the spindle programmed. The actual speed = S instruction value given by the program× the ratio of spindle speed increase<br>② When the speed set by program exceeds the maximum speed of the spindle, the modification speed of the spindle is equal to the maximum speed of the spindle when the speed reaches more than 100%<br>③ When in use, it cooperates with the keys in the spindle control keys |
| 主轴正转 | Name: Spindle Clockwise Rotation<br>(1)After executing a time of S code on the machine, tool, select the manual operation mode, press the clockwise button of the spindle to rotate the spindle clockwise. Spindle rotation speed = the S value of spindle speed previously executed× the gear position where the spindle adjustment knob is located<br>(2)Conditions of use<br>① Only used in "manual" mode, "high-speed" mode, and "inching motion" mode<br>② In automatic mode, when the program executes the spindle clockwise M03 instruction, this button indicator light will be lit<br>(3)When "Spindle Stop Rotation" or "Spindle Anticlockwise Rotation" takes effect, the indicator light will be off<br>(4)The reverse rotation of the spindle may be specified only after the spindle has been stopped |
| 主轴停止 | Name: Spindle Stop Rotation<br>(1)Press this button to stop the spinning spindle, whether it is in clockwise or anticlockwise motion<br>(2)Conditions of use<br>① This button can only be used in "manual" mode, "high-speed" mode, and "inching motion" mode<br>② Invalid for automatic operation<br>(3)The button indicator light will be on when the spindle stops, but if "Spindle Clockwise Rotation"or "Spindle Anticlockwise Rotation" takes effect, the indicator light will be off |
| 主轴反转 | Name: Spindle Anticlockwise Rotation<br>(1)After the machine executes S code for one time, the manual operation mode is selected, and the spindle is rotated counterclockwise by pressing the spindle anticlockwise button. Spindle rotation speed = the S value of spindle speed previously executed× the gear position where the spindle adjustment knob is located<br>(2)Conditions of use<br>① This button can only be used in "manual" mode, "high-speed" mode, and "inching motion" mode<br>② In auto mode, this button lights up when the program executes the rotary-inversion axis clockwise M04 instruction<br>(3)When the spindle anticlockwise, the light inside the button will be on, but if "Spindle Stop Rotation" or "Spindle Clockwise Rotation" takes effect, the indicator will be off<br>(4)The forward rotation of the spindle may be specified only after the spindle has been stopped |

(10) Auxiliary function (see Table 1-2-8)

**Table 1-2-8 Auxiliary function**

| Legend | Description |
| --- | --- |
| | Name: Cooling<br>① In "manual", "high-speed", and "inching motion" mode, press this button to turn the cutting fluid on upon the indicator light on<br>② Press "RESET"button to stop the outflow of coolant, the coolant stops, and the indicator lamp goes out<br>③ Attention should be paid to the direction of the coolant nozzle when the coolant is opened |
| | Name: Manual Tool Changing<br>In "manual", "high-speed", and "inching motion" mode, every time this button is pressed, the tool will rotate one tool position in the direction of addition |

(11) Functional description of axial selection button (see Table 1-2-9)

**Table 1-2-9 Function description of axial selection button**

| Legend | Description |
| --- | --- |
| | Name: + X Control Button<br>+ X button: in JOG mode, press and hold this button, and the $X$-axis moves in the "+" direction (positive direction) of the $X$-axis of the machine tool according to the feed rate/fast rate, and the indicator light of the button is lit at the same time; When the button is released, the axis stops moving in the "+" direction and the button indicator is off<br>In addition, the button can also be a trigger button for $X$-axis zero return |
| | Name: − X Control Button<br>− X button: in JOG mode, press and hold this button, and the $X$-axis moves in the "−" direction (negative direction) of the $X$-axis of the machine tool in according to the feed rate/fast rate, and the indicator light of the button is lit at the same time; When the button is released, the axis stops moving in the "−" direction and the button indicator is off |
| | Name: + Z Control Button<br>+ Z button: in JOG mode, press and hold this button to move the $Z$-axis in the "+" direction (positive direction) of the machine tool in accordance with the feed rate/fast rate, and the indicator light of the button is lit at the same time; When the button is released, the axis stops moving in the "+" direction and the button indicator is off<br>In addition, the button can also be a trigger button for $Z$-axis zero return |
| | Name:-Z Control Button<br>− Z button: in JOG mode, press and hold this button to move the $Z$-axis in the "−" direction (negative direction) of the machine tool in accordance with the feed rate/fast rate, and the indicator light of the button is lit at the same time; When the button is released, the axis stops moving in the "−" direction and the button indicator is off<br>In addition, when the program executes program instruction, of the $Z$-axis negative direction movement, the button indicator light will also be on, when stopping move instruction, the button indicator light will be off |

Continued

| Legend | Description |
| --- | --- |
| | Name: Overrun Release Button<br>① When the travel of each shaft of the machine exceeds the limit of hardware, the machine tool will appear overrun alarm and the action of the machine tool will stop, At this timn-press this button to move the overrun shaft of the machine tool in reverse direction under the hand wheel mode<br>② This button is not suitable for absolute encoder machine tool overrun |
| | Name: Manual Rapid Traverse<br>This function is only valid in manual mode<br>In manual mode, press this button to turn on the indicator light<br>In this mode, the actual fast feed rate = the maximum speed value of G00 set by parameter × rate value % of the fast rate switch |

(12) Feed rate and feed rate adjustment (see Table 1-2-10)

**Table 1-2-10 Feed rate and feed rate adjustment**

| Legend | Description |
| --- | --- |
| | Name: Feed Rate and Feed Rate Adjustment<br>① This knob is located on the operation panel of this machine tool and controls the program to specify the speed of G01. The actual speed of feeding = the given F instruction value × rate value (%) of the feed rate switch<br>② In inching motion mode, the JOG feed rate is controlled at this time, and the actual JOG feed rate = the fixed value set by the parameter × rate value (%) of the feed rate switch<br>③ When in use, it cooperates with the keys in the shaft feed control keys |

(13) Fast rate (see Table 1-2-11)

**Table 1-2-11 Fast rate**

| Legend | Description |
| --- | --- |
| | Name: Fast Rate<br>① This button is located on the operation panel of this machine tool and controls the program to specify the speed of G00. The actual feed speed = the maximum speed value of G00 set by parameter × rate value % of the fast feed rate button<br>② Fast feed mode, at this time control the manual fast feed rate, actual fast feed rate = the maximum speed value of G00 set by parameter × rate value % of the fast feed rate switch. Fast movement rate can be adjusted at F0, 25%, 50% and 100%<br>③ When in use, it cooperates with the keys in the shaft feed control keys |

(14) Manual pulse (see Table 1-2-12, Table 1-2-13)

**Table 1-2-12 The relationship between shaft selection, rate selection and servo axial movement per frame**

| Rate Selection | ×1 | ×10 | ×100 |
|---|---|---|---|
| Metric Movement | 0.001mm/frame | 0.01mm/frame | 0.1mm/frame |
| British-system Movement | 0.0001 inch/frame | 0.001 inch/frame | 0.01 inch/frame |

**Table 1-2-13 Manual pulse**

| Legend | Description |
|---|---|
|  | Name: Axis-based Switch<br>① The knob is located on the hand-held unit and used in conjunction with the hand-wheel feed rate of ×1, ×10 and ×100<br>② This knob is used for "hand wheel" mode<br>③ No axis is selected when the knob in the "0" position, choose the $X$ axis when in the "X" position, choose the $Y$ axis when in the "Y" position, choose the $Z$ axis when in the "Z" position, and choose the 4 th axis when in the "4" position |
|  | Name: Rate-based Button<br>① The knob is located on the hand-held unit and used in conjunction with the hand-wheel feed rate ×1, ×10 and ×100<br>② This knob is used for "hand wheel" mode<br>③ Choose the hand wheel feed rate of 0.001mm/frame when the knob in the "×1" position, choose the 0.01mm/frame when in the "×10", and choose the 0.1mm/frame when in the "×100" |
|  | Name: Hand wheel MPG (Manual Pulse Generator)<br>① This code wheel is only available in "hand wheel" mode and used to operate the direction and speed of the feed shaft<br>② The rotation direction of the MPG is positive in a clockwise direction (i.e. the servo shaft moves in the positive direction after forward rotation) and negative in a counter-clockwise direction (i.e. servo shaft moves in the negative direction after reverse rotation)<br>Warning<br>The hand wheel rotates at a speed not greater than 5 revolutions per second. If the hand wheel rotates more than 5 revolutions per second, the cutters may not stop when the hand wheel stops spinning or the distance the cutting moves is not in accordance with the scale in which the hand wheel rotates<br>③ The hand-held unit should be handled gently with protection |

## Ⅲ. Program Input Practices

(1) Edit program

Switch the MODE SELECT knob on the operation panel to EDIT, press the PRGRM key on the MDI keyboard to enter the EDIT page, select a NC program to displayed it on the CRT interface, which and can edit the NC program.

① Move the cursor  Press PAGE ↓ or ↑ to turn the page, press CURSOR ↓ or ↑ to move the cursor.

② Insert character  First, move the cursor to the desired location, click the number/

alphabet key on the MDI keyboard, enter the code into the input field, press key to insert the contents of the input field behind the code where the cursor is located.

③ Delete data from the input field  Press key to delete the data in the input field;

④ Delete character  First, move the cursor to the place where the characters need to be deleted, press the key to delete the code where the cursor is located.

⑤ Search  Enter the required letter or code; Press CURSOR to start searching at the position where is located after the cursor's position in the current NC program. (The code can be: a letter or a complete set of code. Such as: "N0010", "M".) If there is a code which requires for searching in this NC program, the cursor stays where it is found; If there is no such code after the position where cursor is located in this NC program, the cursor stays where it is.

⑥ Replacement  Move the cursor to the place where the replacement character is required, and the replaced character will be entered through the MDI keyboard into the input field, press the key to replace the content of the input field with the code where the cursor is.

(2) Display NC program directory

Switch the MODE SELECT knob on the operation panel to EDIT, press the key on the MDI keyboard to enter the EDIT page, and then press the soft key . The NC program name will be displayed on the CRT interface.

Select a NC program: Switch the MODE SELECT knob on the operation panel to EDIT or AUTO, press on the MDI keyboard to enter the EDIT page, press to enter the letter "O"; Press the numeric key to enter the search number: XXXX; (The search number is the program number shown in the NC program directory) Start the search by CURSOR . When found, "OXXXX" is displayed in the upper right corner of the screen, and NC program is displayed on the screen.

(3) Delete a NC program

Switch the MODE SELECT knob on the operation panel to EDIT, (press the on MDI keyboard to enter the EDIT page, press to enter the letter "O"; Press the numeric key to enter the number of the program to be deleted: XXXX; Press the key and the program will be deleted.

Create a new NC program: Switch the MODE SELECT knob on the operation panel to EDIT, press the key on the MDI buttonboard to enter the EDIT page, and press to enter the letter "O"; Press the numeric key to enter the program number which should not repeat with the existing program number; Press key to start program input; Press for each code input, the contents of the input field is displayed on the CRT interface, use the key to change the line after the end of the input line.

Note: In terms of the number/letter key on the MDI keyboard, it is alphabet pressed for the first time and number pressed later. To enter the alphabet again, operator firstly display what has already shown in the input domain on the CRT interface (press the

key to display what is in the input domain on the CRT interface).

(4) Delete all NC programs

Switch the MODE SELECT knob  on the operation panel to EDIT, press the key on the MDI keyboard to enter the EDIT page, and press the key to enter letter "O"; Press to enter " – " ; Press to enter "9999"; Press key.

(5) Program simulation practice

After inputting NC program, the running track can be checked. First, move the cursor to the program header, press the three keys  at the same time, then switch the MODE SELECT knob  on the operation panel to AUTO, click the command in the control panel to turn to check the running track mode; Click the key on the operation panel to observe the running of the NC program. By program simulation, we can verify whether the program we input is correct or not. In program simulation, pause operation, stop operation, single block execution and so on are also effective.

## Ⅳ. Manual Operation of Machine Tool

When the work-piece is automatically processed by machine tool according to the machining procedure, the operation of the machine tool is basically done automatically, while in other cases, it is done manually.

(1) Manually return to the reference point of machine tool

Because of the incremental measurement system, once the power of machine tool is cut off, the CNC system on it loses the memory of the reference point coordinates. When the power supply of CNC system is connected again, the operator must firstly carry on the operation of returning to reference point. In addition, when the machine tools encounter a sudden-stop signal or an overrun alarm signal in the process of operation, after troubleshooting, it is also necessary to carry out the operation of returning to the reference point when recovering operation.

The specific operational steps are as follows: Put the "MODE" switch at ZERO RETUEN. Remind the operator to pay attention to: when the distance between the baffle block on the operating slide and the reference point switch is less than 30mm, first use the "JOG" button to move the operating slide towards the negative direction of the reference point until the distance is greater than 30mm to stop inching, and then return to the reference point.

Press the "JOG" button of $X$-axis and $Z$-axis respectively to make the operating slide move forward to the reference point along $X$-axis or $Z$-axis. In this process, the operator should hold down the "JOG" button until the reference point returns and the indicator light is on, and then release the button. Automatic deceleration movement occurs when the operating slide moves near a reference point of the two axes.

(2) Manually feed of operating slide

When manually adjusting the machine tool, or when the cutter is required to move quickly near or away from the work-piece, it is necessary to manually operate the the operating slide for feeding. There are two kinds of manual operation for feeding the operating

slide, one is to use "JOG" button to make the operating slide move quickly, and the other is to use the hand wheel to move it.

(3) Fast moving

When changing cutters or operating manually, the cutter is required to move quickly close to or away from the work piece. The operation is as follows: Firstly, place the "MODE" switch in the RAPID MODE; Use the "APIDOVERRIDE" switch to select the fast moving speed of the operating slide; Press "JOG" button to move the tool rest to the predetermined position quickly.

(4) Feed by hand wheel

When adjusting the cutter manually, use the hand wheel to determine the correct position of the tool nose, or when trying to cut, use the hand wheel to fine-tune the feed speed while observing the cutting. Its operational steps are as follows: Turn "MODE" switch to "HANDLE" position (3 positions can be selected) to select the movement of the operating slide for each rotation of a hand wheel, turn the "MODE" switch to ×1, the hand wheel can move 0.001mm for each rotation, turn to ×10 to move 0.01mm, and turn to ×100 to move 0.1mm; Make the X axis and Z axis on the left side of the handwheel select the coordinate axis of the switch to move towards the operating slide; Turn the MPG to enable that the tool rest moves in the specified direction and speed.

(5) Operation of spindle

The operation of the spindle mainly includes the start and stop of the spindle and its inching.

① Spindle start and stop The start and stop of the spindle is used to adjust the cutter or debug the machine tool. The specific operational steps are as follows:

Set the "MODE" switch to any position in the manual MODE (MANU). Use the "FWD-RVS" switch in the spindle function button to determine the direction of spindle rotation. Switch to "FWD" position for forward rotation and switch to "RVS" position, for reverse rotation. Rotate spindle "SPEED" to low speed zone to prevent spindle from sudden acceleration. Press the "START" button to rotate the spindle. During the rotation of the spindle, the speed of the spindle can be changed through the "SPEED" knob, and the actual speed of the spindle is displayed on the CRT display. Press the STOP button to make the spindle stop turning.

② Inching of spindle The inching of the spindle is used to rotate the spindle to a position where it is convenient to load and unload the jack catch or to check the clamping position of the work-piece. Its operation method is: Set the "MODE" switch to any position in AUTO mode. Point the spindle "FWD-RVS" switch in the desired direction of rotation. Press "START" button to rotate the spindle; Release the button to stop the spindle turning.

(6) Index of tool rest

Loading and unloading the cutters, measuring the cutter position and testing cutting of work piece should be carried out in MDI state. Its operation steps are as follows: Set "MODE" switch to "MDI" MODE; Press the function key "PRGRM" and enter T10, T20, T30 and T40, then press the START button.

(7) Operation of manual tailstock

The operation of the manual tailstock includes the movement of the tailstock body and the tailstock sleeve.

① For the movement of tailstock body  Manually move the tailstock to make it forward or backward, mainly used to adjust the position of the tailstock when machining the shaft parts, or move the tailstock back to a suitable position when machining the short shaft and plate parts. Its operational steps are as follows: Place the "MODE" switch at any position in the "MANU" MODE; Press "TALL STOCK INTERLOCK" button to loosen tailstock, the indicator above the button is on; Move the operating slide drive the tailstock to a predetermined position; Press "TAIL STOK INTERLOCK" button again, the tailstock is locked and the indicator is off.

② For the movement of tailstock sleeve  Move the tailstock sleeve out or back is used to tighten center or loosen the work piece when processing the shaft parts. The operating methods are as follows: First place the "MODE" switch at any position in the "MANU" MODE. Press the "QUILL" button, the tailstock sleeve returns with the center, and the indicator is off.

(8) Clamping and releasing operation of chuck

When the machine is operated manually or automatically, the clamping and releasing of the chuck are realized by the foot switch. The operation steps are as follows: Pull the switch of the forword and reverse chuck in the electric box and select the forward and reverse chuck of the chuck; The first step is to release the switch chuck and the second step is to clamp the switch chuck.

## Ⅴ. Emergency Stop of Machine Tool

Whether the machine tool is in manual or automatic operation, in case of abnormal situations requiring emergency stop of machine tool, the following operation can be used to achieve.

(1) Press the emergency stop button

After pressing the "EMERG STOP" button, in addition to the lubricating oil pump, the operation and various functions of the machine tool are immediately stopped. At the same time, a CNC NOT READY alarm appears on the CRT screen. After troubleshooting, rotate the button clockwise, the button pressed will jump up, then the emergency stop state is lifted. However, in order to resume the work of the machine tool at this time, the operation of returning to the reference point of the machine tool must be carried out.

(2) Press the reset button (RESET)

In the process of automatic operation of machine tool, the machine tool's all operations will stop upon pressing this button, so the operator can use this button to complete the sudden-stop operation. Press power supply disconnect button of the NC device, press the "OFF" of NC to stop the machine tool. When pressing the FEED HOLD, the machine tool is running under the automatic operation, the opering slide can stop, but other functions of the machine tool are still valid. When the machine tool needs to resume running, after pressing "CYCLE START" button, and the machine will continue to execute the following procedure from the current position.

(3) Press feed hold (FEED HOLD)

When the machine tool is running automatically, press the "FEEK HOLD" button to stop the operating slide, but other functions of the machine are still valid. When the machine tool needs to run, after pressing the "CYCLE START" button, and the machine will continue the program from the current position.

## Ⅵ. Tool Setting (Geometric and Wear Compensation of Cutter)

(1) Method for cutter geometric compensation

① Move to make the $X$-axis and $Z$-axis return to reference point, confirm that the reutrn indicator light is on.

② Switch of digital mode select chooses the manual feed.

③ "OFFST" key to show OFFSET/SETTING on CRT.

④ Selector switch selects the required cutter or calls the required cutter in MDI mode.

⑤ Move the cutter so that it is close to the right end of the work piece and smooth the end face without moving the $Z$-axis.

⑥ Move the cursor to the $Z$-axis of the geometric compensation number of the corresponding tool and enter $Z0$, then press the soft key [measurement].

⑦ Move the cutter to measure the outside circle of the workpiece or the inner hole, it would be ready when see the light, measure the outer circle diameter or aperture of the work piece, move the cursor to the $X$-axis of the geometric compensation number of the corresponding tool, input the XD directly (the outside diameter value or the inner diameter value of the work piece) and press the softkey (measurement), as shown in Figure 1-2-3.

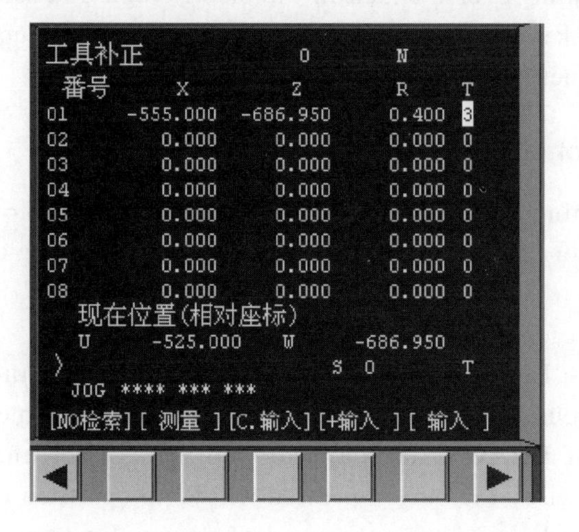


Figure 1-2-3　Geometric Compensation

⑧ Compensation of the other required cutters is similar to that of the first cutter, however, do not cut, but gently lean against the end face or the outer circle of the workpiece. The input data is the same as that of the first cutting.

⑨ Identify the arc radius dimension of tool nose and enter the corresponding tool compensation number in the database.

⑩ Confirm the number of the imaginary position of the tool nose arc, for example, right-deviated cylindrical turning tool is 1, the boring cutter is 3, and input the corresponding tool compensation number in the tool database.

(2) Wear compensation

① After pressing the "OFFSET" key, press the [wear] soft key to show wear compensation picture on the CRT.

② Move the cursor to the cutter compensation number where the wear compensation is

required.

For example, the outer diameter of work piece machined with T11 cutter is $\phi$45.03mm, the length is 20.05mm, while the specified diameter is $\phi$45mm, and the length is 20mm, meaning that the measured diameter is larger 0.03 than the required value, the length is larger than 0.05mm, so the wear compensation should be done: namely, move the cursor to X, press the soft key after typing −0.03; Move the cursor to Z, press the [+ input] soft key after typing −0.05, the X value is changed to the value being added with −0.03mm on the basis of the previous value, and the Z value is changed to the value being added with −0.05mm on the previous basis, as shown in Figure 1-2-4.

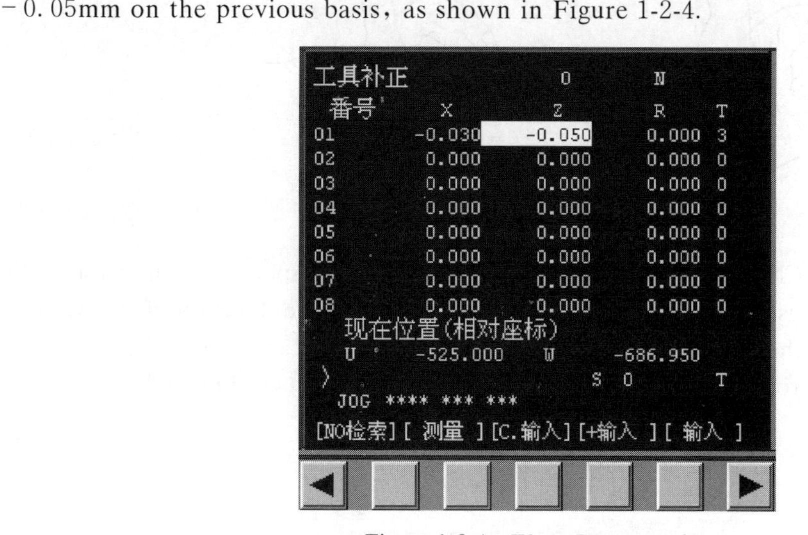


Figure 1-2-4 Wear Compensation

**Exercise**

1. In general, where is the machine origin position of CNC lathe?
2. What is the reference point of machine tool? What is the purpose of returning to the reference point?
3. What should be paid attention to at the tool changing point?
4. What are the reasons for the classification of machining into rough machining, semi-finish machining and finish machining?
5. What are the steps of powering on operation?
6. What are the characteristics of CNC lathe machining?
7. Please introduce the programming characteristics of CNC lathe.
8. How to operate the machine tool in case of a failure?

# Task Ⅲ Basic Components of the Program

## Ⅰ. Coordinate Axis and Movement Direction of CNC Lathe

In order to simplify the programming and guarantee the universality of the program, we establish a unified standard for the coordinate axis and direction naming of NC machine tools, and define the linear feed coordinate axis. The basic coordinate axis is expressed by

$X$, $Y$ and $Z$, and its relation is determined by the right hand rule.

**1. Naming principle of coordinate axis and direction of CNC lathe**

① Adopt the principle of cartesian coordinates (see Figure 1-3-1).

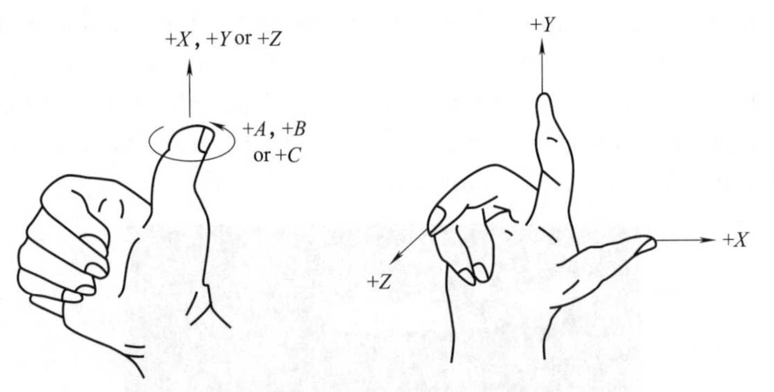


Figure 1-3-1 Rectangular coordinate system

② The $Z$-axis of a CNC lathe is defined as the direction of the spindle which transmits the cutting power; The $X$-axis is specified as the horizontal direction, the direction of $X$-axis is in the radial direction of the work piece, parallel to the transverse operating slide, and the center direction away from the chuck is in the positive direction (see Figure 1-3-2).

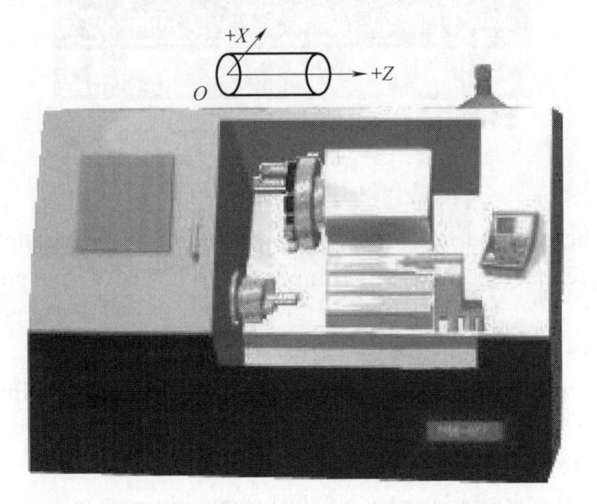


Figure 1-3-2 CNC lathe coordinate axis and its direction

**2. Determination and application of coordinate system of CNC lathe**

① Fix the $Z$-axis first, because the $Z$-axis is the principal axis which transmits cutting power, and then fix the $X$-axis.

② The origin of CNC lathe coordinate system is defined as the intersection point between the center line of the chuck and the middle end.

## Ⅱ. CNC Lathe Coordinate System

**1. CNC lathe coordinate system**

(1) Machine tool coordinate system

In accordance with the Cartesian rule, the CNC lathes manufacturer establishes a rec-

tangular coordinate system of the $Z$-axis and the $X$-axis, which is called the machine tool coordinate system.

(2) Machine tool origin

The zero point of the machine tool coordinate system is called the original point of the machine tool, which is a fixed point on the machine tool and generally defined as the intersection or reference point between the principal axis rotation center line and the end face of the lathe.

(3) Reference point

The reference point is a fixed point on the machine tool, which is determined by the position defined by the mechanical block or system in the $X$ and $Z$ directions and is generally set at the maximum position forward direction of the $Z$-axis and $X$-axis, the position is defined by the manufacturer.

**2. Programming coordinate system**

The programmer selects a known point on the work piece drawing as the origin (also known as the program origin) to establish a new coordinate system which is called the programming coordinate system.

Select programming origin: Theoretically, a programming origin can be selected at any point on the part, but in fact, a reasonable programming origin should be chosen in order to convert the dimensions as easily as possible and to reduce the calculation error. The $X$-direction zero of the programming origin of the turning part should be selected in the return center of the part. $Z$-direction zeros should generally be selected in the right end of the part design benchmark or symmetrical plane. The selection of programming origin for turning parts is shown in Figure 1-3-3.

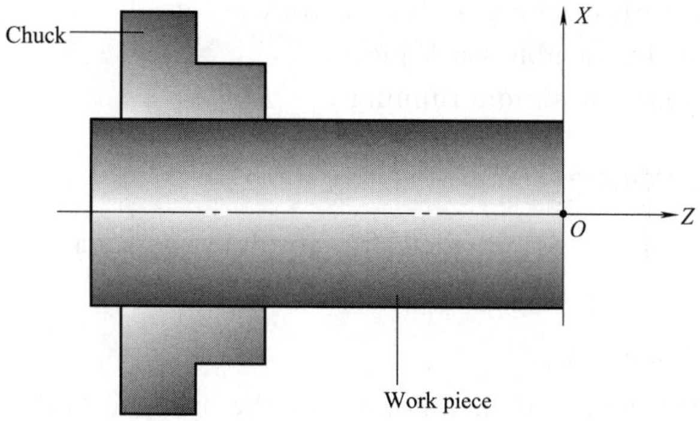

Figure 1-3-3　Programming coordinate origin of turning parts

**3. Workpiece coordinate system**

The operator moves the origin of the programming coordinate system to a CNC lathe by means of tool setting, etc. The coordinate system established on the CNC lathe is called the work piece coordinate system, whose origin is usually chosen at the intersection of the axis with the right end face, the left end face or other positions of the workpiece, the $Z$-axis of the work piece coordinate system generally coincides with the axis of the principal axis. The selection of work piece coordinate origin for turning parts is shown in Figure 1-3-4.

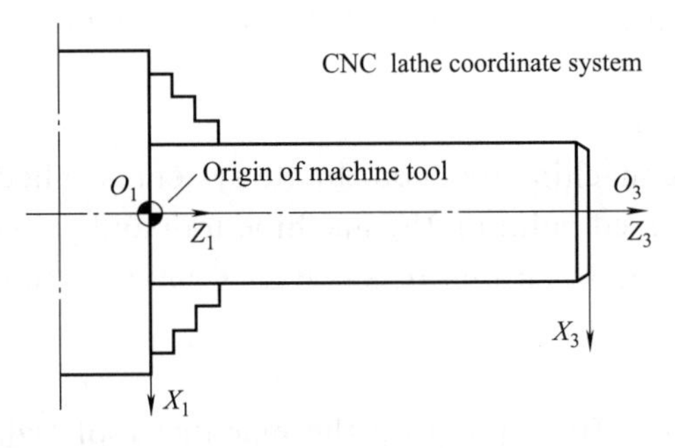


Figure 1-3-4 Coordinate origin for turning parts

#### 4. Point for tool setting, starting, tool Changing

The origin point of the programming coordinate system is transformed into the known point of the machine tool coordinate system to become the origin point of the work piece coordinate system, which is called the tool setting point. The starting point of cutter is the starting point of parts processing. Due to the fact that automatic tool changing is required in the process of part turning, so an automatic tool changing point must be set, which should be separated from the work piece to prevent the cutter from colliding with the work piece when the tool rest is turned back. The tool changing point is usually divided into two types including fixed changing point and self-defined changing point.

The following points should be noted for tool changing position:

① Convenient for mathematical computing and simplifying programming;
② Easy to find correct tool setting;
③ Convenient for process inspection;
④ Small machining error caused;
⑤ Do not collide with machine tools and work pieces;
⑥ Convenient for disassemble work piece;
⑦ Don't be too long for an idle running.

## Ⅲ. Programming Coordinates

Programming coordinates are divided into absolute coordinates ($X$, $Z$), relative coordinates ($U$, $W$) and hybrid coordinates ($X/Z$, $U/W$).

#### 1. Absolute coordinates ($X$, $Z$)

The coordinate parameters of each point take the distance to the origin as the parameter value.

#### 2. Relative coordinates ($U$, $W$)

A coordinate parameter at one point takes the distance to another point as a parameter value, i. e. the distance from the previous position to the next as a parameter value.

#### 3. Hybrid coordinates ($X/Z$, $U/W$)

Absolute coordinates and relative coordinates are used simultaneously, that is, in the same segment, $X$ or $U$, $Z$ or $W$ can be used.

#### 4. Diametral coordinates

The $X$-coordinate parameter is the diameter.

### 5. Radius programming

The $X$-coordinate parameter is radius value

## Ⅳ. Basic Knowledge of CNC lathe Programming

### 1. Initial state and modal state

(1) Initial state

It refers to the system programming state before the program runs, i. e. , the state that has been set in the machine and can be entered upon starting, e. g. G98, G00.

(2) Modal state

It refers to a continuously valid instruction. Once the value of the corresponding field has been set, it remains valid until a program fragment has been reset to that field. After setting, the same group can also use the same function, there is no need to enter the field again.

```
For example:N30 G90 X32. 0Z0 F80;
            N40 X30. 0:
            ......
            N…G02 X30. 0 Z-30. 0 R5. 0 F50:
            N…G01 Z-30. 0 F30;
```

### 2. Program composition

In view of machining parts on a CNC lathe, it is essential to firstly program and then use the program to control the movement of the machine tool. The set of NC instructions is called a program. These instructions are written in the program according to the actual movement order of the machine tool.

(1) Program structure

A complete NC machining program consists of the beginning part of the program, several program segments and the end part of the program. A program segment consists of program segment numbers and several "words", and a "word" consists of address characters and figures.

The following is a complete NC machining program, which starts with program number and ends with M30.

```
Program                          Specification
O1234                            Start of Program
N10 T0101 G97 G99 M03 S500;      Program Segment 1
N20 G00 X100. 0 Z100. 0;         Program Segment 2
N30 G00 X26. 0 Z0;               Program Segment 3
N40 G01 X0 F0. 1;                Program Segment 4
N50 Z1. 0;                       Program Segment 5
N60 G00 X100. 0;                 Program Segment 6
N70 Z100. 0;                     Program Segment 7
N80 M30;                         End of Program
```

(2) Program number

The beginning part of a part program generally consists of the program starting symbol % (or O) followed by 1～4 digits (0000～9999), such as %123, O1234, etc.

(3) Format and composition of program segment

The format of program segment can be divided into address format, segmental address format, fixed program segment format and variable program segment format, among which variable program segment format is the most widely used and means that the length of the program segment is variable.

e. g. :

| N10 | G01 | X40. 0 | Z-30. 0 | F200; |
|---|---|---|---|---|
| Program segment No. ; | Function words; | Coordinate word; | Function words with given speed; | End of program segment |

(4) "Word"

The composition of a "word" is as follows:

| Z | - | 30. 0 |
|---|---|---|
| Address characters; | Sign (positive, negative); | Data word (number) |

A program segment can be formed by adding a number of program words to the program segment number. A letter representing an address in a program segment can be divided into two types: address and non-dimension address. The 18 letters indicating the dimension address are X, Y, Z, U, V, W, P, Q, I, J, K, A, B, C, DERH, and 8 letters indicating the non-dimension address are N, G, F, S, T, M, L, O.

**3. Function word S for spindle speed**

The address character of the spindle speed function word is S, also known as the S function or S instruction, which is used to specify the spindle speed, with the unit of r/min. For CNC lathes with a constant linear speed function, the S instruction in the program is used to specify the number of linear speeds in the turning process.

With gearbox: use S1 (first gear), S2 (second gear)

Without gearbox: directly input speed, such as S100, S210, S500, etc.

**4. Feed function word (cutting speed) F**

The address character of the feed function word is F, also known as the F function or F instruction, which is used to specify the feed speed of the cutting. For the lathe, F can be divided into two types: feed per minute and feed per rev of spindle. For other CNC machine tools, feed per minute is generally used only. F instruction is often used to instruct thread lead in thread cutting program segment.

Units: G98 is feed per minute, mm/ min; G99 is feed per rev, mm/ r.

G00 is fast positioning, without F value, the speed is controlled by the override.

The cutting feed speed should have F value, and the speed of F value can also be controlled in the feed rate.

For example:
```
G00 X32. 0 Z2. 0;
G90 X24. 0 Z-20. 0 F50;
G90 X20. 0 Z-15. 0 F60;
```

**5. Function word T of cutter**

The address character of the tool function word is T, also known as a T function or T instruction, which is used to specify the number of tools used in machining. For CNC

lathes, the figures also specify the function of tool length compensation and tool nose radius compensation. The first and second places after T are the cutter number, and the third and fourth places are the compensation number of the cutter.

Example: T0100 T0200 T0300 T0400 without cutter compensation, e. g. T0200 is No. 2 cutter without cutter compensation.

T0101 T0202 T0303 T0404 has cutter compensation, e. g. T0202 is No. 2 cutter to execute No. 2 cutter compensation.

### 6. Auxiliary function word M

The address character of auxiliary function word is M, and its following number is usually 1～3 digits of positive integer, also called M function or M instruction, which is used to specify the on-off action of auxiliary device of NC machine tool, see Table 1-3-1.

**Table 1-3-1 Meaning of M functional word**

| M function word | Meaning | M function word | Meaning |
|---|---|---|---|
| M00 | Program stop | M07 | No. 2 coolant on |
| M01 | Plan stop | M08 | No. 1 coolant on |
| M02 | Program stop | M09 | Coolant off |
| M03 | Clockwise rotation of spindle | M30 | Program stop and return to the start point |
| M04 | Counterclockwise rotation of spindle | M98 | Call subprogram |
| M05 | Spindle rotation stop | M99 | Return to subprogram |
| M06 | Tool changing | | |

### 7. Preparatory function word G

The address character for a function word is G, also known as a G function or G instruction, which is an instruction used to set up the working way for a machine tool or control system. Its following numbers are usually 1-3 positive integers, see Table 1-3-2.

**Table 1-3-2 Meaning of G functional word**

| G function word | Group | Function |
|---|---|---|
| G00 | | Fast move point location |
| G01 | 01 | Linear interpolation (cutting feed) |
| G02 | | Clockwise circular arc interpolation |
| G03 | | Counterclockwise circular arc interpolation |
| G04 | 00 | Pause, accurate stop |
| G28 | 00 | Return to a reference point (mechanical origin) |
| G32 | 01 | Thread cutting |
| G33 | | $Z$-axis trapping thread cycle |
| G34 | | Thread cutting with variable pitch |
| G50 | 00 | Coordinate system setting |
| G65 | 00 | Macro program command |
| G70 | | Finish maching cycle |
| G71 | | External rough cutting cycle |
| G72 | | End surface rough cutting cycle |

Continued

| G function word | Group | Function |
|---|---|---|
| G73 | | Closed cutting cycle |
| G74 | 00 | End surface deep hole drilling cycle |
| G75 | | External and internal grooving cycle |
| G76 | | Compound thread cutting cycle |
| G90 | 00 | External and internal turning cycle |
| G92 | 00 | Thread cutting cycle |
| G94 | | End surface turning cycle |
| G96 | 02 | Constant linear velocity |
| G97 | | Cancel constant line speed |
| G98 | 03 | Feed per minute |
| G99 | | Feed per rev |

The usage of G code:

One-time G code: the code only valid in the program segment being directed. For example: G04 (pause), G50 (coordinate setting), G70～G75 (compound turning fixed cycle).

G code of modal state: it is valid for other code instructions in the same group, namely, G code in group 01 of the table. For example: G00 (location), G01, G02, G03 (interpolation), G90, G92, G94 (single-type fixed cycle).

G Code of initial state: that is, the state that the system has been set up and can enter into upon starting. The initial state is also a modal state. For example; G98, G00.

**Exercise**

1. What is the difference between the application of G02 instruction and G03 instruction?
2. What is the difference between M02 and M30?
3. Please write the formats of G70, G71 and G73.
4. When are G01 and G00 applied to the program?
5. Can G98 and G99 be applied at the same time? Why?
6. What parts can be machined by CNC lathe?
7. What are the three key elements of cutting?
8. What are the important angles of the cutter and what roles do they play?

# Task Ⅳ Optical Shaft Programming and Processing

## Ⅰ. Drawing and Technical Requirements

As shown in Figure 1-4-1, the part with shaft type characteristics is a cylinder bar with material of No. 45 steel, the specification of $\phi$35mm and hardness of 200HB, which has been done normalizing treatment. Shown in Table 1-4-1.

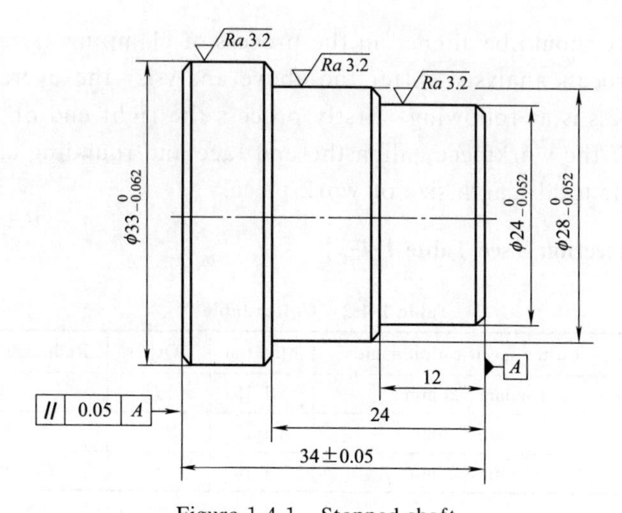


Figure 1-4-1 Stepped shaft

**Table 1-4-1 Scoring form**

| S/N | Project and technical requirements | Partition (IT/$Ra$) | Scoring criteria | Inspection result | Final score |
|---|---|---|---|---|---|
| 1 | External diameter $\phi 33_{-0.062}^{0}$, $Ra3.2$ | 18/4 | Full sore deduction if the result is extreme poor | | |
| 2 | External diameter $\phi 28_{-0.052}^{0}$, $Ra3.2$ | 18/4 | Full sore deduction if the result is extreme poor | | |
| 3 | External diameter $\phi 24_{-0.052}^{0}$, $Ra3.2$ | 18/4 | Full sore deduction if the result is extreme poor | | |
| 4 | Length: 12, 24 | 4 | Full sore deduction if the result is extreme poor | | |
| 5 | Length: 36 ± 0. 05 | 6 | Full sore deduction if the result is extreme poor | | |
| 6 | Rounding chamfer 1×45 (4 pcs) | 4 | Full sore deduction if the result is extreme poor | | |
| Safe and civilized production | | 20 | | | |
| Working hours | | 60 min | | | |

## Ⅱ. Analysis of Drawings

### 1. Parts diagram analysis

The outer outline of the part is composed of three cylinders ($\phi$33mm, $\phi$28mm, $\phi$24mm) with 12mm length, belonging to simple stepped shaft parts. The parts are designed according to the sequence of training and learning.

### 2. Process analysis

① Structural analysis: The structure of the parts is a simple stepped shaft.

② Accuracy analysis: The key dimensions of the parts are all h9 in the external accuracy level, which is a relatively reasonable tolerance size for the training of intermediate pre-work. In addition, length, rounding chamfer and other details of precision also need to pay attention to.

③ Positioning and clamping analysis: This part uses three-jaw self-centering chuck for positioning and clamping. The clamping force of the work piece should be moderate, which should not only prevent the deformation and crush of the work piece, but also prevent the looseness of the work piece in the process. In order to ensure the tolerance of shape and po-

sition, the work piece should be aligned in the process of clamping.

④ Machining process analysis: After the above analysis, the overall arrangement of the parts in the process is as following: firstly process the right end of the part; turn the head after cutting off the workpiece, align the end face and rounding chamfer after turning so as to ensure the total length size of work piece.

### 3. Main cutter selection (see Table 1-4-2)

**Table 1-4-2 Cutter table**

| Cutter name | Cutter specification name | Material | Qty. | Radius of tool nose | Cutter width |
|---|---|---|---|---|---|
| 90° external lathe tool | 25mm×25mm | YT 15 | 1 | 0 | |
| 45° external lathe tool | 25mm×25mm | YT 15 | 1 | 0 | |
| Cutter | 25mm×25mm | YT 15 | 1 | 0 | 4. 5mm |

### 4. Process schedule arrangement (see Table 1-4-3)

**Table 1-4-3 Process table (right end)**

<table>
<tr><td colspan="2">Unit</td><td colspan="2">Product name and model</td><td colspan="2">Part name</td><td>Part drawing number</td></tr>
<tr><td colspan="2"></td><td colspan="2">Task 4</td><td colspan="2">Simple stepped shaft</td><td>Figure 1-4-1</td></tr>
<tr><td>Working procedure</td><td>Program No.</td><td colspan="3">Jig name</td><td>Used equipment</td><td>Work piece material</td></tr>
<tr><td>001</td><td>O0001</td><td colspan="3">three-jaw chuck</td><td>SK50</td><td>No. 45 steel</td></tr>
<tr><td colspan="7">Stepped shaft parts</td></tr>
<tr><td>Working step</td><td>Working step content</td><td>Cutter No.</td><td>Cutting amount</td><td>Remark</td><td colspan="2">Procedure diagram</td></tr>
<tr><td>1</td><td>Radial facing</td><td>T11</td><td>n = 600r/min (Manual feed rate mode position is at×10)</td><td>Manual processing</td><td colspan="2">Jaw<br>(50~60)</td></tr>
<tr><td>2</td><td>Rough turning Set aside 0. 5mm finish machining allowance for the outer circle with φ24, φ28, φ33</td><td rowspan="2">T11</td><td>n = 500r/min<br>f = 0. 2mm/r<br>$a_p$ = 2. 0mm</td><td rowspan="2">Automatic processing</td><td colspan="2" rowspan="2">Jaw<br>φ33 φ28 φ24<br>12 12<br>37</td></tr>
<tr><td>3</td><td>Finish turning φ24, φ28, φ33 outer circle</td><td>n = 800r/min<br>f = 0. 1mm/r<br>$a_p$ = 0. 5mm</td></tr>
</table>

## Ⅲ. Programming

### 1. Fast Positioning (G00)

(1) Programming formats:

```
N10  G00 X(U)__  Z(W)__;
```

Where, the value of X and Z is the end-point coordinate dimension of fast point positioning, which is the absolute value coordinate programming;

The value of U, W is the end-point coordinate dimension of the fast point positioning, and is the relative (incremental) value coordinate programming.

For instance, The procedure segment for fast movement from point $A$ to point $B$ is as follows:

```
N10  G00  X20.0  Y30.0;
Or  N10  G00  U-20.0  W-10.0;
```

(2) G00 feed path

The fast point positioning command controls the tool to move quickly to the target position in a point-controlled manner, and its moving speed is set by the parameters. After the execution of the instruction, the cutter moves along all coordinate directions at the same time at the rate set by the parameters, and reaches the end point by deceleration, as shown in Figure 1-4-2. Note: It is possible not to reach the destination at the same time in all coordinate directions. The tool-moving path is a combination of several lines, not a straight line. In FANUC systems, for example, motion is always moved first along a straight line at an angle of 45° and then to the target point position by one-way on a certain shaft, as shown in Figure 1-4-2(b). The programmer should be aware of the tool movement path of the NC system used to avoid possible collisions in the machining process.

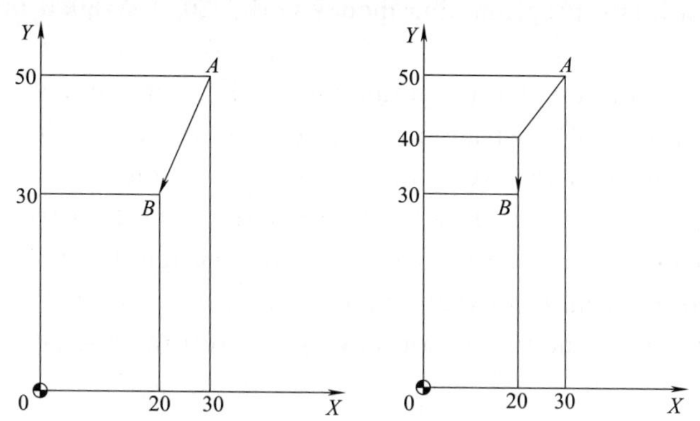


(a) Arrive at the destination at the same time (b) One-way movement to the end

Figure 1-4-2 Fast point positioning

(3) No F Value in G00

Fast moving speed is set by manufacturer and is controlled by fast rate switch (F0, 25%, 50%, 100%). Feeding speed specified by F value is invalid.

### 2. Line Interpolation G01

The G01 enables the cutter to move along a straight line to the target point at the feed

speed of the instruction.

The instruction format is: G01X (U) ____ Z (W) ____ F ____;

Where, X and Z denotes the absolute value coordinates of the target point;

U and W represents the incremental coordinates of the target point relative to the previous point;

F represents the feed, which can be omitted if specified earlier.

Usually, when conducting the machining of the end face of the turning and groove parallel to the *X*-axis, only the X(or U) coordinates need to be specified separately, and when machining the outer circle and inner hole of the turning parallel to the *Z*-axis, only the Z(or W) value is specified separately. Figure 1-4-3 is the case where the two axes are simultaneously instructed to move the turning cone, programmed with G01 as follows:

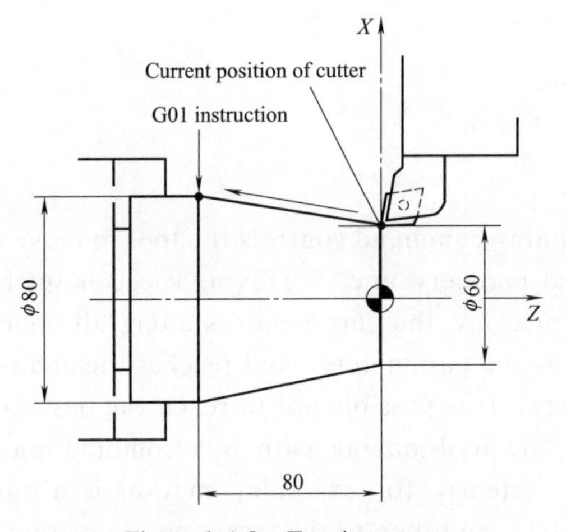


Figure 1-4-3 Turning cone

Absolute coordinate programming mode: G01 X80. 0 Z-80 F0. 25

Incremental coordinate programming mode: G01 U20. 0 W-80. 0 F0. 25

Descriptions:

① Choosing the absolute value programming or the increment value programming of the coordinate value after G01 instruction is decided by the dimension and word address, some CNC lathe is decided by the NC system state at that time.

② The feed speed is determined by the F instruction. The F instruction is also a modal instruction, which can be canceled with the G00 instruction. If the segment prior to the G01 program does not have an F instruction and there is no F instruction in the current G01 program segment, the machine tool is not moving. Therefore, the G01 program must contain F instruction.

### 3. Arc interpolation instruction G02, G03

(1) Instruction format

```
G02/G03  X(U)__ Z(W)__ I __ K __ F __;
G02/G03  X(U)__ Z(W)__ R __ F __;
```

Judgment of clockwise and anticlockwise arc: The arc interpolation instructions are divided into the clockwise arc interpolation instruction G02 and the anticlockwise arc interpolation instruction G03. The clockwise and anticlockwise of arc interpolation can be

judged by the direction given in Figure 1-4-4: seeing along the negative direction of the vertical shaft ( $-Y$ ) of the plane in which the arc is located (such as the $XZ$ plane), the clockwise direction is G02 and counterclockwise direction is G03.

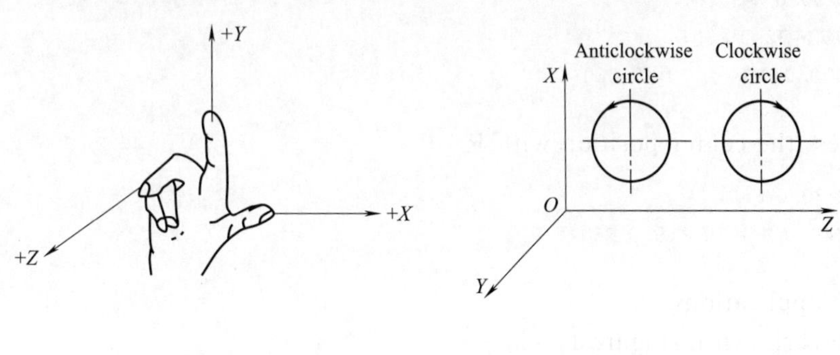


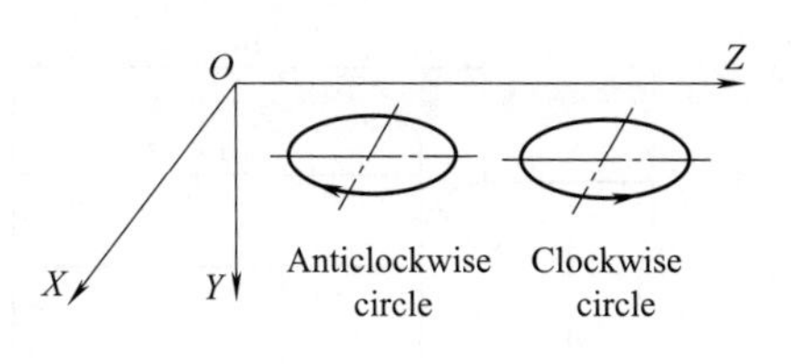


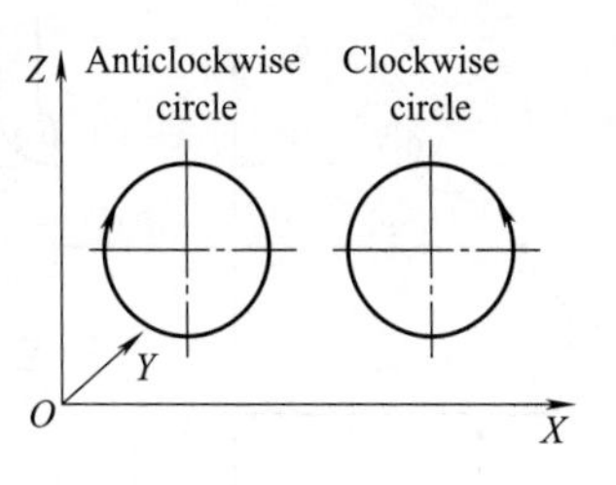


Figure 1-4-4 Judgment of clockwise and anticlockwise arc

(2) Descriptions

① When programming with absolute value, the coordinate of the arc end point is the coordinate value of the arc end point in the workpiece coordinate system which is expressed by X and Z. When programming with incremental value; the coordinate of the end point of the arc is the increment of the arc end point relative to the arc starting point which is expressed by U and W.

② The coordinate I and K of the circle center is the sub-vector (the vector direction points to the center of the circle) of the vector from the beginning of the arc to the center of the arc in the direction of the $X$-axis and $Z$-axis. In this system, I and K are incremental values with "±" sign. When the direction of sub-vector is inconsistent with that of coordinate shaft, "−" sign is taken.

③ When the position of the circle center is only specified by the radius, it is possible to have two arcs from the starting point to the end point of an arc with the same radius. Therefore, to distinguish the two, it is stipulated that "+R" is used when the central angle is $\leqslant 180°$. If the central angle of an arc is $>180°$, it is expressed by "−R".

④ The circle cannot be described when only the center of the circle is specified by the radius.

(3) The G02 application

Example is shown in Figure 1-4-5.

① Express the position of the circle center with I and K, the program with the absolute value is:

```
N03 G00 X20. 0 Z2. 0;
N04 G01 Z-30. 0 F80;
N05 G02 X40. 0 Z-40. 0 IO KO F60;
```

② Express the position of the circle center with I and K, the program with the increment value is:

```
N03 G00 U-80. 0 W-98. 0;
N04 G01 UO W-32. 0 F80;
N05 G02 U20. 0 W-10. 0 I0 K0 F60;
```

③ Express the center position with R

```
N04 G01 Z-30. 0 F80;
N05 G02 X40. 0 Z-40. 0 R10. 0 F60,
```

(4) G03 applications

Example is shown in Figure 1-4-6.

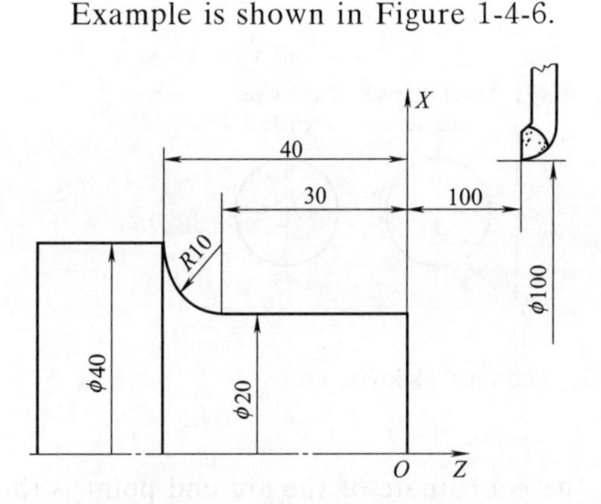


Figure 1-4-5 G02 application example

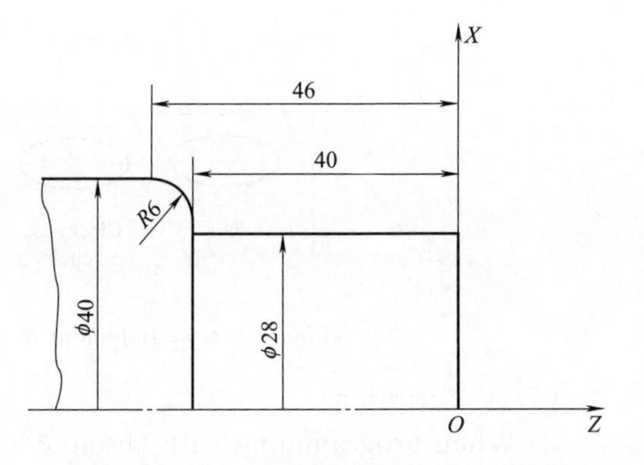


Figure 1-4-6 G03 application example

① Express the position of the circle center with I and K, the program with absolute value is:

```
N04 G00 X28. 0 Z2. 0;
N05 G01 Z-40. 0 F80;
N06 G03 K40. O Z-46. 0 I0 K-6. 0 F60;
```

② Program with incremental value:

```
N04 G00 U-150. 0 W-98. 0;
N05 G01 W-42. 0 F80;
N06 G03 U12. 0 W-6. 0 I0 K-6. 0 F60;
```

③ Express the circle center position with R, the program with absolute value is:

```
N04 G00 X28. 0 Z2. 0;
N05 G01 Z-40. 0 F80;
```

(5) Method of lathing arc with G02/G03

When using G02 (or G03) instruction to lathe arc, if the arc is processed only one time, the cutter penetration would be bite too large and it is easy to hit the cutter. Therefore, when actually lathing arc, multi-cutter is required, remove the majority of the surplus firstly and then get the arc required by lathing. The common machining route of lathing arc is described below.

Figure 1-4-7 is the cutting route of the lathing cone method to lathe the arc. That is, lathe a cone at first and then the arc. However , it is necessary to pay attention to the confirmation of starting and ending point of the cone, if the determination is not good, it may damage the conical surface and may also leave the surplus too large. Method of determination: connect $OC$ to intersect the arc at $D$, and make the tangent $AB$ of the arc thought point $D$.

Figure 1-4-8 represents the concentric arc cutting route of lathing the arc which means to lathe with different radius circle, and finally machine the required arc. For this method, after determining the cutter penetration $a_p$, it is quite easy to determine the coordinates of starting point and ending point of 90° arc, with simple numerical calculation and convenient programming , which is often used. but its idle run time is quite long.

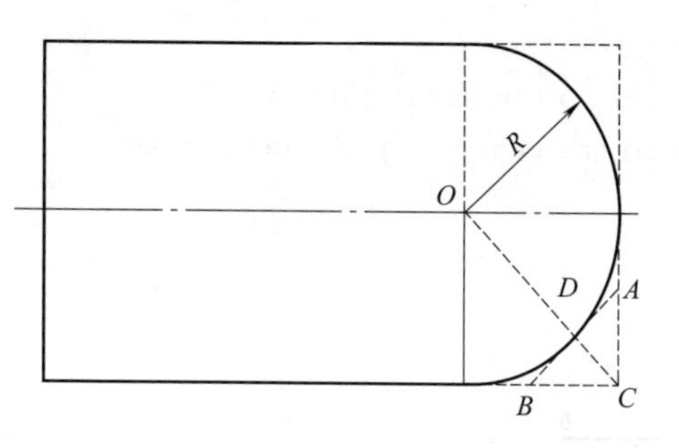

Figure 1-4-7 Lathing cone method

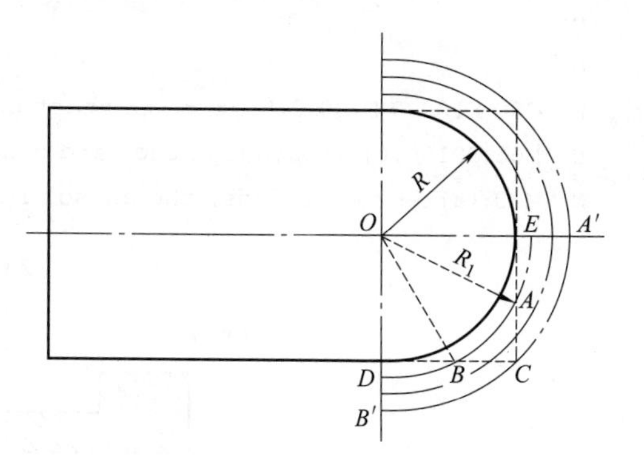

Figure 1-4-8 Lathing circle method

**4. Program the function used at the beginning of the procedure (the programming of first three steps)**

N10 G50 X ____ Z ____; set the coordinate system of parts, that is, the starting point of the cutter (program)

N20 M ____ S ____ T ____; spindle forward and reverse rotation (M03 spindle forward rotation, M04 spindle reverse rotation);

Spindle speed: If there is a gearbox, then S1 is the first gear and S2 is second, input value directly if there is no gearbox;

The used cutter No. (e. g. T0100)

N30 G00 X ____ Z ____; move the cutter quickly to the edge where the workpiece is ready to be machined

**5. Program the function used in end of the procedure (the programming of the last three steps)**

N __ G00 X ____ Z ____; move the tool back quickly to the program's starting point

N __ M05 T ____; the spindle stops and changes back to the datum cutter

N __ M30; the program ends, the cursor returns back to the starting point of the program, ready for the next workpiece processing

**6. Program the functions used in the middle part of the procedure**

According to the requirements of the processing drawing, choose the processing technology and program the middle part of the procedure.

**7. Programming examples**

**【Example 1-4-1】** Workpiece machining is shown in Figure 1-4-9, and please try to

program a NC machining program.

Programming:

```
O0401
N10 G50 X100. 0 Z100. 0;(starting point of cutter program)
N20 M03 S800 T0101;(Spindle forward rotation with a speed of 800r/min with No. 1 datum cutter)
N30 G00 X30. 0 Z2. 0;(quickly locate to the place near the workpiece)
N40 …
…
…        (Cutting process)
…
…
N  G00 X100. 0 Z100. 0;(return to start point quickly)
N  M05 T0100;(The spindle stops and changes back to the datum cutter)
N  M30;(The program ends, the cursor returns to the beginning of the program)
```

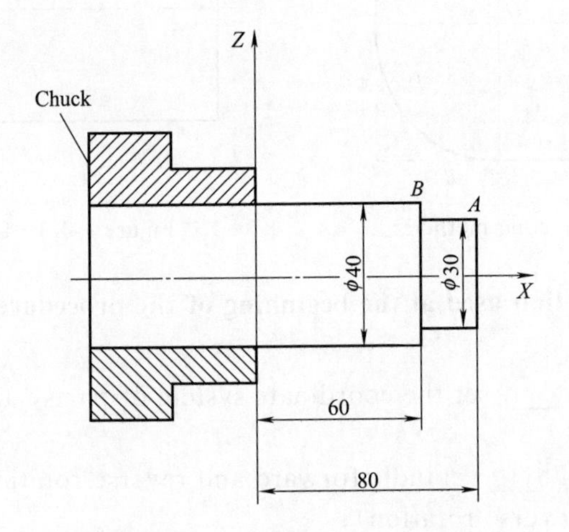


Figure 1-4-9 Programming example Ⅰ

**【Example 1-4-2】** As shown in Figure 1-4-10, the workpiece has done the rough machining and the margin of 0. 2mm is left in each position. Please rewrite the finish machining without cutting off.

Processing program

```
O0402
N10 G99 G50 X150. 0 Z100. 0;
N20 M03 S1000 T0101;
N30 G00 X16. 0 Z2. 0;
N40 G01 X16. 0 Z0 F0. 5;
N50 G01 X20. 0 Z-2. 0 F0. 1;
N60 Z-20. 0;
N70 X40. 0 Z-30. 0;
N80 G00 X150. 0 Z100. 0;
N90 M05 T0100;
N100 M30;
```

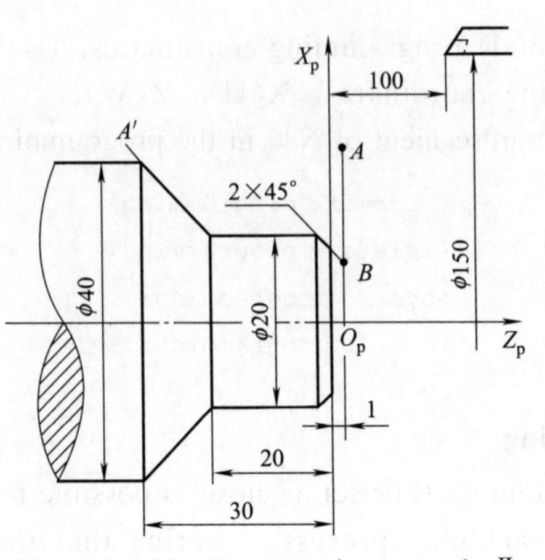

Figure 1-4-10　Programming example Ⅱ

**【Example 1-4-3】** As shown in Figure 1-4-11, the workpiece has done the rough machining and the margin of 0. 2mm is left in each position. Please rewrite the finish machining without cutting off.

Processing program

```
O0403
N10 G99 G50 X100. 0 Z100. 0;
N20 M03 S1000 T0101;
N30 G00 X10. 0 Z2. 0;
N40 G01 Z0 F0. 5;
N50 G03 X12. 0 Z-1. 0 R1. 0 F0. 2;
N60 G01 Z-12. 0;
N70 G02 X18. 0 Z-15. 0 R3. 0;
N80 G03 X22. 0 W-2. 0 R2. 0;
N90 G01 Z-28. 0;
N100 G00 X100. 0 Z100. 0;
N110 M05 T0100;
N120 M30;
```

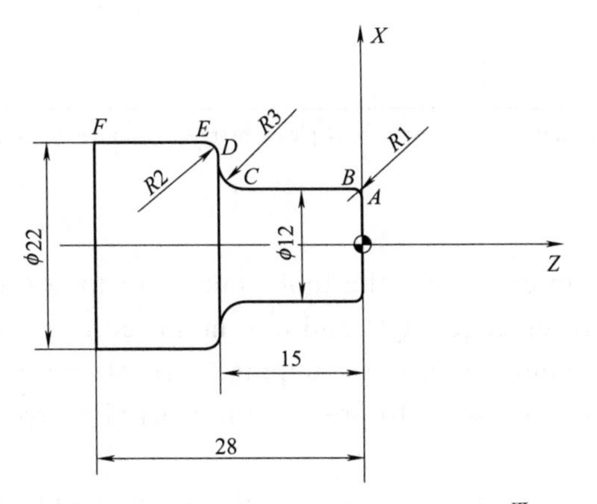

Figure 1-4-11　Programming example Ⅲ

**8. Programming coordinates that may be used in the same program segment**

① Absolute programming coordinates: X, Z.

② Relative (incremental) programming coordinates: U, W.

③ Hybrid programming coordinates: X(U), Z(W).

For example, the program segment of N30 in the programming example is available for:

```
N30 G00 X30. 0 Z2. 0;          (absolute programming)
or  N30 G00 U-70. 0 W-98. 0;   (relative programming)
N30 G00 X30. 0 W-98. 0;        (hybrid programming)
N30 G00 U-70. 0 Z2. 0;         (hybrid programming)
```

**9. Notes in programming**

① The coordinate system shall be set as close as possible to the workpiece according to the requirements of the machining process, ensuring that the workpiece is not touched when changing the cutter.

② The format of first three steps are the same as the last three steps, but the setting range of coordinate system is different, the location of workpiece G00 should be determined according to the machining requirements.

③ A program segment of G00 cannot contain an F value, otherwise it is invalid.

**10. Processing procedures**

Abridgment.

## Ⅳ. Preparation Before Processing

(1) Preparation of machine tool (see Table 1-4-4)

**Table 1-4-4 Machine tool preparation card**

| Project | Mechanical part | | | | Electrical equipment | | CNC system part | | | Auxiliary part | |
|---|---|---|---|---|---|---|---|---|---|---|---|
| Equipment inspection | Spindle part | Feed part | Tool rest part | Tail stock | Main power supply | Cooling fan | Electrical component | Control section | Drive section | Cooling | Lubri cating |
| Check the condition | | | | | | | | | | | |

Note: If this part is in good condition after inspection, please mark "√" under the corresponding item; Report to repair in time if problems arise.

(2) Other notes

① When installing an external lathe tool, take care to control the length of the tool rod and angle of and cutting edge angle and tool minor cutting edge angle.

② Pay attention to controlling the clamping force of the workpiece when the workpiece is rotated for clamping, so as to prevent the workpiece from being clamped.

(3) Parameter setting

① The value of the cutter setting shall be input in the cutter compensation No. corresponding to that of the cutter in the program.

② The value of the radius value of the cutter and the serial number of the false tool nose shall be input in the numerical value of the cutter.

## Ⅴ. Processing of Actual Parts

**1. Teacher's demonstration**

① Input and emulation checking of the program.
② Establishment of cutter setting and cutter compensation.
③ Method of ensuring dimensional tolerances of external circles.

**2. Students' processing training**

During the training, trainers detour to instruct, correct the wrong operation behavior in time, and solve the problems that arise in the student's practice.

## Ⅵ. Parts Measurement

Teaching strategy: teaching method, demonstration method

Emphasis is placed on the selection of measuring tools and the method and matters needing attention of micrometer calipers. Because the students know relatively little about NC machining in the initial stage of learning and training, each link of machining should be carefully explained to the students at the initial stage, so as to lay a good foundation for the students' later improvement. Introduce the factors of error generation and the method of reducing it. After the teaching and demonstration, physical measurement can be conducted in groups to enhance the proficiency of testing and improve the accuracy and stability of measurement.

(1) Check the external circle dimensions $\phi 24_{-0.052}^{0}$mm of the parts and check the surface roughness $Ra3.2\mu m$

Measure the readings directly using a micrometer caliper with 0-25mm external diameter.

Check the surface roughness and conduct the comparison verification using the surface roughness comparison template.

(2) Check the external circle size of the parts of $28_{-0.052}^{0}$mm and $30_{-0.062}^{0}$mm. Check surface roughness of $Ra3.2\mu m$

Measure the readings directly using a a micrometer caliper with 25-50mm external diameter.

Check the surface roughness and conduct comparison verification using the surface roughness comparison template.

(3) Check length and size

Check the length of 36mm ± 0.05mm with slide caliper.

(4) Dimensions of rounding chamfer

Check the rounding chamfer with slide caliper or eye measurements.

**Exercise**

1. What is the difference between G90 and G94?
2. What is the relationship between G40, G41 and G42?
3. Please explain the format and meaning of G04.
4. What basic measuring tools should be mastered when operating CNC lathe?
5. How to delete a single program and all programs on the machine tool?
6. When is the spindle forward rotation M03 and spindle reverse rotation M04 available?
7. How to calculate the minor diameter of thread?
8. Please write the program for returning to the reference point.

# Task Ⅴ　Turning of Connecting Shaft

## Ⅰ. Drawing and Technical Requirements

As shown in Figure 1-5-1, the part with shaft type characteristics is a cylinder bar with material of No. 45 steel, the specification of $\phi$40mm and hardness of 200HB, which has been done normalizing treatment.

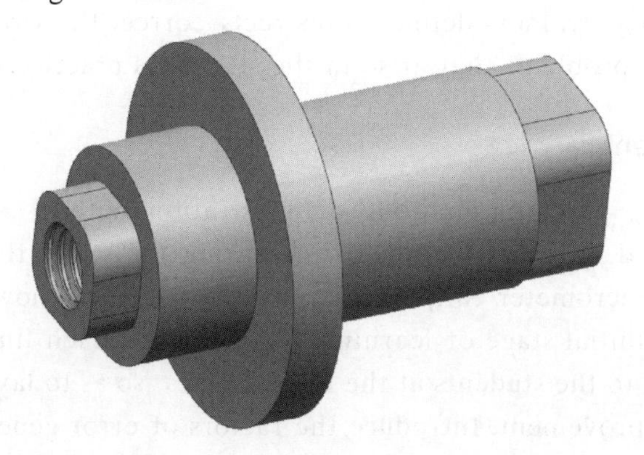

Figure 1-5-1　Physical map of connecting shaft

## Ⅱ. Analysis of Drawings

### 1. Part diagram analysis

The outer outline of the part is composed of three cylinders ($\phi$20mm, $\phi$35mm, $\phi$20mm) with different length, belonging to simple stepped shaft parts, shown in Figure 1-5-2. The parts are designed according to the sequence of training and learning.

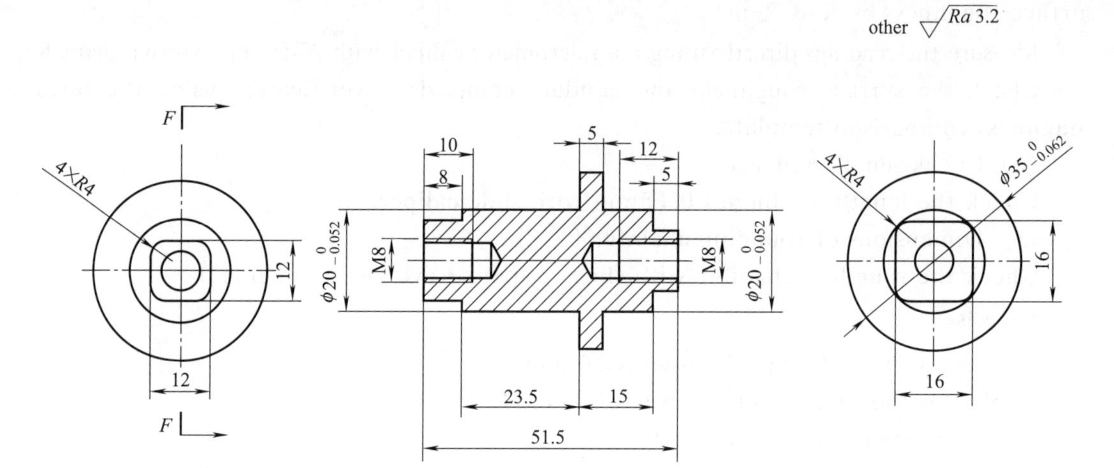


Figure 1-5-2　Figure of connecting shaft parts

### 2. Process analysis

① Structural analysis: The structure of the parts is a simple stepped shaft.

② Accuracy analysis: The key dimensions of the parts are all h9 in the external accuracy level, which is a relatively reasonable tolerance size for the training of intermediate pre-work. In addition, length, rounding chamfer and other details of precision also need to pay attention to.

③ Positioning and clamping analysis: This part uses three-jaw self-centering chuck for positioning and clamping. The clamping force of the work piece should be moderate, which should not only prevent the deformation and crush of the work piece, but also prevent the looseness of the work piece in the process. In order to ensure the tolerance of shape and position, the work piece should be aligned in the process of clamping.

④ Machining process analysis: After the above analysis, the overall arrangement of the parts in the process is as following: firstly process the right end of the part; turn the head after cutting off the workpiece, align the end face and rounding chamfer after turning so as to ensure the total length size of work piece .

### 3. Main cutter selection (see Table 1-5-1)

**Table 1-5-1 Cutter table**

| Cutter name | Cutter specification name | Material | Qty. | Radius of tool nose | Cutter width |
|---|---|---|---|---|---|
| 90° external lathe tool | 25mm×25mm | YT 15 | 1 | 0 | |
| 45 ° external lathe tool | 25mm×25mm | YT 15 | 1 | 0 | |
| Cut-off tool | 25mm×25mm | YT 15 | 1 | 0 | 4mm |

### 4. Process schedule arrangement (see Table 1-5-2)

**Table 1-5-2 Process table (right end)**

<table>
<tr><td colspan="2">Unit</td><td>Product name and model</td><td>Part name</td><td>Part drawing number</td></tr>
<tr><td colspan="2"></td><td>Task Ⅴ</td><td>Simple stepped shaft</td><td>Figure 1-5-2</td></tr>
<tr><td>Working procedure</td><td>Program No.</td><td>Jig name</td><td>Used equipment</td><td>Work piece material</td></tr>
<tr><td>001</td><td>O0001</td><td>three-jaw chuck</td><td>SK50</td><td>No. 45 steel</td></tr>
<tr><td colspan="5">Stepped shaft parts</td></tr>
<tr><td>Working step</td><td>Working step content</td><td>Cutting tool No.</td><td>Cutting engagement</td><td>Remarks</td></tr>
<tr><td>1</td><td>Radial facing</td><td>T11</td><td>$n = 700r/min$<br>(Manual feed rate mode at × 10% position)</td><td>Manual processing</td></tr>
<tr><td>2</td><td>Rough lathe for outer circle of $\phi20$ and $\phi30$ and leave a finish processing margin of 0.5mm</td><td rowspan="2">T11</td><td>$n = 700r/min$<br>$f = 0.2mm/r$<br>$a_p = 3.0mm$</td><td rowspan="2">Automatic processing</td></tr>
<tr><td>3</td><td>Fine lathe for outer circle of $\phi20$ and $\phi35$</td><td>$n = 900r/min$<br>$f = 0.1mm/r$<br>$a_p = 1.0mm$</td></tr>
<tr><td>4</td><td>Manually cut-off work piece and leave a total margin length: 1.0mm or so</td><td>T33</td><td>$n = 500r/min$<br>(feed rate switch mode at × 10% position)</td><td>$B = 4mm$ cut off the work-piece, pay attention to avoid the impact of the workpiece</td></tr>
<tr><td>5</td><td>Turn flat end to ensure total length and chamfer</td><td>T11</td><td>$n = 700r/min$<br>(feed rate switch mode at × 10% position)</td><td>Manual processing</td></tr>
</table>

## Ⅲ. Programming

**1. The turning cycle of external circle and innernal circle**

Function: When the blank margin on the external and internal cylinder (conical surface) of the part is quite large, G90 can be used to remove most of the blank margin.

Linear cutting cycle:

Format: G90 X (U)_ Z (W)_ F;

Where, X, Z represent the absolute value coordinates of the end point;

U, W represent the end-point coordinate size of relative (incremental) value;

F means cutting feed speed.

Its path is shown in Figure 1-5-3 and consists of four steps:

1 (R) represents the rapid movement in the first step;

2 (F) represents that the second step is cutting according to the feed speed;

3 (F) represents that the third step is cutting according to the feed speed;

4 (R) represents the rapid movement in the fourth step.

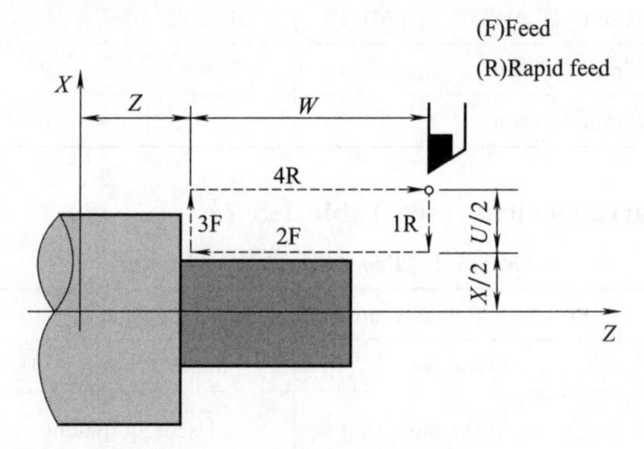


Figure 1-5-3 G90 Rough lathe cycle chart

**2. Cone turning cycle**

(1) Format: G90 X (U)_ Z (W)_ R_ F;

Where, X, Z represent the absolute value coordinates of the end point;

U, W represent the end-point coordinate size of relative (incremental) value ;

R means taper size [$R = (D - d)/2$, $D$ is large end diameter of taper, $d$ is small end diameter of taper]. If the turning of external taper is from small end to large end, the cutting taper R is negative; If the turning of internal taper is from the large end to the small end, internal taper R is positive;

F represents the cutting feed speed.

The path of F cutting feed speed is shown in Fig. 1-5-4, and the positive and negative values of R are related to the tool path.

(2) Examples of programming

See Figure 1-5-5, Figure 1-5-6.

**3. End face turning cycle**

(1) Flat end surface turning cycle

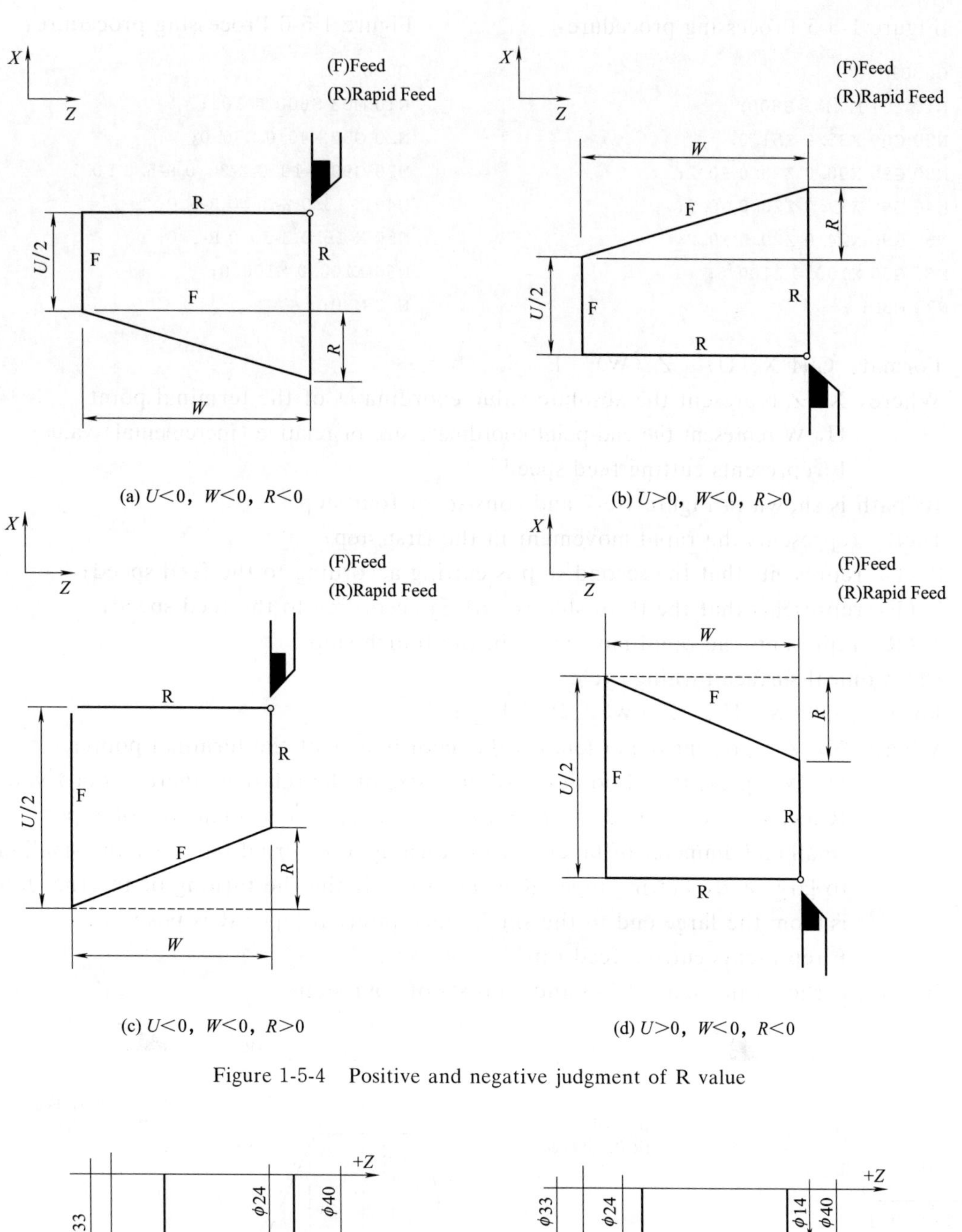


Figure 1-5-4　Positive and negative judgment of R value

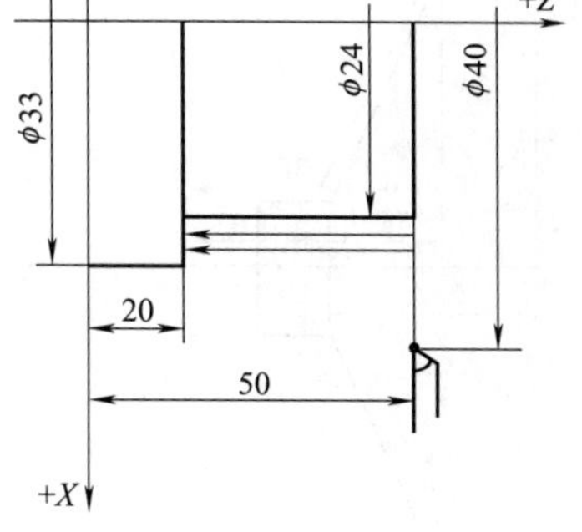


Figure 1-5-5　G90 programming example Ⅰ



Figure 1-5-6　G90 programming example Ⅱ

Figure 1-5-5 Processing procedure:

```
O0501
N10 T0101 M03 S800;
N20 G00 X35. 0 Z51. 0;
N30 G90 X30. 0 Z20. 0 F0. 2;
N40 G90 X27. 0 Z20. 0 F0. 2;
N50 G90 X24. 0 Z20. 0 F0. 2;
N60 G00 X100. 0 Z100. 0;
N70 M30;
```

Figure 1-5-6 Processing procedure:

```
O0502
N10 M03 S600 T0101;
N20 G00 X40. 0 Z50. 0;
N30 G90 X-10. 0 Z-30. 0 R-5. 0 F0. 1;
N40 X-13. 0 Z-30. 0 R-5. 0;
N50 X-16. 0 Z-30. 0 R-5. 0;
N60 X100. 0 Z100. 0;
N70 M30;
```

Format: G94 X (U)__ Z (W)__ F__;

Where, X, Z represent the absolute value coordinates of the terminal point;

U, W represent the end-point coordinate size of relative (incremental) value;

F represents cutting feed speed.

Its path is shown in Figure 1-5-7 and consists of four steps:

1 (R) represents the rapid movement in the first step;

2 (F) represents that the second step is cutting according to the feed speed;

3 (F) represents that the third step is cutting according to the feed speed;

4 (R) represents the rapid movement in the fourth step.

(2) Conical surface turning cycle

Format: G94 X (U)__ Z (W)__ R__ F__;

Where, X, Z represent the absolute value coordinates of the terminal point;

U, W represent end-point coordinate size of the relative (incremental) value;

R means taper size [$R=(D-d)/2$, $D$ is large end diameter of taper, $d$ is small end diameter of taper]. If the turning of external taper is from small end to large end, cutting taper R is negative; if the the turning of internal taper is from the large end to the small end, internal taper R is positive;

F represents cutting feed rate.

Its path is shown in Figure 1-5-8 and consists of four steps.

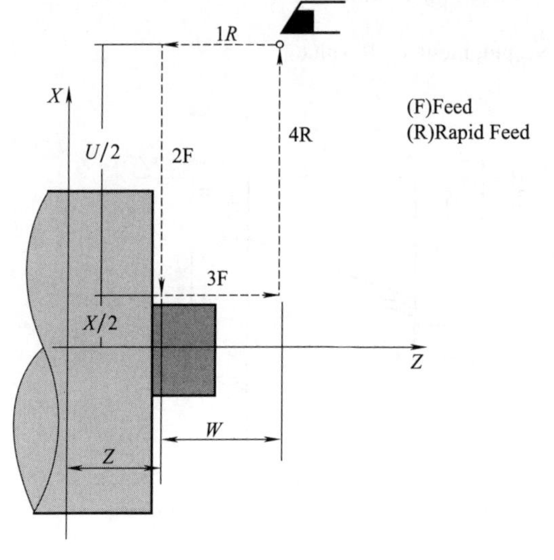


Figure 1-5-7 Flat end surface turning cycle

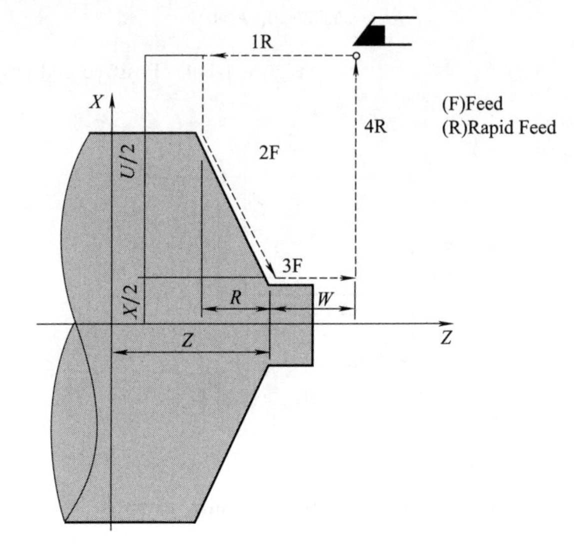


Figure 1-5-8 Conical surface turning cycle

(3) G94 Programming example (see Figure 1-5-9、Figure 1-5-10)

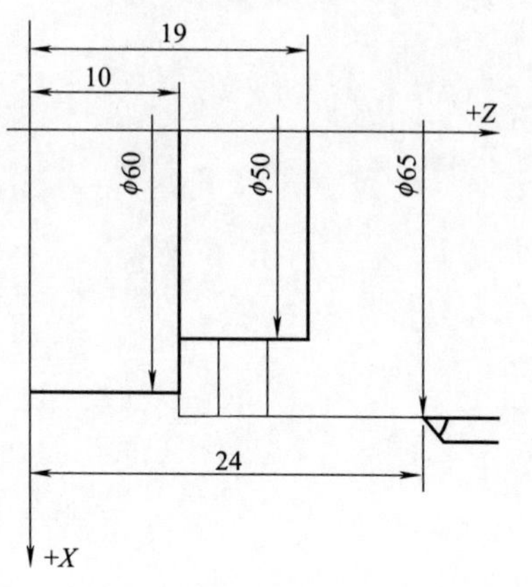


Figure 1-5-9 G94 programming example Ⅰ

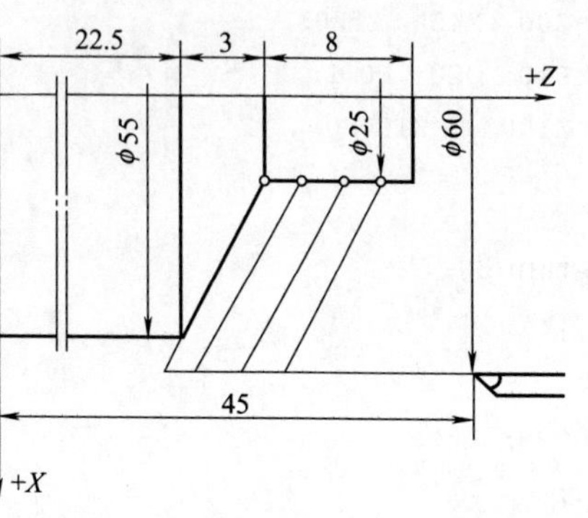


Figure 1-5-10 G94 programming example Ⅱ

Figure 1-5-9 Processing procedure:

```
O0503
N10 M03 S600 T0202;
N20 G00 X65. 0 Z24. 0;
N30 G94 X-15. 0 Z-8. 0 F0. 1;
N40 X-15. 0 Z-11. 0;
N50 X-15. 0 Z-14. 0;
N60 G00 X100. 0 Z100. 0;
N70 M30;
```

Figure 1-5-10 Processing procedure:

```
O0504
N10 M03 S700 T0101;
N20 G00 X60. 0 Z45. 0;
N30 G94 X25. 0 Z31. 5 R-3. 5 F0. 15;
N40 X25. 0 Z29. 5 R-3. 5;
N50 X25. 0 Z27. 5 R-3. 5;
N60 X25. 0 Z25. 5 R-3. 5;
N70 G00 X100. 0 Z100. 0;
N80 M30;
```

**4. Reference program**

Program 1

```
O0001;
N1;
G97  G99;
T0101;
M03  S700;
G40  G00  X45. 0  Z5. 0;
G71  U1. 5  R0. 5;
G71  P10  Q20  U0. 5  W0  F0. 2;
N10  G00  X0;
G01  Z0;
X20. 0,C0. 5;
Z-15. 0;
X35. 0,C0. 5;
W-6. 0;
N20  X45. 0;
G00  X100. 0 Z100. 0;
M05;
M00;
```

```
N2;
T0202;
M03  S900;
G42  G00  X45.0 Z5.0;
G70  P10  Q20  F0.1;
G00  X100.0  Z100.0;
M30;
```

Program 2

```
O0001;
N1;
G97  G99;
T0101;
M03 S700;
G40  G00  X45.0 Z5.0;
G71  U1.5  R0.5;
G71  P10  Q20  U0.5  W0  F0.2;
N10  G00  X0;
G01  Z0;
X20.0,C0.5;
Z-31.5;
X35.0,C0.5;
W-5.5;
N20  X45.0;
G00  X100.0 Z100.0;
M05;
M00;
N2;
T0202;
M03  S900;
G42  G00  X45.0 Z5.0;
G70  P10  Q20  F0.1;
G00  X100.0  Z100.0;
M30;
```

## Ⅳ. Preparation before Processing

(1) Preparation of machine tool (see Table 1-5-3)

**Table 1-5-3 Machine tool preparation card**

| Project | Mechanical part | | | | Electrical equipment | | CNC system part | | | Auxiliary part | |
|---|---|---|---|---|---|---|---|---|---|---|---|
| Equipment inspection | Spindle part | Feed part | Tool rest part | Tail stock | Main power supply | Cooling fan | Electrical component | Control section | Drive section | Cooling | Lubricating |
| Check the condition | | | | | | | | | | | |

Note: If this part is in good condition after inspection, please mark "√" under the corresponding item; Report to repair in time if problems arise.

(2) Other notes

① When installing an external lathe tool, take care to control the length of the tool rod and angle of and cutting edge angle and tool minor cutting edge angle;

② Pay attention to controlling the clamping force of the workpiece when the workpiece is rotated for clamping, so as to prevent the workpiece from being clamped.

(3) Parameter setting

① The value of the cutter setting shall be input in the cutter compensation No. corresponding to that of the cutter in the program;

② The value of the radius value of the cutter and the serial number of the false tool nose shall be input in the numerical value of the cutter.

## Ⅴ. Processing of Actual Parts

**1. Teacher's demonstration**

① Input and emulation checking of the program;

② Establishment of cutter setting and cutter compensation;

③ Method of ensuring dimensional tolerances of external circles.

**2. Students' processing training**

During the training, trainers detour to instruct, correct the wrong operation behavior in time, and solve the problems that arise in the student's practice.

## Ⅵ. Parts Measurement

Teaching strategy: teaching method, demonstration method.

Emphasis is placed on the selection of measuring tools and the method and matters needing attention of micrometer calipers. Because the students know relatively little about NC machining in the initial stage of learning and training, each link of machining should be carefully explained to the students at the initial stage, so as to lay a good foundation for the students' later improvement. Introduce the factors of error generation and the method of reducing it. After the teaching and demonstration, physical measurement can be conducted in groups to enhance the proficiency of testing and improve the accuracy and stability of measurement.

(1) Check the external circle dimensions $\phi 20_{-0.052}^{0}$ mm of the parts and check the surface roughness $Ra3.2\mu m$

Measure the readings directly using a micrometer caliper with 0-25mm external diameter.

Check the surface roughness and conduct the comparison verification using the surface roughness comparison template.

(2) Check the external circle size of the parts of $\phi 35_{-0.062}^{0}$ mm. Check surface roughness of $Ra3.2\mu m$

Measure the readings directly using a micrometer caliper with 25-50mm external diameter.

Check the surface roughness and conduct comparison verification using the surface roughness comparison template.

(3) Check length and size

Check the length of 51.5mm ± 0.05mm with slide caliper.

(4) Dimensions of rounding chamfer

Check the rounding chamfer with slide caliper or eye measurements.

## Ⅶ. Processing Precautions

① The influence of cutter center height on machining.

② The workpiece coordinate system set by the cutter setting should be corrected and verified.

**Exercise**

1. When is the positive value for arc radius used? And when is the negative value used?
2. Please briefly describe the basic parameters of the cone.
3. Given an outer cone, the diameter of its large end is $D = 80$mm, the diameter of its small end is $d = 60$mm, and the half angle of its cone is 30°, please calculate the length of the cone part.
4. Please write a cycle machining program. It is known that: $d = 40$mm, $d_1 = 30$mm, $d_2 = 20$mm, the depth for each section is 10mm, and the cutter safety point is ($a$, $b$).
5. Write the thread program of M20×2 with self-designed cycle point.
6. What is the tool name for verifying the thread? And explain it.
7. What are the functions of cutting fluid?
8. Draw the position maps of the front tool rest and rear tool rest.

# Task Ⅵ Thread Shaft Machining

## Ⅰ. Drawing and Technical Requirements

As shown in Figure 1-6-1, the part with thread type characteristics is a cylinder bar with material of No. 45 steel, the specification of $\phi$10mm and hardness of 200HB, which has been done normalizing treatment.

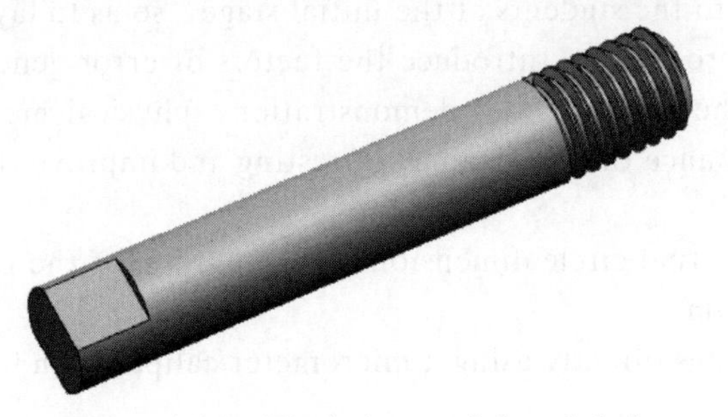

Figure 1-6-1 Physical map of thread shaft

## Ⅱ. Analysis of Drawings

Teaching strategies: group discussion, group report and summary of teachers

### 1. Parts diagram analysis

The outer contour of the part is a cylindrical surface of $\phi$10mm which are composed of external thread of M8×1.25, which is included in simple thread-like parts, shown in Figure 1-6-2. The parts are designed according to the sequence of training and learning. Score form is shown in Table 1-6-1.

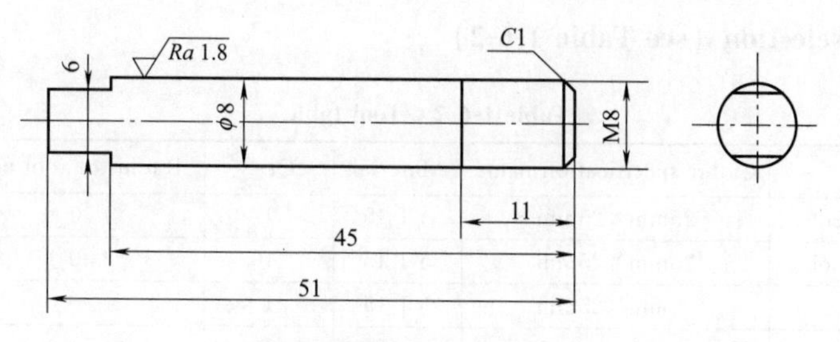


Figure 1-6-2 Thread shaft

**Table 1-6-1 Score form**

| No. | Project and technical requirements | Partition(IT/*Ra*) | Scoring standard | Inspection result | Real score |
|---|---|---|---|---|---|
| 1 | External Diameter: $\phi 8_{-0.039}^{0}$, $Ra1.6_{-0.039}^{0}$ | 16/2 | Full sore deduction if the result is extreme poor | | |
| 2 | External Diameter Scan 6, $Ra3.2_{-0.052}^{0}$ | 16/2 | Full sore deduction if the result is extreme poor | | |
| 3 | Length: $51 \pm 0.1$ | 16 | Full sore deduction if the result is extreme poor | | |
| 4 | Length: $45_{-0.06}^{0}$ | 16 | Full sore deduction if the result is extreme poor | | |
| 5 | Length: $11_{0}^{+0.1}$ | 16 | Full sore deduction if the result is extreme poor | | |
| 6 | Rounding Chamfer *C*1. (4) | 8 | Full sore deduction if the result is extreme poor | | |
| 7 | Shape outline is complete | 3 | No score if not finished | | |
| Safe and civilized production | | 5 | | | |
| Working hours | | 60min | | | |

## 2. Process analysis

① Structure analysis: the structure of the part is simple thread shaft.

② Accuracy analysis: The key dimension of the part is h7 in the external circle accuracy level and 6g in the middle (the diameter between the external and internal) tolerance level of the thread, which is a relatively reasonable tolerance dimension for the training of this module. Especially in programming, pay attention to the length of the thread. In addition, length, rounding chamfer and other details accuracy of the problems are also needed to pay attention to.

③ Positioning and clamping analysis: this part uses three-jaw self-centering chuck for positioning and clamping. The clamping force of the workpiece should be moderate which can not only prevent the deformation and clamping of the workpiece, but also prevent the looseness of the workpiece in the process. In order to ensure the tolerance of shape and position, the workpiece should be straightened in the process of clamping.

④ Machining process analysis: After the above analysis, the overall arrangement of the parts in the process is as following: firstly process the right end of the part; turn the head after cutting off the workpiece, align the end face and rounding chamfer after turning so as to ensure the total length size of work piece.

## 3. Main tool selection (see Table 1-6-2)

Table 1-6-2 Tool table

| Cutter name | Cutter specification name | Material | Qty. | Radius of tool nose | Cutter width |
|---|---|---|---|---|---|
| 90° external lathe tool | 25mm×25mm | YT 15 | 1 | 0 | |
| 45° external lathe tool | 25mm×25mm | YT 15 | 1 | 0 | |
| Cutter | 25mm×25mm | YT 15 | 1 | 0 | 4mm |
| Thread cutter | 25mm×25mm with tool included angle of 60° | YT 15 | 1 | 0 | |

## 4. Process schedule arrangement (see Table 1-6-3)

Table 1-6-3 Process table (Right End)

<table>
<tr><td colspan="2">Unit</td><td colspan="2">Product name and model</td><td>Part name</td><td>Part drawing number</td></tr>
<tr><td colspan="2"></td><td colspan="2">Task Ⅵ</td><td>Thread shaft</td><td>Figure 1-6-2</td></tr>
<tr><td>Working procedure</td><td>Program No.</td><td colspan="2">Jig name</td><td>Used equipment</td><td>Work piece material</td></tr>
<tr><td>001</td><td>O0001</td><td colspan="2">three-jaw chuck</td><td>SK50</td><td>No. 45 steel</td></tr>
<tr><td colspan="6">Piece 1</td></tr>
<tr><td>Working step</td><td>Working step content</td><td>Cutter No.</td><td colspan="2">Cutting engagement</td><td>Remarks</td></tr>
<tr><td>1</td><td>Rough machining</td><td>T11</td><td colspan="2">$n = 700$r/min<br>(Feed ratio switch at×10% position)</td><td>Automatic processing</td></tr>
<tr><td>2</td><td>Finish machining</td><td>T22</td><td colspan="2">$n = 900$r/min<br>(Feed ratio switch at×10% position)</td><td>Automatic processing</td></tr>
<tr><td>3</td><td>Rough machining M8×1.25 thread of the outer circle of $\phi 8$</td><td>T44</td><td colspan="2">$n = 500$r/min<br>$f = 1.25$mm/r<br>$a_p = 0.2$mm</td><td>Automatic processing</td></tr>
<tr><td>4</td><td>Finish machining M8×1.25 thread</td><td>T44</td><td colspan="2">$n = 500$r/min<br>$f = 1.25$mm/r<br>$a_p = 0.1$mm</td><td>Automatic processing</td></tr>
<tr><td>5</td><td>Manually cut-off work piece<br>And leave a total of margin of 1mm or so</td><td>T33</td><td colspan="2">$n = 500$r/min<br>(Feed ratio switch at×10% position)</td><td>Manually processing $B = 4$mm cut off the workpiece, pay attention to avoid the impact of the workpiece</td></tr>
<tr><td>6</td><td>Turn flat end to ensure total length and chamfer</td><td>T11</td><td colspan="2">$n = 700$r/min<br>(Feed ratio switch at×10% position)</td><td>Automatic processing</td></tr>
</table>

# Ⅲ. Programming

## 1. Basic knowledge of thread processing

Two problems that should be paid attention to when threading

① There must be a cut-in section $\delta_1$ and a cut section $\delta_2$ in the thread turning. See Figure 1-6-3.

② Thread machining generally requires multiple cuts, and the depth of each cut shall

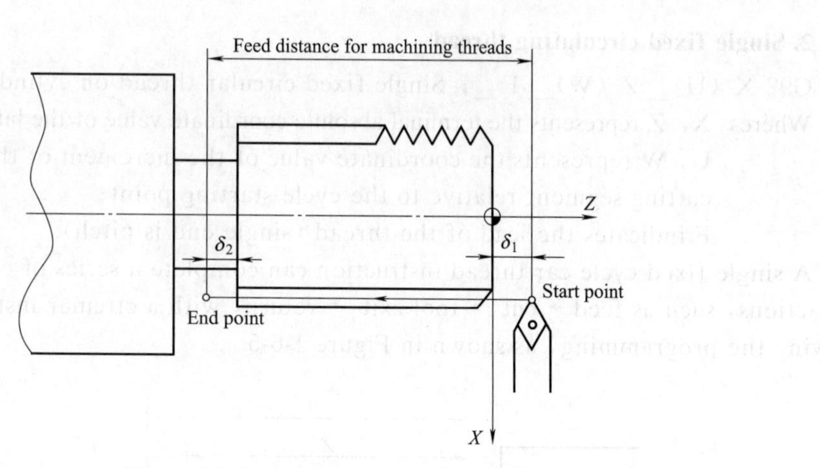


Figure 1-6-3 Cut-in section $\delta_1$ and $\delta_2$

be assigned in accordance with the law of decreasing, as shown in Figure 1-6-4.

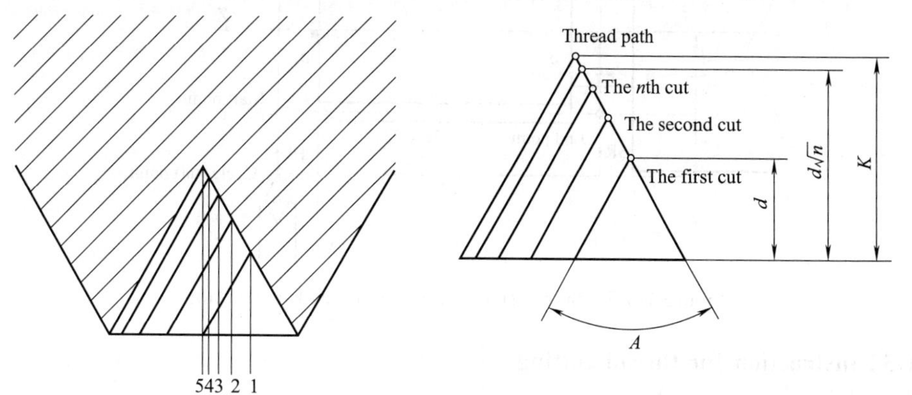


Figure 1-6-4 Multi-feed route

Table 1-6-4 shows feeding times and depth of layered cutting for a common metric thread.

**Table 1-6-4 Feeding times of common metric threads**

| Common Metric Threads | | | | | | | | |
|---|---|---|---|---|---|---|---|---|
| Pitch/mm | | 1.0 | 1.5 | 20 | 2.5 | 3.0 | 3.5 | 4.0 |
| Thread height/mm | | 0.649 | 0.977 | 1.299 | 1.624 | 1.949 | 2.273 | 2.598 |
| Feeding times and depth of layered cutting(diameter value)/mm | One time | 0.7 | 0.8 | 0.9 | 1.0 | 1.2 | 1.5 | 1.5 |
| | Twice | 0.4 | 0.6 | 0.6 | 0.7 | 0.7 | 0.7 | 0.8 |
| | Three times | 0.2 | 0.4 | 0.6 | 0.6 | 0.6 | 0.6 | 0.6 |
| | Four times | | 0.16 | 0.4 | 0.4 | 0.4 | 0.6 | 0.6 |
| | Five times | | | 0.1 | 0.4 | 0.4 | 0.4 | 0.4 |
| | Six times | | | | 0.15 | 0.4 | 0.4 | 0.4 |
| | Seven times | | | | | 0.2 | 0.2 | 0.4 |
| | Eight times | | | | | | 0.15 | 0.3 |
| | Nine tines | | | | | | | 0.2 |

**2. Single fixed circulating thread**

G92 X (U)__ Z (W)__ F __; Single fixed circular thread on cylindrical surface.

Where, X, Z represents the terminal absolute coordinate value of the lathe thread segment;

U, W represents the coordinate value of the increment of the end point of the cutting segment relative to the cycle starting point;

F indicates the lead of the thread (single end is pitch).

A single fixed cycle car thread instruction can complete a series of continuous processing actions, such as feed→ cut → tool exit → return, with a circular instruction, thus simplifying the programming, as shown in Figure 1-6-5.

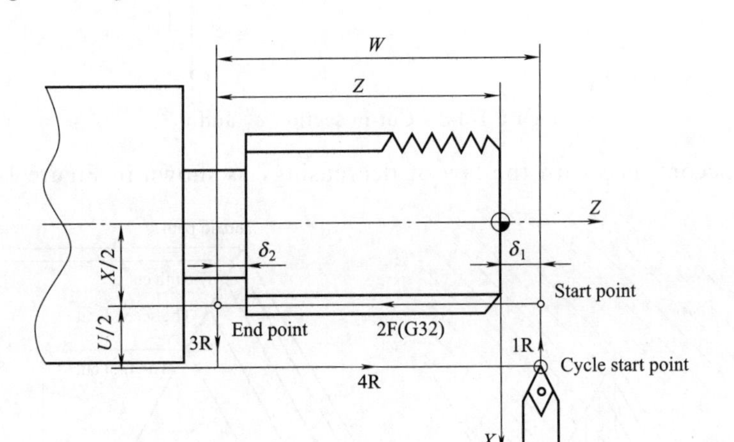


Figure 1-6-5 Single fixed cycle car thread instruction

**3. G32 instruction for thread cutting**

Format: G32 X (U)__ Z (W)__ F__

Where, X, Z: For absolute programming, the coordinates of the effective thread end point in the workpiece coordinate system;

U, W: For incremental programming, the displacement of the end point of the effective thread relative to the starting point of the thread cutting;

F: Thread lead, i. e. the feed value of the cutter relative to the workpiece per turn of the spindle.

Description:

① From thread rough machining to finish machining, the speed of spindle must keep constant.

② It is very dangerous to stop the cutting of thread without stopping the spindle; Therefore, the FEED HOLD function is not effective when cutting threads. If operator press the FEED HOLD button, the tool stops moving after completing the machining.

③ Do not use the constant linear speed control function in the thread processing.

④ Sufficient up-speed feed segment $\delta$ and down-speed return segment $\delta'$ in the thread processing path shall be set to eliminate pitch error caused by servo lag.

**4. G92 instruction for thread cutting cycle**

Format: G92 X (U)__ Z (W)__ F __;

Where, X, Z: for absolute value programming, the coordinates of the thread end point in the workpiece coordinate system;

U, W: for incremental programming, the directed distance of the end point of the thread relative to the starting point of the cycle;

F: Thread lead.

### 5. Reference program

```
O0001;
N1;
G97  G99;
T0101;
M03  S700;
G40  G00  X35.0  Z5.0;
G71  U1.5  R0.5;
G71  P10  Q20  U0.5  W0  F0.2;
N10  G00  X0;
G01  Z0;
X7.8  ,  C1.0;
Z-54.0;
N20  X35.0;
G00  X100.0  Z100.0;
M05;
M00;
N2;
T0202;
M03  S900;
G42  G00  X35.0  Z5.0;
G70  P10  Q20  F0.1;
G00  X100.0  Z100.0;
M05;
M00;
N3;
T0404;
M03  S500;
G40  G00  X35.0  Z5.0;
G92  X7.4  Z-10.5  F1.25;
X7.;
X6.7;
X6.5;
X6.3;
X6.3;
G00  X100.0  Z100.0;
M30;
```

## Ⅳ. Preparation Before Processing

### 1. Machine tool preparation (see Table 1-6-5)

**Table 1-6-5 Machine tool preparation card**

| Project | Mechanical part | | | | Electrical equipment | | CNC system part | | | Auxiliary part | |
|---|---|---|---|---|---|---|---|---|---|---|---|
| Equipment inspection | Spindle part | Feed part | Tool rest part | Tail stock | Main power supply | Cooling fan | Electrical component | Control section | Drive section | Cooling | Lubricating |
| Check the condition | | | | | | | | | | | |

Note: If this part is in good condition after inspection, please mark "√" under the corresponding item; Report to repair in time if problems arise.

**2. Other notes**

① When installing an external lathe tool, take care to control the length of the tool rod and angle of and cutting edge angle and tool minor cutting edge angle;

② Pay attention to controlling the clamping force of the workpiece when the workpiece is rotated for clamping, so as to prevent the workpiece from being clamped.

**3. Parameter setting**

① The value of the cutter setting shall be input in the cutter compensation No. corresponding to that of the cutter in the program;

② The value of the radius value of the cutter and the serial number of the false tool nose shall be input in the numerical value of the cutter.

## Ⅴ. Processing of Actual Parts

**1. Teacher's demonstration**

① Input and emulation checking of the program.

② Establishment of cutter setting and cutter compensation.

③ Manual turning, thread machining and dimensional accuracy assurance of thread cutter exit groove.

**2. Students' processing training**

During the training, trainers detour to instruct, correct the wrong operation behavior in time, and solve the problems that arise in the student's practice.

## Ⅵ. Parts Measurement

Teaching strategy: teaching method, demonstration method.

Emphasis is placed on the selection of measuring tools and the method and matters needing attention of micrometer calipers. Because the students know relatively little about NC machining in the initial stage of learning and training, each link of machining should be carefully explained to the students at the initial stage, so as to lay a good foundation for the students' later improvement. Introduce the factors of error generation and the method of reducing it. After the teaching and demonstration, physical measurement can be conducted in groups to enhance the proficiency of testing and improve the accuracy and stability of measurement.

**1. Reference detection technology**

(1) Check the external circle dimensions $\phi$8mm of the parts and check the surface roughness $Ra3.2\mu m$.

Measure the readings directly using a micrometer caliper with 0 ~ 25mm external dremeter.

Check the surface roughness and conduct the comparison verification using the surface roughness comparison template.

(2) Check the external thread M8×1.25

Use the go gauge and no go gauges of M8×1.25 thread ring gauge to measure whether the thread is qualified.

(3) Check length and size

Check the length of 51mm with wilde caliper.

(4) Dimensions of rounding chamfer

Check the fouding chamfer with slide caliper or eye measurements.

**2. Test and fill in a record form**

Teaching strategies: group mutual inspection, individual verification and teacher selective examination.

First of all, students are divided into groups, and conduct mutual examination within the group, mutual examination result will be given by the test student according to the score sheet; Then the individual carries on the self-examination for his/her workpiece, and compares the result with the mutual examination result to find out the problem size and find out the reasons of different results, and corrects the wrong ring size section. At last, the teacher conducts a selective examination to students' parts, explains the causes of measurement errors centrally according to the problems, and puts forward the improved methods.

## Ⅶ. Notes for thread processing

① The speed can not be changed when processing the thread.

② The speed for thread processing should not be too high.

**Exercise**

1. What are the notes for installing the external thread cutter?
2. What are the coarse thread pitches of M10, M12, M16, M20 and M24 threads?
3. How to determine the positive and negative value of program margin for inner hole machining? What is the method?
4. What is the difference between automatic operation and NDI?
5. How to use the computer transmission program?
6. What are the consequences if the cutter is lower than the workpiece center?
7. How to set the feed speed F when machining the thread?

# Task Ⅶ  Handle Processing

## Ⅰ. Drawing and Technical Requirements

As shown in Figure 1-7-1-Figure 1-7-3, the part is a cylinder bar with material of

No. 45 steel, the specification of $\phi$30mm, and hardness of 200HB, which has been done normalizing treatment.

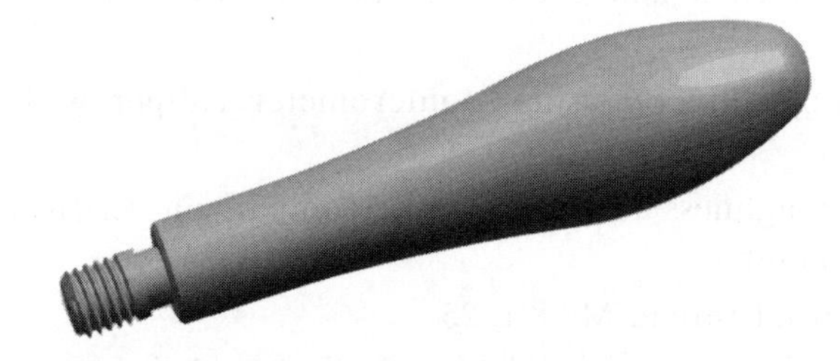

Figure 1-7-1　Handle physical diagram

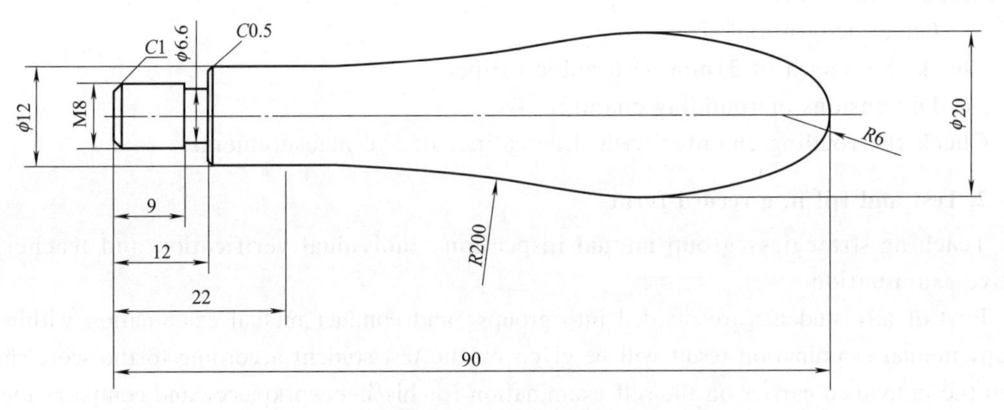


Figure 1-7-2　Handle parts

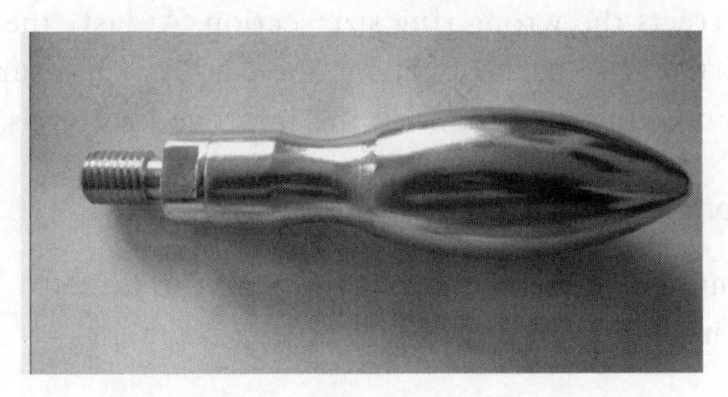

Figure 1-7-3　Handle physical diagram

## Ⅱ. Analysis of Drawings

### 1. Parts diagram analysis

As shown in Figure 1-7-2, the part consists of an external circle, wide groove and external thread, the surface roughness is $Ra1.6\mu m$, $Ra3.2\mu m$ and $Ra6.3\mu m$.

### 2. Process analysis

① Structure analysis: The structure of this part is relatively complex. Rigidity, programming instruction, tool working angle, cutting angle and other problems should be considered in machining.

② Accuracy analysis: It is necessary to ensure part size and form and position tolerance requirements, such as coaxiality, run-out and verticality, and ensure the surface roughness requirements, so the machining rigidity of the workpiece, cutter rigidity, pro-

cessing technology and other problems shall be paid attention to when machining.

③ Positioning and clamping analysis: Three-jaw self-centering chuck is used for positioning and clamping of the part. Especially, the clamping force should be moderate when the workpiece turns direction for clamping, which should not only prevent the workpiece from deforming and damaging, but also prevent the workpiece from loosing during machining. During the clamping process of the workpiece, the workpiece should be aligned to ensure the tolerance of each form and position.

④ Processing process analysis: According to the drawing analysis, parts processing difficulty is in the wide groove, because the cutting groove radial force is relatively large, when clamping the right end of the processing part for the first time, the wide groove should be processed. Also because the protruded section of the part is longer, so when machining the thread, the speed should not be too high, otherwise it is easy to produce the taper thread.

When machining the part after turning direction, it is necessary to align the part to ensure the coaxial requirement and ensure the length size of 80mm ± 0. 03mm.

### 3. Main tool selection (see Table 1-7-1)

Table 1-7-1 Tool table

| Cutter name | Cutter specification name | Material | Qty. | Radius of tool nose | Cutter width |
|---|---|---|---|---|---|
| 90° external circular cutter | 25mm × 25mm | YT 15 | 1 | 0 | |
| 45° circular cutter | 25mm × 25mm | YT 15 | 1 | 0 | |
| 35° external circular cutter | 25mm × 25mm | YT 15 | 1 | 0. 4mm | |
| Cut-off tool | 25mm × 25mm | YT 15 | 1 | 0 | 3mm |
| Thread cutter | 25mm × 25mm with tool included angle of 60° | YT 15 | 1 | 0 | |

### 4. Process schedule arrangement (see Table 1-7-2)

Table 1-7-2 Process table

<table>
<tr><td colspan="3">Unit</td><td colspan="2">Product name and model</td><td>Part name</td><td>Part drawing number</td></tr>
<tr><td colspan="3"></td><td colspan="2">Task Ⅶ</td><td>Handle diagram</td><td>Figure 1-7-2</td></tr>
<tr><td colspan="2">Process</td><td>Program number</td><td colspan="2">Jig name</td><td>Used equipment</td><td>Process</td></tr>
<tr><td colspan="2">001</td><td>O0001, O0002</td><td colspan="2">Three-jaw chuck</td><td>SK50</td><td>001</td></tr>
<tr><td>Working step</td><td colspan="3">Working step content</td><td>Cutter No.</td><td>Cutting amount</td><td>Remarks</td></tr>
<tr><td>1</td><td colspan="3">Radial facing</td><td>T11</td><td>$n = 700$r/min</td><td>Manual processing</td></tr>
<tr><td>2</td><td colspan="3">Outer circle of the rough turning</td><td>T11</td><td>$n = 700$r/min<br>$f = 0.2$mm/r<br>$a_p = 2$mm</td><td>Automatic processing</td></tr>
<tr><td>3</td><td colspan="3">Contours of finish turning</td><td>T22</td><td>$n = 900$r/min<br>$f = 0.1$mm/r<br>$a_p = 0.5$mm</td><td>Automatic processing</td></tr>
</table>

Continued

| Working step | Working step content | Cutter No. | Cutting amount | Remarks |
|---|---|---|---|---|
| 4 | Cutter groove | T33 | $n$ = 500r/min<br>$f$ = 0. 07mm/r | Automatic processing |
| 5 | Threading | T44 | $n$ = 500r/min<br>$P$ = 1. 25mm | Automatic processing |
| 6 | Cut-off | T33 | $n$ = 300r/min<br>(Feed ratio switch is at × 10 position) | Manual processing |
| 7 | Processing contours after turning the part around | T11 | $n$ = 700r/min<br>$f$ = 0. 15mm/r<br>$a_p$ = 1mm | Automatic processing |
| 8 | Contours of finish turning | T22 | $n$ = 900r/min<br>$f$ = 0. 1mm/r<br>$a_p$ = 0. 5mm | Automatic processing |

## Ⅲ. Compound Cycle Instruction

### 1. G71 axial multiple cycles of rough turning

(1) G71 instruction format

The compound cycle instruction of inernal and external circles of the rough turning is applicable to rough machining of shaft sleeve parts that require multiple cutting on internal and external cylinders, as shown in Figure 1-7-4 and Figure 1-7-5.

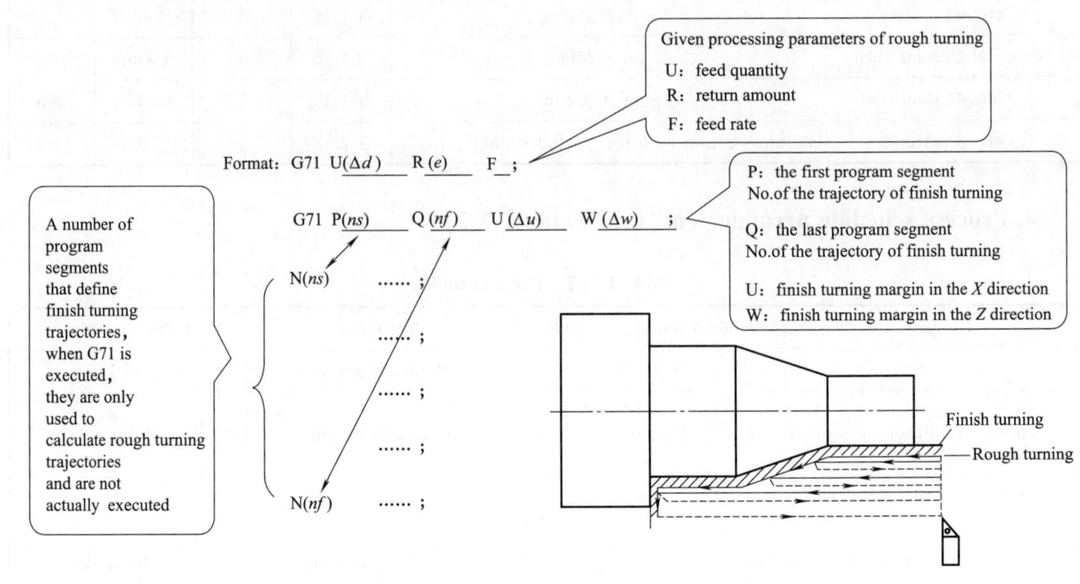


Figure 1-7-4 G71 Instruction format

(2) Examples of programming

**【Example 1-7-1】** The machining procedure of the parts shown in Figures 1-7-6 drawing of compound cycle for rough machining of outer diameter: the cycle start point should be $A$ (46, 1), the cutting depth should be 1. 5mm (radius). the cutter exit amount should be 1mm, the finish machining allowance in $X$ direction is 0. 4mm, the finish machining allowance in $Z$ direction is 0. 1mm. Wherein the double dot-dash section is a workpiece blank.

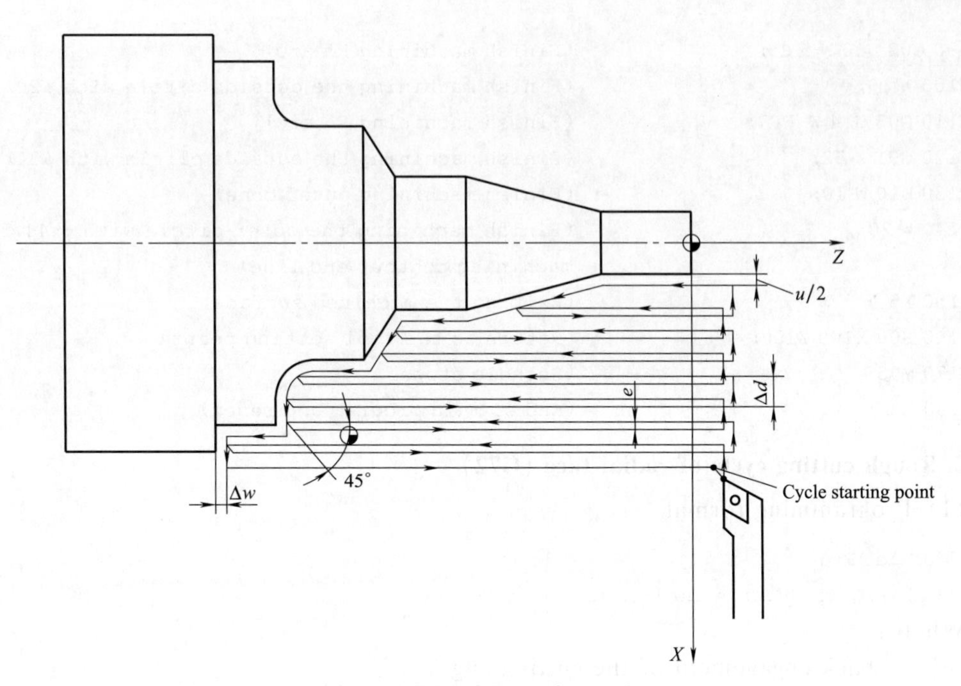


Figure 1-7-5 G71 cutting path

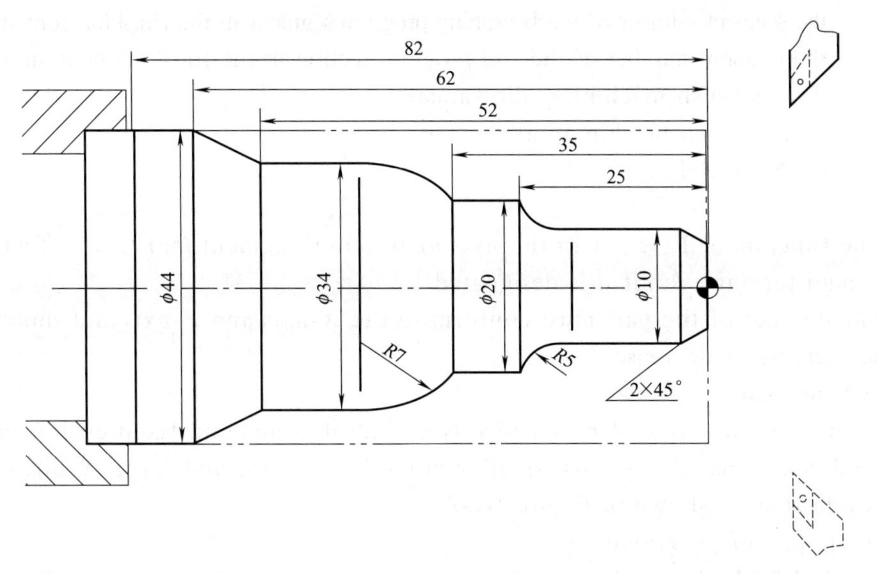


Figure 1-7-6 Examples of programming

O0701 (See Figure 1-7-6)

```
N10 T0101 G00 X100 Z100 G95;      (Select cutter to the program starting point)
N20 M03 S800;                     (Spindle rotates forward at 800r/min)
N30 G00 X46 Z2;                   (Cutter arrives to cycle start point)
N40 G71 U1. 5 R0. 5;              (Rough cut:1. 5mm)
N50 G71 P60Q150 U0. 4 W0. 1 F0. 25; (Fine cut:X0. 4mm Z0. 1mm)
N60 G00 X4 Z2;                    (Finish contour starting line,to rounding chamfer
                                  extension line)
N70 G01 X10 Z-2 F0. 1;            (Finish machining 2 × 45° rounding chamfer)
N80 Z-20;                         (Finish machining the outside circle with ϕ10)
```

```
N90 G02 U10 W-5 R5;        (Finish machining R5 arc)
N100 W-10;                 (Finish machining the outside circle with ϕ20)
N110 G03 U14 W-7 R7;       (Finish machining R7 arc)
N120 G01 Z-52;             (Finish machining the outside circle with ϕ34)
N130 U10 W-10;             (Finish machining outer cone)
N140 W-20;                 (Finish machining the outer circle with ϕ 44,finish
                           machining contour end line)
N150 X50;                  (Exiting the machined surface)
N160 G00 X100 Z100;        (Return to the tool setting point)
N170 M05;                  (Spindle stops)
N180 M30;                  (End of main program and reset)
```

**2. Rough cutting cycle of radial face (G72)**

(1) Programming format

```
G72 U(Δd )R(e );
G72 P(ns)Q(nf )U(Δu )W(Δw )F(f )S(s )T(t );
```

Where:

$\Delta d$——back engagement of the cutting edge;

$e$——cutter exit amount;

$ns$——the segment number of the beginning program segment in the finishing contour segment;

$nf$——the segment number of the end program section in the finishing contour section;

$\Delta u$——$X$-axis finish machining allowance;

$\Delta w$——$Z$-axis finish machining allowance;

$f, s, t$——F、S、T code.

Note:

① The function of F、S、T in the $ns \rightarrow nf$ program segment makes no effect on the cycle of rough turning even if it is designated.

② The contour of the part must conform to the $X$-axis and $Z$-axis and simultaneously monotone increase or decrease.

(2) Cycle routes

The rough cutting cycle of radial face is a kind of compound fixed cycle, and is suitable for rod rough machining with small $Z$-direction margin and large $X$-direction margin. The cycle route is shown in Figure 1-7-7.

(3) Examples of programming

**【Example 1-7-2】** Write the rough cutting procedure of radial face according to the dimensions shown in Figure 1-7-7.

```
O0702
N10 G50 X200 Z200 T0101;
N20 M03 S800;
N30 G90 G00 G41 X176 Z2 M08;
N40 G96 S120;
N50 G72 U3 R0. 5;
N60 G72 P70 Q120 U2 W0. 5 F0. 2;
N70 G00 X160 Z60;//ns
```

```
N80 G01 X120 Z70 F0.15;
N90 Z80;
N100 X80 Z90;
N110 Z110;
N120 X36 Z132;//nf
N130 G00 G40 X200 Z200;
N140 M30;
```

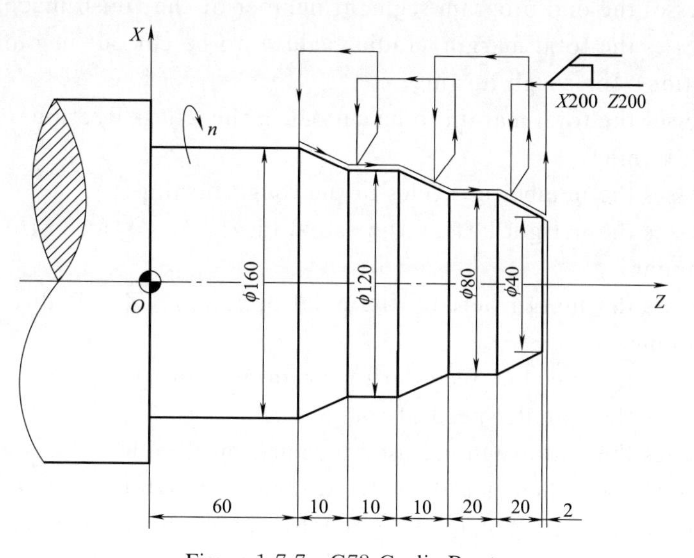


Figure 1-7-7　G72 Cyclic Route

**3. G73 instruction for fixed-shape rough-turning compound cycle**

(1) Instruction format

Compound fixed cycle instruction of closed rough turning is suitable for castings, rough machining of forgings, as shown in Figure 1-7-8.

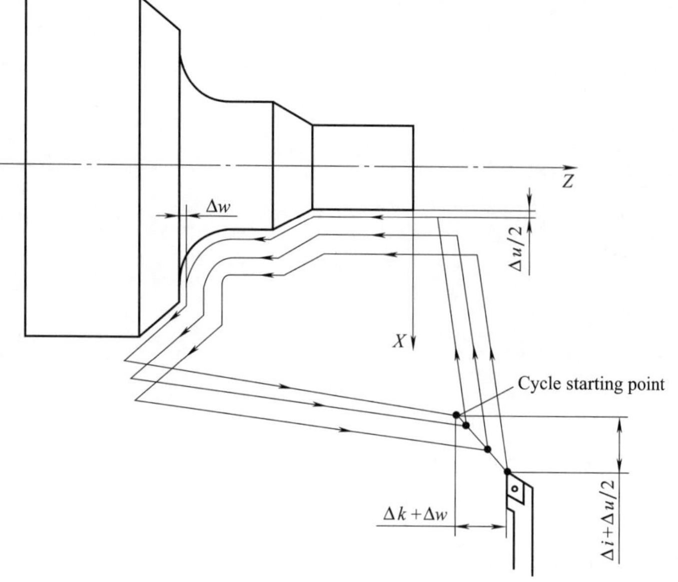


Figure 1-7-8　G73 instruction for fixed-shape rough-turning compound cycle

Programming format: G73U ($\Delta i$) W ($\Delta k$) R ($d$)
G73P ($ns$) Q ($nf$) U ($\Delta u$) W ($\Delta w$) F ($f$) S ($s$) T ($t$)
N$ns$
……
N$nf$

Where:

$ns$—expresses the beginning program segment number of the finish machining program;
$nf$—expresses the end program segment number of the finish machining program;
$\Delta i$—expresses the total margin (radius value) to be cut out in radial direction ($X$ direction) for rough turning;
$\Delta k$—expresses the total margin to be cut out in the axial direction ($Z$ direction) for rough turning;
$d$—expresses the number of cycles in the rough turning;
$\Delta u$—expresses the margin left by the radial direction ($X$-axis direction) for finish machining;
$\Delta w$—expresses the margin left by the axial direction ($Z$-axis direction) for finish machining;
$f$—expresses the speed of feed during rough machining;
$s$—expresses the spindle speed of rough machining;
$t$—expresses the cutter number used in rough machining.

Explanation: The so-called closed (or fixed-shaped) rough turning compound fixed cycle is to follow a certain cutting shape to gradually approach the final shape, it is suitable for rough turning of contour shape of a blank similar to part contour shape. Therefore, this kind of machining is an efficient rough machining method for casting or forgings.

(2) Examples of programming

**【Example 1-7-3】** Make a cutting cycle machining program according to the dimensions shown in Figure 1-7-9.

```
O0703
N10 G99 G50 X200 Z200 T0101;
N20 M03 S2000;
N30 G00 G42 X140 Z40 M08;
N40 G96 S150;
N50 G73 U9.5 W9.5 R3;
N60 G73 P70 Q130 U1 W0.5 F0.3;
N70 G00 X20 Z0;//ns
N80 G01 Z-20 F0.15;
N90 X40 Z-30;
N100 Z-50;
N110 G02 X80 Z-70 R20;
N120 G01 X100 Z-80;
N130 X105;//nf
N140 G00 X200 Z200 G40;
N150 M30;
```

**4. G70 finishing machining cycle**

Completing rough machining by G71, G72 and G73, finish machining can be done with G70.

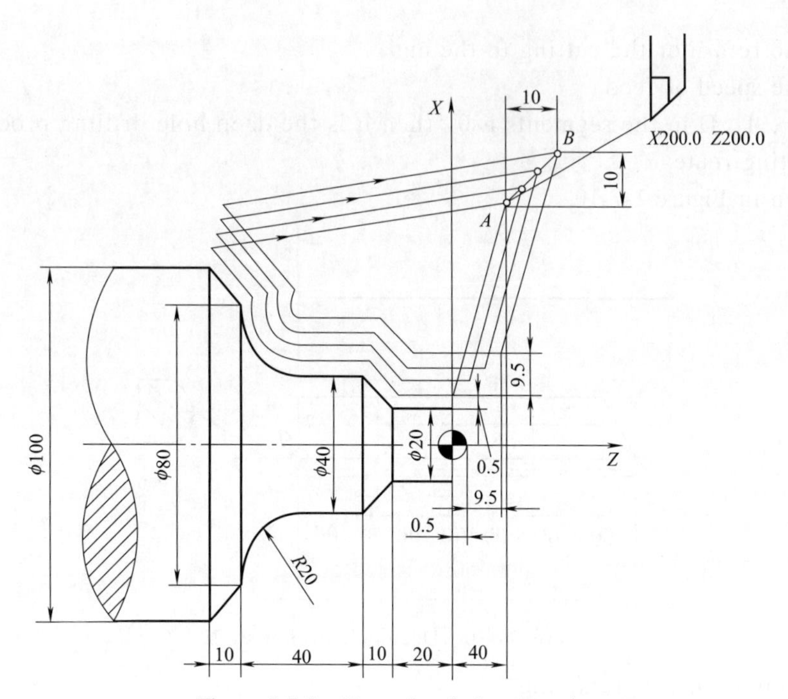


Figure 1-7-9 Example of closed cutting

(1) Programming format

```
G70 P(ns)Q(nf)
```

Where:

*ns*—The segment number of the beginning program segment in finish machining contour segment;

*nf*—The segment number at the end program segment in finish machining contour segment.

During finish machining, F、S、T instructions in G71, G72, and G73 sections are invalid. Only F. S. T instructions in *ns*-*nf* section are valid.

(2) Examples of programming

For example: Add "G70 P*n*s Q*nf*" to the nf section of G71, G72, and G73 applications, and add F、S、T suitable for finish machining in *ns*-*nf* program segment, then the whole process from rough machining to fine machining can be completed.

**5. G74 for deep hole drilling cycle function**

Suitable for deep hole drilling, as shown in Figure 1-7-10.

(1) Programming format

```
G74 R(e )
G74 X(U)__ Z(W)__ I __K __D __F __
```

Where:

X—the *X*-coordinate of the *B*-point;

U—the increment of $A \rightarrow B$;

Z—the *Z* coordinates of the point C;

W—the increment of $A \rightarrow C$;

I—the amount of movement in the direction of *x* (without sign designation);

K—the cut in the *Z* direction (without sign designation);

D——the return of the cutting to the end;

F——the speed of feed.

If X(U)、I、D in the segments is 0, then it is the deep hole drilling processing.

(2) Cutting route

As shown in Figure 1-7-10.

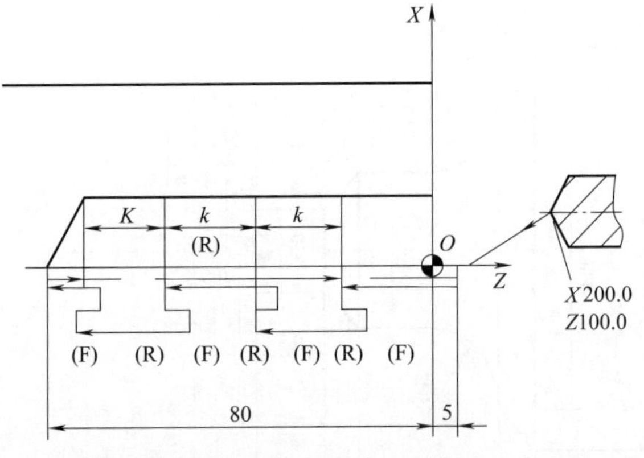


Figure 1-7-10 Deep hole drilling cycle

(3) Examples of programming

**【Example 1-7-4】** As shown in Figure 1-7-11, a deep hole with a diameter of 10mm and a depth of 100mm shall be drilled on the lathe. The procedure is as follows:

```
N01 G50 X50. 0Z100. 0;           Establishing the coordinate system of workpiece
N02 G00 X0 Z680;                 Rapid approach of bit
N03 G74 Z 8. 0 K5. 0 F0. 1 S800; Drilling cycle by G74 instruction
N04 G00 X50. 0 Z 100. 0;         Tool retreats to reference point rapidly
```

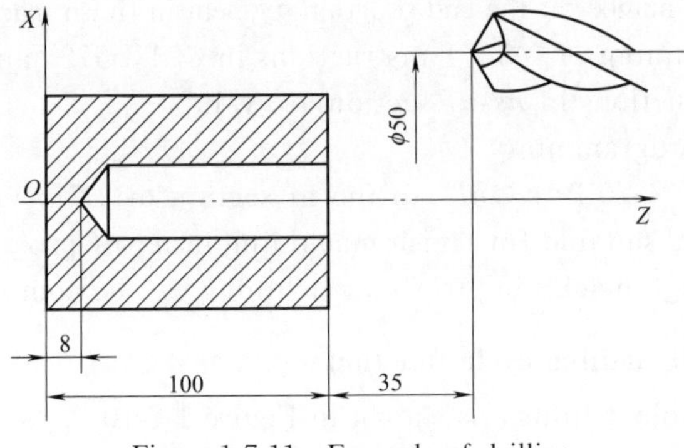


Figure 1-7-11 Example of drilling

## Ⅳ. Processing Program (Reference)

Program 1

```
O0001;
N1;
G97  G99;
T0101;
M03  S700;
G40  G00  X35. 0  Z5. 0;
G71  U1. 5  R0. 5;
```

```
G71  P10  Q20  U0.5  W0  F0.2;
N10  G00  X0;
G01  Z0;
X7.8  ,  C1.0;
Z-12.0;
X12.0,C0.5;
Z-13.0;
N20  X35.0;
G00  X100.0  Z100.0;
M05;
M00;
N2;
T0202;
M03  S900;
G42  G00  X35.0  Z5.0;
G70  P10  Q20  F0.1;
G00  X100.0  Z100.0;
M05;
M00;
N3;
T0303;
M03  S400;
G40  G00  X35.0  Z5.0 F0.07;
G00  Z-12.0;
G01  X6.6;
G04  X1.0;
G01  X35.0;
G00  X100.0  Z100.0;
M05;
M00;
N4;
T0404;
M03  S500;
G40  G00  X35.0  Z5.0;
G92  X7.4  Z-10.5  F1.25;
X7.0;
X6.7;
X6.5;
X6.3;
X6.3;
G00  X100.0  Z100.0;
M30;
```

Program 2

```
O0001;
N1;
```

```
G97  G99;
T0202;
M03  S700;
G40  G00  X35.0  Z5.0;
G73  U15.  W0  R15;
G73  P10  Q20  U0.5  W0  F0.15;
N10  G00  X0;
G01  Z0;
G03  X10.83  Z-3.42  R6.0;
G03  X18.47  W-28.73  R47.16;
G02  X12.0  W-35.85  R200.0;
G01  W-12.0;
N20  X35.0;
G00 X100.0  Z100.0;
M05;
M00;
N2;
T0202;
M03  S900;
G42  G00  X35.0  Z5.0;
G70  P10  Q20  F0.1;
G00  X100.0  Z100.0;
M30;
```

## Ⅴ. Preparation Before Processing

### 1. Machine tool preparation (see Table 1-7-3)

**Table 1-7-3 Machine tool preparation card**

| Project | Mechanical part | | | | Electrical equipment | | CNC system part | | | Auxiliary part | |
|---|---|---|---|---|---|---|---|---|---|---|---|
| Equipment inspection | Spindle part | Feed part | Tool rest part | Tail stock | Main power supply | Cooling fan | Electrical component | Control section | Drive section | Cooling | Lubricating |
| Check the condition | | | | | | | | | | | |

Note: If this part is in good condition after inspection, please mark "√" under the corresponding item. Report to repair in time if problems arise.

### 2. Other notes

① The cutting edge angle is $90^\circ - 93^\circ$ when installing the external circle tool.

② When installing the cut-off tool, the tool major cutting edge shall be parallel to the axis of the spindle.

### 3. Parameter setting

① The value of the cutter setting shall be input in the cutter compensation No. corresponding to that of the cutter in the program;

② The value of the radius value of the cutter and the serial number of the false tool

nose shall be input in the numerical value of the cutter.

## Ⅵ. Processing of Actual Parts

### 1. Teacher's demonstration

① Clamping and alignment of workpiece.

② Measuring methods of workpiece.

### 2. Students' processing training

During the training, trainers detour to instruct, correct the wrong operation behavior in time, and solve the problems that arise in the student's practice.

## Ⅶ. Parts Measurement

(1) Check the outer circle size and length of the part, check the surface roughness of $Ra1.6\mu m$

The dimension of each outer circle is measured by a first-class precision external micrometer caliper. According to the measurement result and the tolerance requirement of the tested outside circle, whether the outside circle is qualified or not is determined. Then the measurement is conducted after rotating the spindle at 90°. When measuring, attention should be paid to the usage of micrometer caliper: the measuring head of the micrometer caliper should lightly touch the surface of the circle being measured, operator rotates the fine-tuning ratchet of the micrometer caliper two or three times, and while rotating the fine-tuning ratchet, swing the movable measuring head of the micrometer caliper along the surface of the outer circle at the same time, the maximum size of the circle being measured will be get.

Check the surface roughness and conduct the comparison verification using the surface roughness comparison template.

(2) Check the size of the wide groove and check the surface roughness of $Ra3.2\mu m$

The width of the groove can be measured with an internal micrometer caliper. Dimension $3\pm0.03$ can be measured by micrometer caliper with common normal line. The measured results and the tolerance requirement of the shaft being measured can be used to determine whether the shaft are qualified or not.

Check the surface roughness and conduct the comparison verification using the surface roughness comparison template.

(3) Testing of external thread

The outer cone is detected by thread ring gauge.

## Ⅷ. Machining Error Analysis and Follow-up Processing

Teaching strategies: student feedback, teaching method and questioning method.

In view of the students' processing errors and timely feedback, teacher make centralized summary and use the teaching method to guide the students to understand the reasons for the problems frequently happened. For cases with little or no chance of occurrence, teachers use the method of asking questions to guide students to analyze the causes of processing errors by themselves.

There are many kinds of machining errors which are often encountered in NC lathe. The reasons for their occurrence and measures for preventing and eliminating them

are shown in Table 1-7-4.

**Table 1-7-4 Machining errors and follow-up processing**

| Problem | Cause | Prevention and elimination |
|---|---|---|
| Vibration in cutting process | 1. The workpiece is not clamped correctly<br>2. The cutter installation is incorrect<br>3. The cutting parameters are incorrect | 1. Check the installation of the workpiece and increase the rigidity of the installation<br>2. Adjust the cutter installation position<br>3. Improve or decrease the cutting speed |
| Poor surface quality | 1. The cutting speed is improper<br>2. The tool center is too low<br>3. The chip control is poor<br>4. The cutter nose produces a built-up edge<br>5. The selection of cutting fluid is unreasonable | 1. Adjust the speed of spindle<br>2. Adjust the cutter center height<br>3. Choose a reasonable tool rake angle, feed method and cutting depth<br>4. Choose the right cutting fluid and spray it fully |
| Large axial dimension error when turning the parts around | 1. There are clearance in spindle and with axial runout<br>2. The workpiece is not chucked properly | 1. Adjust the spindle clearance<br>2. Align the workpiece |

**Exercise**

1. What are the two key technical problems to be solved for the lathe hole?

2. As hole machining is much more difficult than turning the outer circle, what are the main characteristics for hole machining?

3. What should be paid attention to when programming the inner hole with G71 instruction?

4. How to open the Morse taper-shank?

5. How to operate the tailstock when drilling with a twist drill?

6. How to find out the problem when the machine tool alarms?

7. How to do a good maintenance work for the machine tool?

# Project Ⅱ

## CNC Milling Machining

【Project Description】

Milling, as one of the most commonly used methods in mechanical processing, mainly includes planar milling and contour milling, which can also carry out machining processes to parts such as drilling, expanding, reaming, boring, countersinking and thread turning.

【Capability Goals】

1. To master the classification, function and cutting tools of NC milling machine, and master the process analysis of NC boring and milling.
2. To be familiar with operation panel and button function of typical FANUC 0i MC NC milling system.
3. To master the use method of rapid positioning instruction G00 and linear interpolation instruction G01 of NC milling machine.
4. To master the use method of the circular interpolation instruction G02 and G03 of NC milling machine.
5. To master the application method of tool setting instruction G92 and defining coordinate system G54-G59 of NC milling machine.
6. To master the use method of radius compensation and length compensation of NC milling machine.
7. To master the use method of subprogram call of NC milling machine.
8. To master the characteristics of climb milling and reverse milling.
9. To master the using skill of each key function of the CNC milling machine.
10. To master the skills of machining complex parts by CNC milling machine.

## Task Ⅰ Machining Support Frame

### Ⅰ. Drawing and Technical Requirement

Shown in Figure 2-1-1、Figure 2-1-2、Table 2-1-1、Table 2-1-2。

Table 2-1-1 List of training equipment

| No. | Type | Name | Specification | Quantity | Note |
|---|---|---|---|---|---|
| 1 | Material | 6061 | 100mm×42mm×25mm | 2 | |
| 2 | Cutting tools | High-speed steel end mill | $\phi$12mm | 1 | |
| | | Twist drill | $\phi$6, $\phi$6.8mm | 1 for each | |
| | | Centre drill | A3 | 1 | |
| | | Screw tap | M8 | 1 | |
| 3 | Fixture | Precision plain vice | 0-300mm | 1 set | |

Continued

| No. | Type | Name | Specification | Quantity | Note |
|---|---|---|---|---|---|
| 4 | Tools | Milling chuck | 0-13 | 1 | |
| | | Tapping wrench | | 1 | |
| | | Clamping bush | $\phi$12mm | 1 for each | Matching with cutting tools |
| | | Parallel clamp | | 1 pair | |
| | | Oil stone | | 1 | |

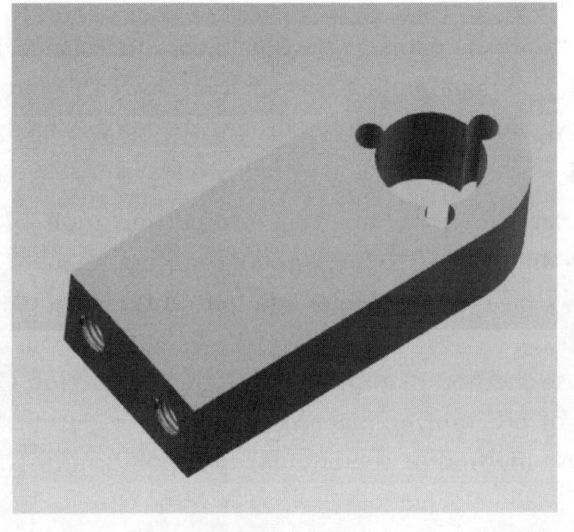

Figure 2-1-1 Physical drawing of support frame

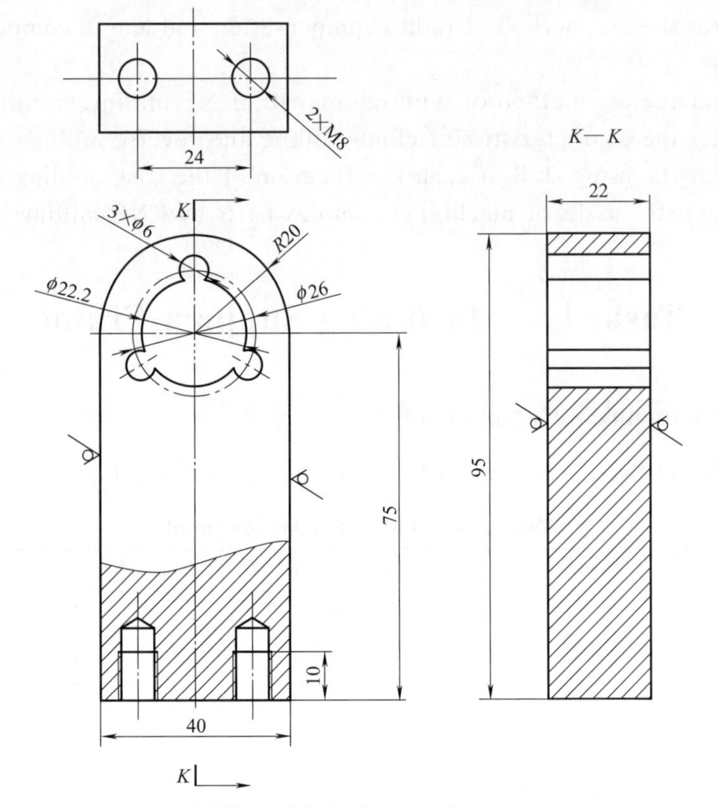


Figure 2-1-2 Support frame

Table 2-1-2 Part inspection items and score sheet (100 points) Score ______

| S/N | Assessment Items | Assessment content and accuracy requirements | Score distribution | Scoring criteria | Measured results | Score |
|---|---|---|---|---|---|---|
| 1 | Overall dimensions | 95±0.1 | 15 | No score when it is out of tolerance | | |
| 2 | | 40±0.05 | 15 | No score when it is out of tolerance | | |
| 3 | | 22±0.05 | 18 | No score when it is out of tolerance | | |
| 4 | | $R20$ | 12 | No score when it is out of tolerance | | |
| 5 | | $\phi22.2$ | 10 | No score when it is out of tolerance | | |
| 6 | | $3\times\phi6$ | 10 | No score when it is out of tolerance | | |
| 7 | Others | Surface roughness3.2 | 5 | Points are deducted when the surface roughness is out of tolerance | | |
| 8 | | Bluntness of edges | 2 | All points are deducted if it is out of tolerance | | |
| 9 | | Graphic integrity | 5 | All points are deducted if it is incomplete | | |
| 10 | | Civilized production | 8 | All points are deducted if there is violation operation | | |
| 11 | | Man-hour | | 5 points are deducted for per 15 minutes overtime | | |

## Ⅱ. Drawing Analysis

Through the Figure 2-1-1、Figure 2-1-2, the part has the main boundary dimension of 95mm×40mm×22mm, the upper part is a through hole of $\phi22^{+0.03}_{0}$ mm, the bottom has 2 deep threaded holes of M8mm with the depth of 10mm.

The shape of the part is simple with a maximum tolerance requirement of 0.03mm.

### 1. Selection of cutting tools

The choice and rational application of materials of cutting tools is very important, at present, the materials of cutting tools commonly used in cutting are mainly high-speed steel and cemented carbide. The material of the workpiece in this example is hard aluminum, and the cutting tool selects high-speed steel solid end mill with sharp edge, good straightness accuracy and high precision.

According to the drawing, considering the structure of the part, the cutting tools are selected, see Table 2-1-3.

Table 2-1-3 Card of cutting tools

| Name of Cutting Tools | Specifications of Cutting Tools | Material | Quantity | Use of Cutting Tools | Note |
|---|---|---|---|---|---|
| End Mill | $\phi12$mm | High-speed Steel | 1 | Plane processing, rough and finish machining of contour | |
| Center Drill | $\phi3$mm | High-speed Steel | 1 | Drilling center holes | |
| Twist Drill | $\phi6$mm | High-speed Steel | 1 | Drilling | |
| Twist Drill | $\phi6.8$mm | High-speed Steel | 1 | Drilling thread bottom holes | |

## 2. Selection of cutting parameters

According to the materials of the machining objects, the materials and specifications of the cutting tools, and by reading the book of metal cutting parameters to find the cutting speed of the cutter and per tooth feed rate, the speed and feed speed of the cutter selected are determined, reference cutting parameters see Table 2-1-4.

**Table 2-1-4 Card of cutting parameters**

| Cutting tool | Cutting speed $v$/(m/min) | Feed rate per cut $f$/(mm/cut) | Speed of spindle $S$/(r/min) | Feed rate $F$/(mm/min) | Note |
|---|---|---|---|---|---|
| $\phi$12m end mill | 50 | 0.04 | 1300 | 200 | Rough machining |
| | 80 | 0.03 | 2100 | 240 | Finish machining |
| $\phi$3mm centre drill | 30 | 0.03 | 3200 | 200 | |
| $\phi$6mm drill | 25 | 0.05 | 1400 | 140 | |
| $\phi$6.8mm drill | 25 | 0.05 | 1400 | 140 | |

## 3. Depth of cutting $a_p$

The depth of cutting is mainly limited by machine and tool rigidity during rough machining, and in general, the cutting depth is taken $(0.6\sim0.8)D_{cut}$, when the radial cutting volume is large. Otherwise the depth of cutting can be larger.

The contour processing volume of the part is not large, processing depth for each contour can be processed according to the drawing dimension, no layered processing is needed.

## 4. Process procedure planning

The each contouring process of the part is planned as follows Table 2-1-5.

**Table 2-1-5 Card of part working process**

| Unit | | Product name and Model | Part Name | Part Chart Number |
|---|---|---|---|---|
| | | | Simple parts | |
| Process | Working procedure number | Fixture name | The device used | Materials of workpiece |
| 1 | O0001 | Precision plain vice | VMC850 | LY12 |
| Step | Content | Number of cutting tool | Tool and cutting volume | Note |
| 1 | Milling the upper surface | T01 | $\phi$12mm End Mill<br>$n$ = 2100r/min<br>$F$ = 240mm/min<br>$a_p$ = 0.3mm | Milling by finish machining |
| 2 | The processing origin is set at the center of the upper surface of the workpiece | T01 | | Using the test cutting method |
| 3 | Roughing 95 × 40 as the the outer profile of the main size, leaving a margin of 0.2mm | T01 | $\phi$12mm End Mill<br>$n$ = 1300r/min<br>$F$ = 200mm/min<br>$a_p$ = 6.8mm | |

Continued

| Process | Working procedure number | Fixture name | The device used | Materials of workpiece |
|---|---|---|---|---|
| 4 | Roughing the through holes of $\phi$22, leaving a margin of 0.2mm | T01 | $\phi$12mm End Mill<br>$n$ = 1300r/min<br>$F$ = 200mm/min<br>$a_p$ = 14.8mm | |
| 5 | Finishing 95 × 40 as the outer profile of the main size to the specified dimension | T01 | $\phi$12mm End Mill<br>$n$ = 1300r/min<br>$F$ = 200mm/min<br>$a_p$ = 4.8mm | |
| 6 | Finishing the through hole of $\phi$22 to the specified dimension. | T01 | $\phi$12mm End Mill<br>$n$ = 2100r/min<br>$F$ = 240mm/min<br>$a_p$ = 7mm | |
| 7 | Drilling the positioning center holes of the three $\phi$6mm holes | T02 | $\phi$3mm Centre Drill<br>$n$ = 3200r/min<br>$F$ = 200mm/min<br>$a_p$ = 3mm | |
| 8 | Drilling three holes of $\phi$6mm with a depth of 22mm | T03 | $\phi$5.8mm Drill<br>$n$ = 1400r/min<br>$F$ = 140mm/min<br>$a_p$ = 2mm | |
| 9 | Drilling the positioning center holes of two holes of $\phi$6.8mm | T02 | $\phi$3mm Centre Drill<br>$n$ = 3200r/min<br>$F$ = 200mm/min<br>$a_p$ = 3mm | Bottom facing up, re-clamping and presetting cutter |
| 10 | Drilling 2 14mm-deep thread bottom holes of $\phi$6.8mm | T03 | $\phi$5.8mm Drill<br>$n$ = 1400r/min<br>$F$ = 140mm/min<br>$a_p$ = 2mm | |

## 5. Processing procedures

```
O0001(main program)
G69  G40;
G28  G91  Z0;
G54  G90  G00  X0  Y0;
Z100;
M03  S1500;
X0  Y0;
Z5;
G01  Z-5   F100;
G01  G41  D01  X-11.1  Y0;
G02  X-11.1  Y0  I11.1  J0;
G40  X0  Y0;
M98  P0002  L3;
G01  Z-22  F100;
G01  G41  D01  X-11.1  Y0;
G02  X-11.1  Y0;
```

```
G40  X0  Y0;
G00  Z100;
M05;
M30;
O0002(subprogram)
Z5;
G01  Z-5  F100;
G01  G41  D01  X-11.1  Y0;
G02  X-11.1    Y0  I11.1  J0;
G40  X0  Y0;
Z5;
M99;
```

## Ⅲ. Basic Knowledge of NC Milling Machine

### 1. Classification of NC milling machine

NC milling machine can be divided into NC vertical milling machine, NC horizontal milling machine, NC compound milling machine and gantry NC milling machine.

(1) NC vertical milling machine

The spindle axis of the NC vertical milling machine is perpendicular to the horizontal plane, as shown in Figure 2-1-3, which is mainly used in machining planes, internal and external contours, holes, and tapping thread of parts, as well as various types of molds. At present, three-coordinate NC vertical milling machine occupies a large proportion, which can generally perform three-coordinate processing.

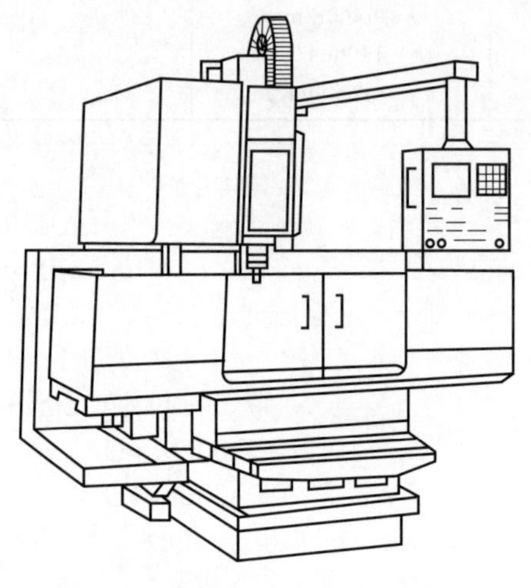

Figure 2-1-3 Three-axis NC vertical milling machine

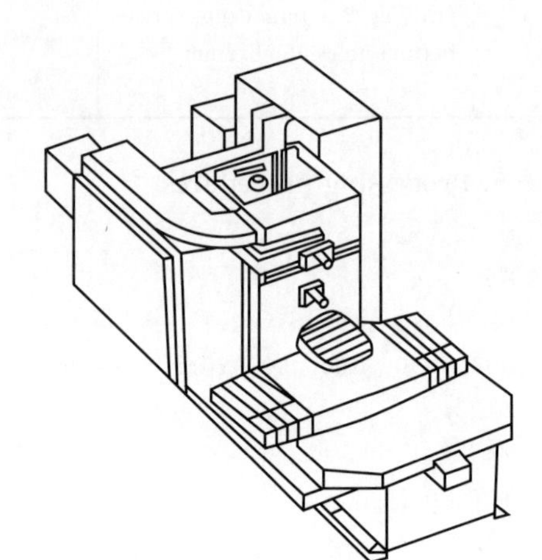

Figure 2-1-4 NC horizontal milling machine

(2) NC horizontal milling machine

The spindle axis of the NC horizontal milling machine is parallel to the horizontal plane, as shown in Figure 2-1-4. In order to expand the range and function of machining, the NC horizontal milling machine usually adopts the method of adding NC rotary table (universal NC rotary table) to realize four-coordinate and five-coordinate machining. In this way, it can not only process the continuous rotary profile on the side of the work-

piece, but also change the position through the rotary table in an installation, so as to carry out "four-side machining". NC horizontal milling machine is mainly suitable for machining machine parts such as box type.

(3) NC compound milling machine

NC compound milling machine refers to a NC milling machine has both vertical and horizontal spindle, or its spindle can be rotated at 90° and provides the functions of both vertical and horizontal milling machine. A NC compound milling machine with vertical and horizontal spindle is shown in Figure 2-1-5.

NC compound milling machine is mainly used for processing box type parts and various types of molds.

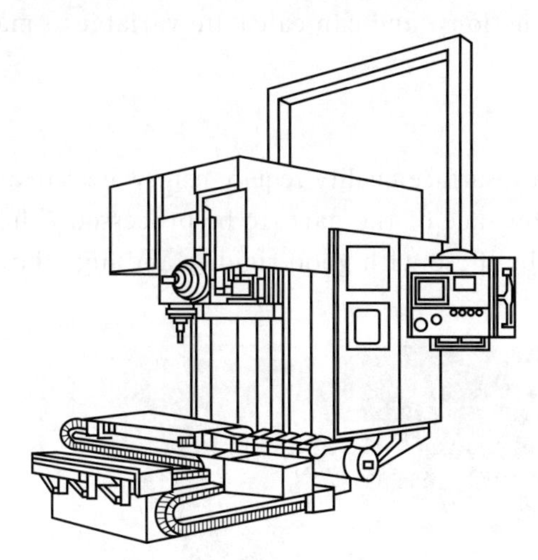

Figure 2-1-5 NC compound milling machine

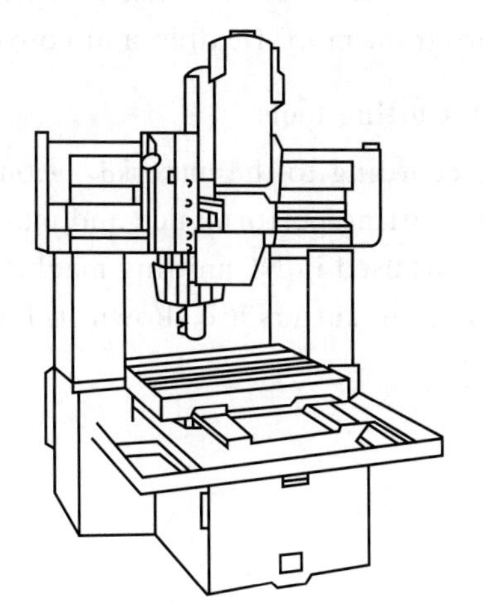

Figure 2-1-6 Gantry NC milling machine

(4) Gantry NC milling machine

The spindle of the gantry NC milling machine is fixed on the gantry, as shown in Figure 2-1-6. The gantry NC milling machine is mainly used for machining large-scale mechanical parts and molds.

**2. Main function of NC milling machine**

Although the CNC systems of various types of NC milling machines are different, except for some special functions, the main functions of various CNC systems are basically the same.

① Point-to-point control function This function can realize hole machining with high requirement of position accuracy.

② Continuous contour control function It can realize the linear interpolation and circular interpolation as well as the machining of non-circular curve.

③ Tool radius compensation function It can be programme according to the dimension of the part drawing without considering the actual radius of the tool used, thus reducing the complicated numerical calculation in programming.

④ Tool length compensation function This function can automatically compensate tool length to meet the requirement of tool length adjustment in machining.

⑤ Proportion and mirror processing function The function can be implemented by changing the coordinates of the programmed program according to the specified propor-

tion. Mirror processing is also called axisymmetric processing. When the shape of a part is axially symmetrical with respect to coordinates, only one or two quadrants need to be programmed, while the contours of other quadrants can be achieved by mirror processing.

⑥ Rotation function　This function can be implemented by rotating the programmed processing program at any angle in the processing plane.

⑦ Subprogram call function　It is necessary for some parts to repeatedly process the same contour shape at different positions. The processing program of this contour shape can be regarded as a subprogram, which can be called repeatedly at the required position to complete the processing of the part.

⑧ Macro program function　This function can use a general instruction to represent a series of instructions for achieving a certain function, and can calculate variables, making the program more flexible and convenient.

**3. Cutting tools**

According to the material, geometric shape, surface quality requirements, heat treatment status, cutting performance and machining allowance of the parts to be processed, the cutting tools used in NC milling machines should be these with good rigidity and high durability. Common cutters are shown in Figure 2-1-7.

Figure 2-1-7　Common cutting tools

(1) The selection of milling cutter type

The geometric shape of the machined parts is the main basis for choosing tool types.

① When machining curved surface parts, in order to ensure that the cutting edge of the tool and the processing contour are tangential at the cutting point so as to avoid interference between the cutting edge and the contour of the workpiece, generally, ball-end cutters, two-edge milling cutters for rough machining, four-edge milling cutters for semi-finish machining and finish machining are used, as shown in Figure 2-1-8.

② When milling a larger plane, in order to improve production efficiency and processing surface roughness, the blade mosaic disc milling cutter is generally used, as shown in Figure 2-1-9.

③ Universal milling cutters are generally used when milling small planes or stepped

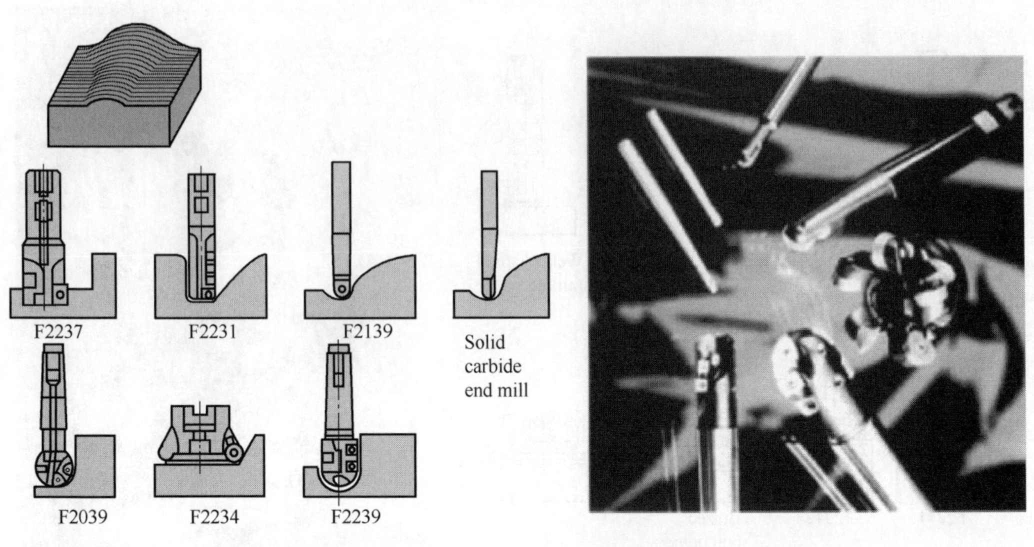


Figure 2-1-8 Milling cutters for machining curved surface

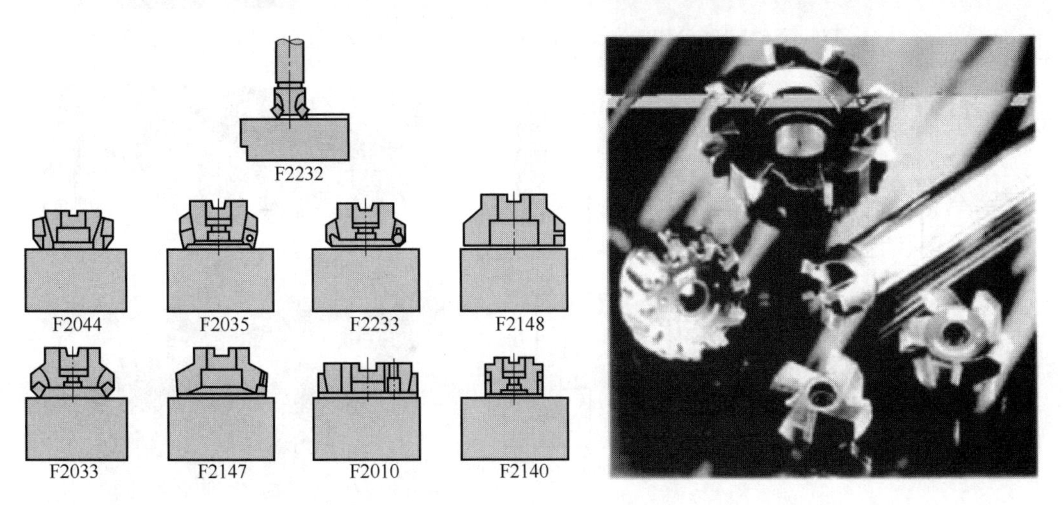


Figure 2-1-9 Milling cutters for machining large plane

surfaces, as shown in Figure 2-1-10.

④ In order to ensure the dimension accuracy of the key way when milling the key way, a two-edge key way milling cutter is usually used, as shown in Figure 2-1-11.

⑤ When machining holes, the tools such as drilling bit, boring cutter can be used, as shown in Figure 2-1-12.

(2) Selection of milling cutter structure

The milling cutter generally consists of a blade, a positioning element, a clamping element and a cutter body. Due to the fact that there are many methods to position and clamp the blade on the cutter body, and the structure of the blade positioning element has different types, there are many kinds of structure forms of the milling cutter, with many classification methods. Milling cutter can be selected mainly according to the blade arrangement. Blade arrangement can be divided into two categories: horizontal-mounted structure and vertical-mounted structure.

① Horizontal-mounted structure (radial arrangement of blades) The horizontal-mounted structure milling cutter (as shown in Figure 2-1-13) has a good structure and workmanship

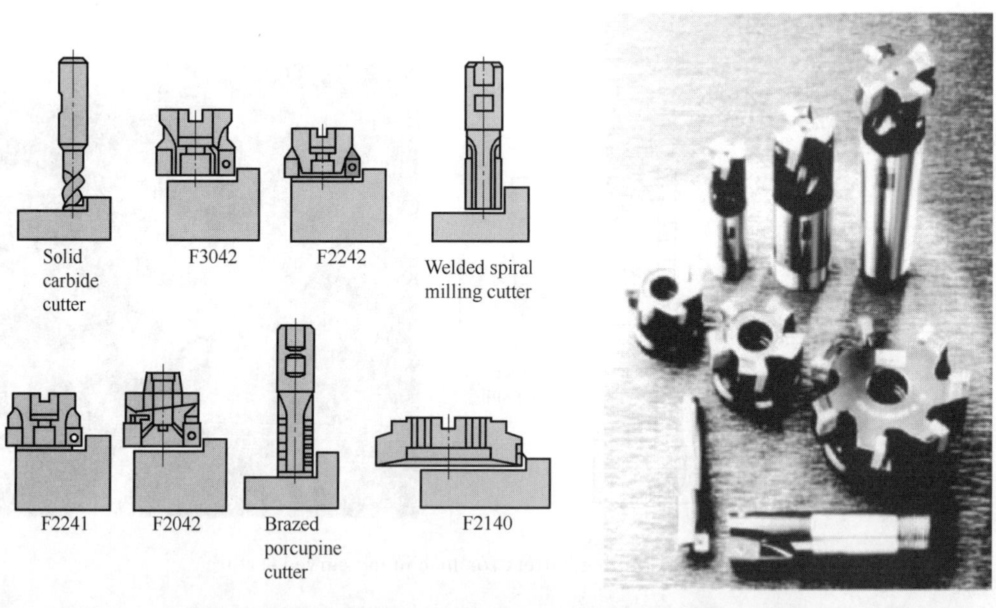


Figure 2-1-10 Milling cutters for machining stepped surface

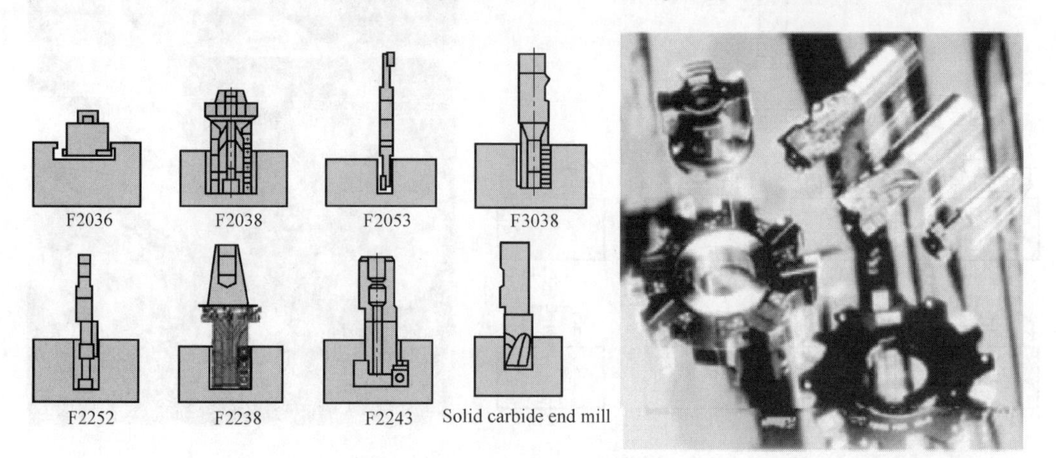


Figure 2-1-11 Milling cutters for machining slot

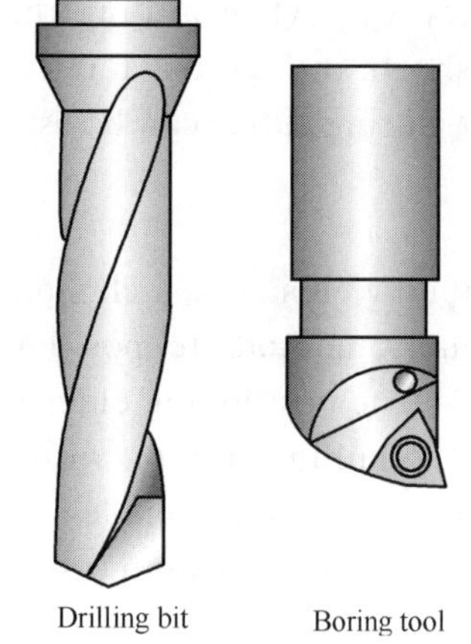


Figure 2-1-12 Milling cutters for machining hole

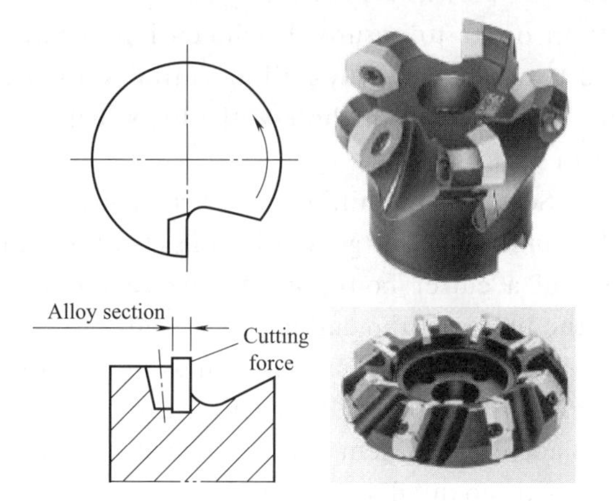


Figure 2-1-13 Horizontal-mounted structure milling cutter

and can be used for machining easily. And non-porous blades can be adopted (whose blade is low in price and can be reground). Due to the use of clamping element, a part of the blade is covered, which makes the space for filling chip smaller and the cemented carbide section smaller in the direction of cutting force. Therefore, the horizontal-mounted milling cutters are generally used for light and medium-sized milling.

② Vertical-mounted structure (tangential arrangement of blade) The blade of the vertical-mounted structure milling cutter (as shown in Figure 2-1-14) is fixed on the tool slot with only one screw, with simple structure and convenient transposition. Although there are fewer tool parts, it is difficult to process parts with the tool, which usually needs the tool to process parts in a five-coordinate machining center. Because the blade is clamped by cutting force, the clamping force increases with the increase of cutting force, the clamping element can be omitted and the chip space can be increased. Due to the tangential installation of the blade, the cemented carbide cross-section in the direction of cutting force is large, so it can be used for deep cutting and large cutting feed. This kind of milling cutter is suitable for heavy and medium-sized milling machining.

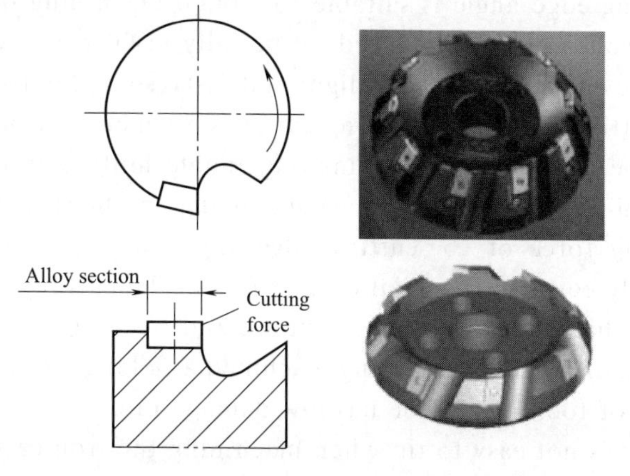


Figure 2-1-14 Vertical-mounted structure milling cutter

(3) The angle selection of milling cutter

The angle of milling cutter includes the rake angle, the back angle, the cutting edge angle, the end cutting edge angle, the cutting edge inclination angle and so on. To meet different processing needs, there are various angle combinations. The most important in the various angles are the cutting edge angle and the rake angle (In general, the cutting edge angle and the rake angle of the tool are clearly stated in the product samples of the manufacturer).

① The cutting edge angle $\kappa_r$ The cutting edge angle is the included angle between the cutting edge and the cutting plane, as shown in Figure 2-1-15. The cutting edge angle of the milling cutter has 90°, 88°, 75°, 70°, 60°, 45° and so on.

The cutting edge angle has a great influence on the radial cutting force and cutting depth. The size of radial cutting force directly affects cutting power and tool vibration resistance. The smaller the cutting edge angle is, the smaller the radial cutting force is and the better the vibration resistance will be, but the cutting depth decreases accordingly.

The 90° cutting edge angle is selected when milling the plane with convex shoulder, which is not generally used for simple plane processing. This kind of cutter has good versatility (that is to say, it can process stepped surface and plane surface) and is selected in single

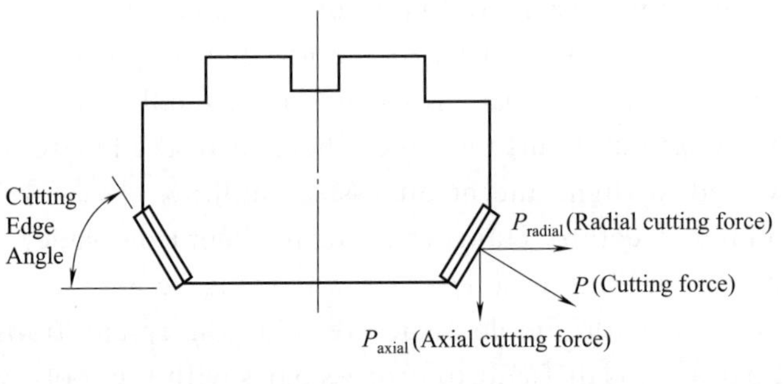


Figure 2-1-15 Cutting edge angle

piece and small batch processing. Because the radial cutting force of this kind of cutting tool is equal to the cutting force, with large feed resistance and easy to vibrate, the machine tool is required to have high power and enough rigidity. When machining the plane with convex shoulder, the 88°cutting edge angle milling cutter can also be selected, whose cutting performance is improved compared with 90°milling cutter.

The 60°-75° cutting edge angle is suitable for rough machining of plane milling. Since the radial cutting force is obviously reduced (especially at 60°), its vibration resistance is improved greatly, the cutting is smooth, light and fast, so it should be preferred in plane machining. The 75°cutting edge angle milling cutter is a general-purpose cutter, which has wide application scope; while the 60°cutting edge angle milling cutter is mainly used for rough milling and semi-finish milling on boring and milling machines and machining centers.

The radial cutting force of 45° cutting edge angle milling cutter decreases greatly, which is approximately equal to the axial cutting force. The cutting load distributes on the longer cutting edge. It has good vibration resistance and is suitable for the machining occasion where the spindle of boring and milling machine has a long overhang. When machining planes with this kind of tool, the blade has low damage rate and high durability, and the edge of the workpiece is not easy to tip when machining the iron casting.

② Rake angle $\gamma$ The rake angle of milling cutter can be decomposed into radial rake angle $\gamma_f$ [Figure 2-1-16(a)] and axial rake angle $\gamma_p$ [Figure 2-1-16(b)]. The rake angle $\gamma_f$ mainly affects cutting power, while the rake angle $\gamma_p$ mainly affects chip formation and direction of axial force. When the $\gamma_p$ is positive, the chip will fly away from the machining surface. The positive and negative discrimination between radial rake angle $\gamma_f$ and axial rake angle $\gamma_p$ is shown in Figure 2-1-16. The common combination of rake angle is as follows:

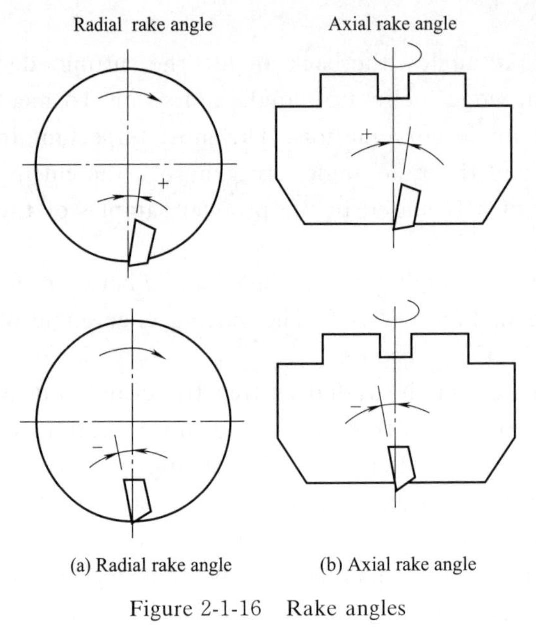


Figure 2-1-16 Rake angles

a. Double negative rake angle, whose milling cutter usually adopts square (or rectangular) blade without rear angle, the number of cutting edges of the cutter is many (generally 8), with high strength, good impact resistance, and it is suitable for rough

machining of cast steel and iron casting. As a result of large chip compression ratio, larger cutting force is needed, so the machine tool is required to have higher power and higher rigidity. Due to the negative axial rake angle, the cuttings can't flow out automatically, it is easy to produce built-up edge and tool vibration when cutting ductile material.

In order to make full use of and save the cutting blade, it is recommended that the double-negative rake angle milling cutter should be selected first when the double-negative rake angle cutting tool is available. If tipping is produced when double-positive rake angle milling cutter is used (i. e. the impact load is large), the double-negative rake angle milling cutter should be preferred under the conditions allowed by the machine tool.

b. Double positive rake angle, whose milling cutter is provided with a blade with a back-angle, which has a small wedge angle and a sharp cutting edge. As a result of small chip compression ratio and the low cutting power consumption, the chip is spirally discharged, and it is not easy to form built-up edge. This kind of milling cutter is suitable for cutting soft material, stainless steel, heat-resistant steel, etc. For machine tools with poor rigidity (e. g. boring and milling machine with longer spindle overhang), low power and welding structural parts, double positive rake angle milling cutters should also be preferred.

c. The positive-negative rake angle (axial positive rake angle and radial negative rake angle), whose milling cutter combines the advantages of double positive rake angle milling cutter and double negative rake angle milling cutter. The positive axial rake angle is beneficial to the formation and discharge of chips, while the negative radial rake angle can improve the edge strength and impact resistance. This milling tool has the characteristics of smooth cutting, smooth chip removal and high metal removal rate, which is suitable for milling with large allowance. The tangential tooth placement heavy cutting milling cutter F2265 produced by WALTER Company is a kind of milling cutter with positive axial rake angle and negative radial rake angle.

(4) Selection of teeth number (pitch) of milling cutter

The milling cutter with more teeth can improve the production efficiency, but the number of teeth of milling cutters with different diameters is regulated due to the limitation of chip space, strength of cutter teeth, power and rigidity of machine tools. In order to meet the needs of different users, the milling cutter with the same diameter has three types: coarse tooth, medium-dense tooth and dense tooth.

① Coarse tooth milling cutter is suitable for large margin rough machining of general machine tools and milling of soft material or large cutting width. When the power of machine tool is low, coarse tooth milling cutter is often used to make the cutting stable.

② Medium-dense tooth milling cutter is an universal series, widely used, with high metal removal rate and cutting stability.

③ Dense-tooth milling cutter is mainly used for the machining of iron casting, aluminum alloy and non-ferrous metal with large feed speed. In specialized production (e. g. assembly line processing), for making full use of the power of equipment and meeting the requirements of production rhythm, the dense tooth milling cutter is often chosen (at this time, it is a special non-standard milling cutter).

In order to prevent the resonance of the process system and make the cutting smooth, there is also a kind of milling cutter with unequal pitch. For example, NOVEX series milling cutter by WALTERadopts unequal pitch technology. In the large margin rough machining of steel and iron casting, it is suggested that the milling cutter with unequal pitch

should be preferred.

(5) The diameter selection of milling cutter

The selection of milling cutter diameter varies greatly according to the different products and production batches, and the selection of cutter diameter mainly depends on the specifications of the equipment and the processing size of the workpiece.

① Face milling cutter  When choosing the diameter of face milling cutter, it is necessary to consider that the power of cutting tool should be within the power range of machine tool, and the diameter of spindle of machine tool can also be taken as the basis of selection. The diameter of face milling cutter can be selected according to $D=1.5d$ ($d$ is spindle diameter). In mass production, the cutter diameter can also be selected according to the 1.6 times of the cutting width of the workpiece.

② End milling cutter  The diameter selection of the end milling cutter should consider the machining dimension of the workpiece and ensure that the power needed for the cutter is within the rated power range of the machine tool. If it is a small diameter end milling cutter, it should be considered whether the maximum revolution of the machine tool can reach the minimum cutting speed of the cutter (60 m/min).

③ Slot milling cutter  The diameter and width of slot milling cutter shall be selected according to the size of the workpiece and the cutting power shall be within the allowable power range of the machine tool.

(6) Maximum cutting depth of the milling cutter

Different series of indexable face milling cutters have different maximum cutting depth. The larger the maximum cutting depth of the cutter is, the larger the blade is and the higher the price is, so from the point of saving expense and reducing cost, when selecting the tool, the appropriate specifications should be chosen according to the maximum margin of machining and the maximum cutting depth of the tool. Moreover, the rated power and rigidity of the machine tool should also be considered to meet the needs of the maximum cutting depth of the tool.

(7) Selection of blade grade

The properties of processed materials and cemented carbide are the main basis for the selection of blade grade. Generally, when choosing milling cutters, carbide blades of corresponding grades can be equipped according to the material and processing conditions provided by the tool manufacturer.

Due to the different components and properties of cemented carbides with the same use produced by different factories, the methods for expressing the cemented carbide types are also different. For the convenience of users, the International Organization for Standardization stipulates that the cemented carbides for machining are divided into three categories according to the type of chip removal and processed materials: category P, category M and category K. In line with the processed materials and applicable processing conditions, each category is divided into several groups, which are represented by two digit numbers. The greater the number is in each class, the lower the wear resistance is and the higher the toughness is.

P alloys (including cermet) are used to process metal materials with long chips such as steel, cast steel, malleable cast iron, stainless steel, heat resistant steel, etc. The larger the group number is, the greater the feed and cutting depth may choose, and the smaller the cutting speed is.

M alloys are used to process ferrous or non-ferrous metals with long chips and short chips such as steel, cast steel, austenitic stainless steel, heat resistant steel, malleable

cast iron, alloy cast iron, etc. The larger the group number is, the greater the feed and cutting depth may choose, and the smaller the cutting speed is.

K alloys are used to process ferrous metals, non-ferrous metals and non-metallic materials with short chips, such as cast iron, aluminum alloy, copper alloy, plastic, hard bake lite, etc. The larger the group number is, the greater the feed and cutting depth may choose, and the smaller the cutting speed is.

The selection criteria for the above three categories are shown in Table 2-1-6.

**Table 2-1-6 Selection of Cutting Parameters for P, M and K Alloys**

| P | P01 | P05 | P10 | P15 | P20 | P25 | P30 | P40 | P50 |
|---|---|---|---|---|---|---|---|---|---|
| M | M10 | M20 | M30 | M40 | | | | | |
| K | K01 | K10 | K20 | K30 | K40 | | | | |
| Feed | → | | | | | | | | |
| Depth of cutting | → | | | | | | | | |
| Cutting speed | ← | | | | | | | | |

Although the cemented carbide produced by each factory has its own grade, it has the classification number corresponding to the international standard, which is very convenient to choose.

**4. Machining range of NC milling machine**

Milling, as one of the most commonly used machining methods in machining, mainly includes face milling and contour milling, which can also carry out machining processes to parts such as drilling, expanding, reaming, boring, countersinking and thread turning. NC milling is mainly suitable for processing the following parts.

① Planar type parts, as shown in Figure 2-1-17.

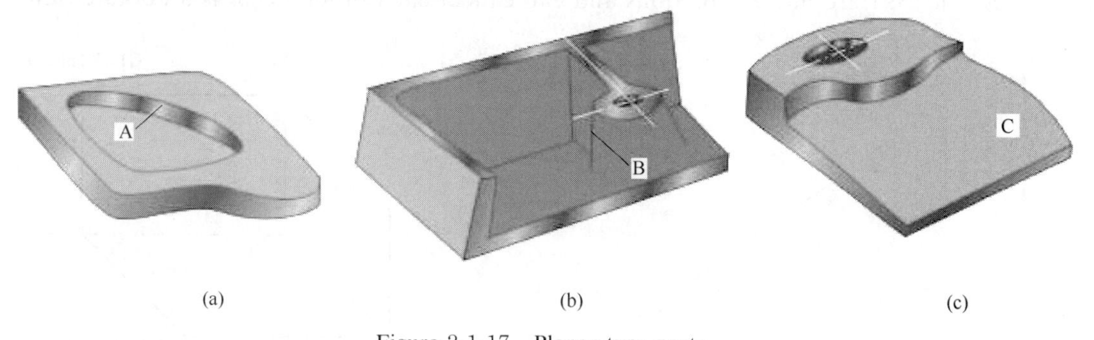


Figure 2-1-17 Planar type parts

② Ruled surface parts, as shown in Figure 2-1-18.

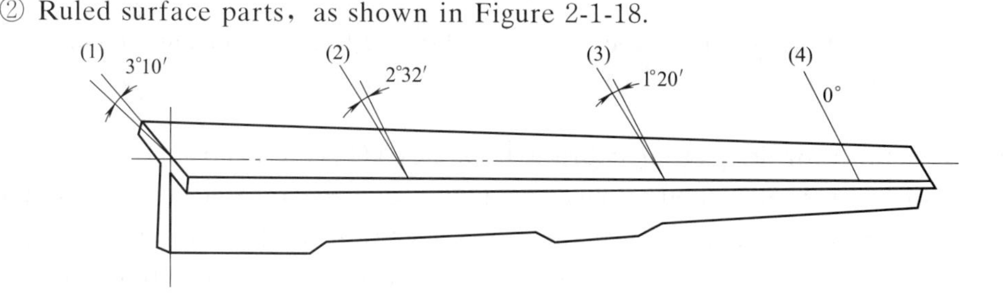


Figure 2-1-18 Ruled surface part

③ Three-dimensional curved parts, as shown in Figure 2-1-19.

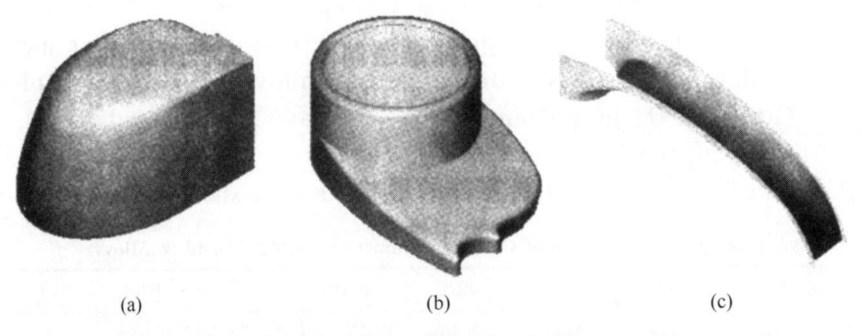


Figure 2-1-19 Three-dimensional curved parts

## Ⅳ. Setting of Workpiece Coordinate System

(1) Absolute value instruction and relative value instruction

G90 is an absolute value programming instruction, indicating that the coordinate dimension of tool motion given in the program section is an absolute coordinate value, that is, the given coordinate value is relative to the coordinate origin.

G91 is a relative programming instruction, indicating that the coordinate dimension of tool motion given in the program section is an incremental coordinate, that is, it is relative to the previous one, as shown in Figure 2-1-20, if the tool straightly moves from point A to point B, then:

When programming in absolute value, the program segments are as follows: G90 G01 X10. 0 Y20. 0;

When programming in incremental values, the program segments are as follows: G91 G01 X-20. 0 Y15. 0;

G90 and G91 are modal functions and can cancel each other, G90 is a default value.

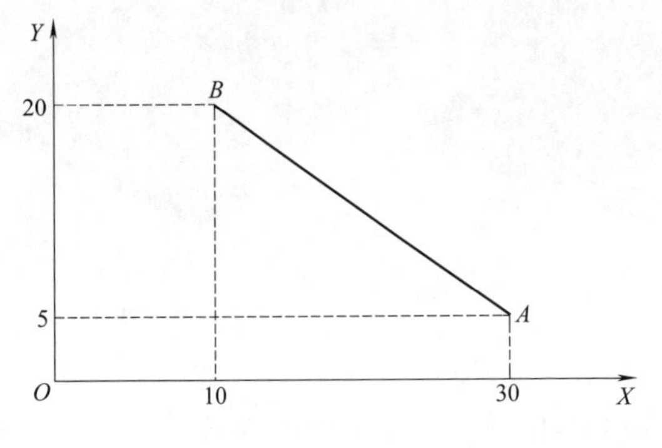


Figure 2-1-20 Functions of G90 and G91 instructions

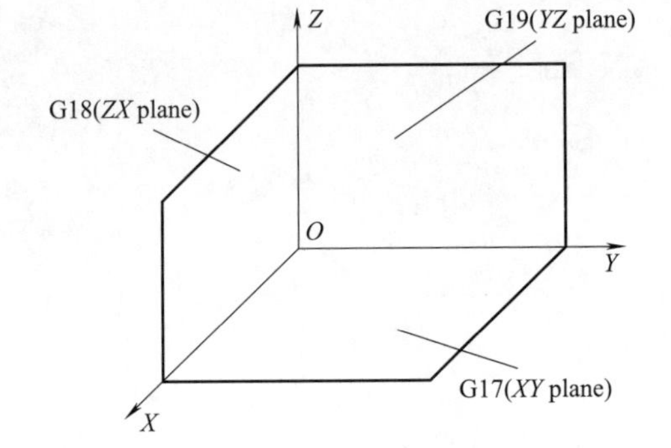


Figure 2-1-21 Selection of coordinate plane

(2) Instructions for designating coordinate planes (G17, G18, G19)

This set of instructions is used to select the plane for circular interpolation and tool radius compensation. G17 specifies XY plane, G18 specifies ZX plane, and G19 specifies YZ plane, as shown in Figure 2-1-21. G17, G18, and G19 are modal functions, which can cancel each other, G17 is a default value. Therefore, this group of instructions for NC

vertical milling machine (including NC machining center) can be implied and not written.

In addition, it is important to note that the line movement instruction has nothing to do with the plane selection. For example, when the instruction G17 G01 Z10. 0 is executed, the $Z$-axis movement is not affected.

(3) Selection instruction of origin setting for workpiece coordinate system selection (G54-G59)

The workpiece can set origins of six working coordinates from G54 to G59, as shown in Figure 2-1-22. The working origin data value can be pre-input into the offset register of the machine tool after the operation of the tool setting, which is not reflected in the programming.

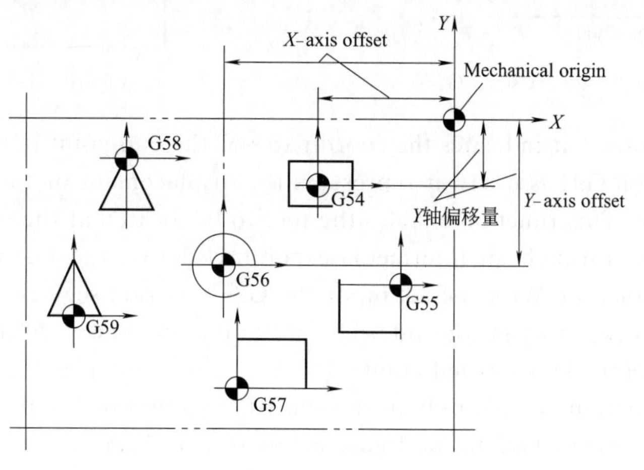


Figure 2-1-22 Setting working coordinate system

(4) Control instruction for returning to reference point

① Format: G90/G91 G27 X __ Y __ Z __;

Among them, X, Y, Z are the coordinate values of machine tool reference points in the workpiece coordinate system.

The function of the instruction is to check the correctness of the origin of the workpiece after the machine runs continuously for a long time so as to improve the machining reliability and ensure the accuracy of the size of the workpiece.

② Automatically return to reference point -G28

Format: G90/G91 G28 X __ Y __ Z __

Among them, X, Y, Z are the end position of the instruction.

The end point of the instruction is called the "intermediate point" instead of the reference point. It indicates the coordinates of the end point in the workpiece coordinate system when G90 is used, and the displacement of the end point relative to the starting point is indicated in G91. The axis specified by this instruction can be automatically located at the reference point.

Example: The usage of G28 is shown in Figure 2-1-23 as following:

G91 G28 X100. Y150. ;

G90 G28 X300. Y250. ;

③ Automatically return from reference point -G29

Format: G90/G91 G29 X __ Y __ Z __

Among them, X, Y, Z are the locating end points of instructions.

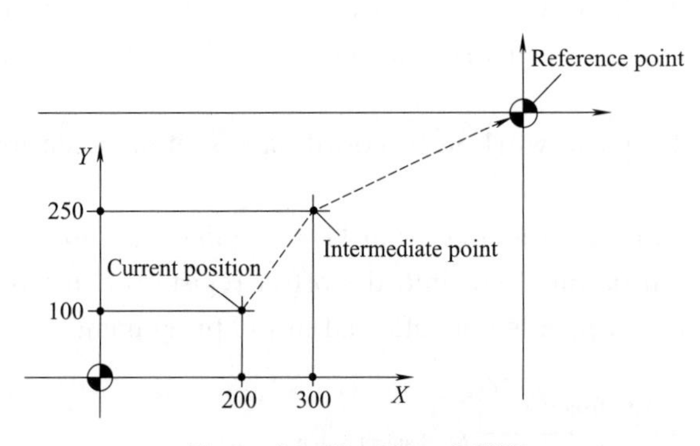


Figure 2-1-23 Use of G28

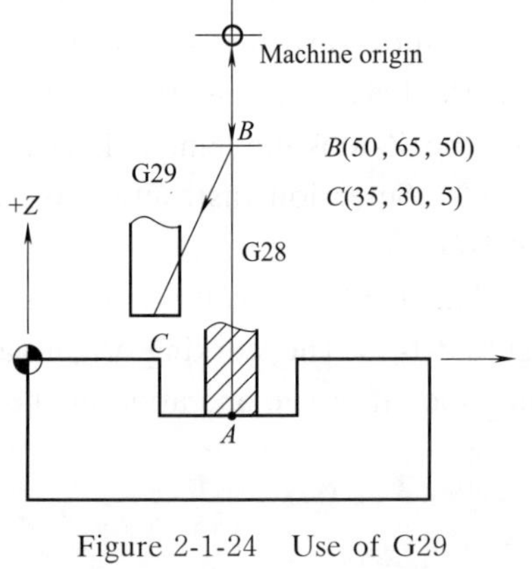

Figure 2-1-24 Use of G29

When G90 is used, it indicates the coordinates of the end point in the workpiece coordinate system; when G91 is used, it represents the displacement of the end point relative to the middle point. This function enables the tool to be located at the specified point from the reference point through an intermediate point. Usually, the instruction immediately follows a G28 instruction. With the action of the G29 program segment, all axes being instructed are fed quickly to pass the intermediate point previously defined by the G28 instruction, and then to the specified point.

The G29 instruction is valid only in the specified section of the program.

Example: The use of G29 in the figure below is as follows:

```
M06 T02;
……
G90 G28 Z50. 0;
M06 T03;
G29 X35. Y30. Z5. ;
```

## Ⅴ. Types of Common Fixture

The commonly used fixture includes universal fixture, modular fixture, special fixture and adjustable fixture.

### 1. Universal fixture

Universal fixtures are generally machine accessories or universal fixtures that can clamp various parts, such as jaw vice, dividing head, disc chuck and three-jaw chuck. Figure 2-1-25, shows an uniaxial indexing disc that can machine the workpiece from all sides by installing it once. Figure 2-1-26, shows an uniaxial NC rotary table (bed) that can install the workpiece once to process it from all sides, and also form five-axis linkage processing, such as processing the spatial forming surface and plane cam of cylindrical cam. A two-axis indexing disc is shown in Figure 2-1-27, which can be used to process holes arranged at different angles on the surface, and can process in five directions and inclined directions.

Figure 2-1-28, shows a two-axis NC rotary table (bed) which can be process in five directions and in five-axis linkage. Figure 2-1-29 shows a two-axis tiltable NC turntable with pallet exchange. The pallet size of the turntable is 1,000mm×1,000mm, the maximum load is

Figure 2-1-25 Uniaxial indexing disc

Figure 2-1-26 Uniaxial NC rotary table (bed)

Figure 2-1-27 Two-axis indexing disc

Figure 2-1-28 Two-axis NC rotary table (bed)

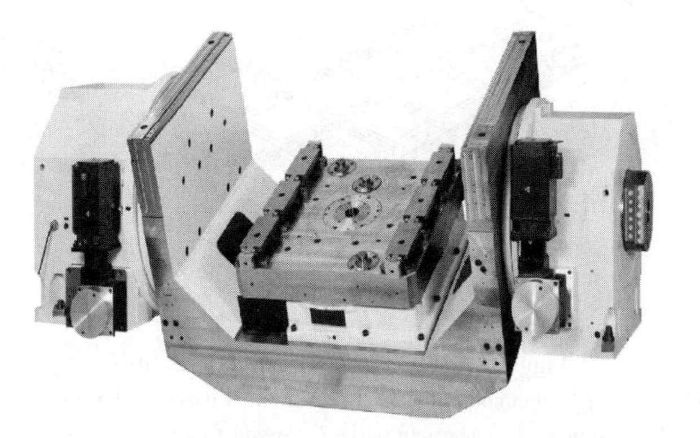

Figure 2-1-29 Two-axis tiltable NC turntable with pallet exchange

6,000kg, the rotation inertia $J$ is 2,000kg · $m^2$, the range of A-axis inclination angle is −115°(CCW)-5°(CW), the maximum rotation speed is 6r/min, the coaxiality is 0.01mm, the axial runout is 0.02mm, and the positioning accuracy is ±10. "B axis (rotating axis) table size is 1 000mm × 1 000mm, maximum rotation speed is 6 r/min, coaxiality is 0.01mm, axial run-out is 0.015mm, positioning accuracy is ±7 ".

**2. Modular fixture**

The modular fixture is composed of a set of general-purpose components and parts whose structures and sizes have been standardized, which can form fixtures with various functions according to the processing needs of workpieces. The modular fixture can be divided into the hole system modular fixture and the groove system modular fixture. Figure 2-

1-30 shows the groove modular fixture. The modular fixture can generally meet the requirements of standardization, serialization and universalization, with the advantages of combination, adjustability, simulation, flexibility, emergency, cost-effectiveness, and long service life, and can meet the requirements of short cycle and low cost in product processing. It is suitable for using in machining center.

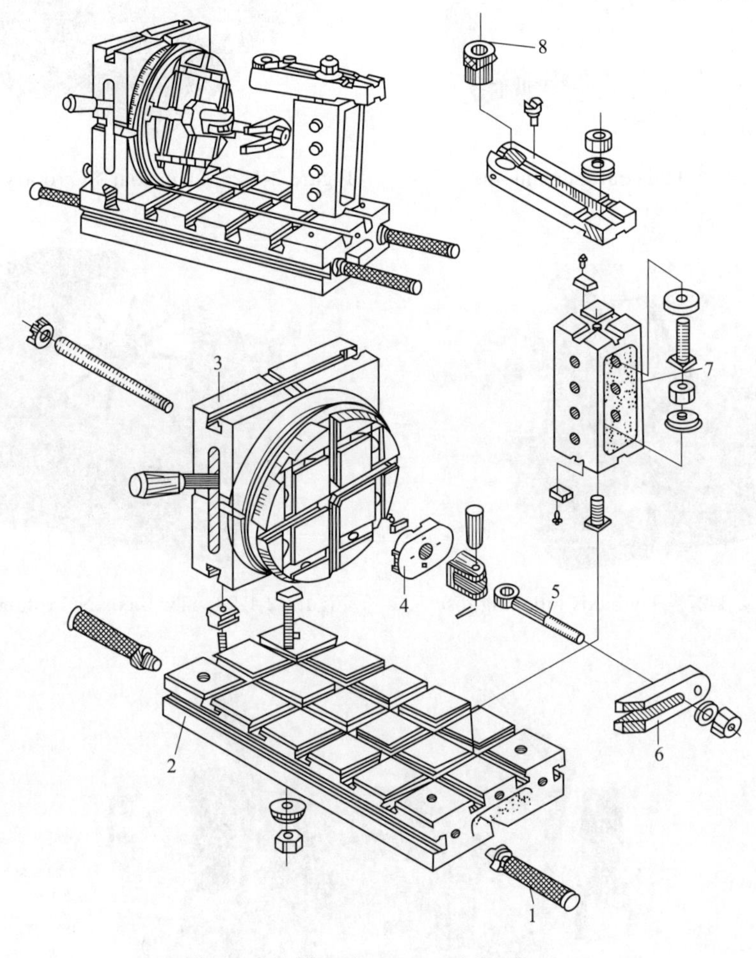


Figure 2-1-30 Groove modular fixture

1—Other part; 2—Foundation part; 3—Combined part; 4—Positioning part; 5—Fastener; 6—Clamping part; 7—Supporting part; 8—Guide

However, the modular fixture is made up of all kinds of universal standard components. There are many links between the components, the accuracy and rigidity of the fixtures are not as good as those of the special fixtures, especially the stiffness of the joint surface of the components, which has a great influence on the processing accuracy. Generally, the machining precision of the modular fixture can only reach IT9 and IT8 levels. Furthermore, the modular fixture has some shortcomings such as bulky overall appearance and inconvenient chip removal.

**3. Special fixture**

It is specially designed and manufactured for one or several similar processes, with reasonable structure, strong rigidity, stable and reliable clamping, convenient operation,

high installation precision and fast clamping speed. After processing by using this kind of fixture, the size of a batch of workpiece is relatively stable, with good interchangeability, which can greatly improve the productivity. However, this kind of fixture is only used to process one kind of part and is incompatible with the situation of product varieties constantly changing and updating. Especially, the special fixture also has the shortcomings of long design and manufacturing cycle, large labor cost and poor economy of machining simple parts. Generally, leading products of a factory and key parts with large batches and high precision requirements can choose special fixture when processing on the processing center. The clamping mechanism in special clamp usually adopts pneumatic or hydraulic clamping mechanism, which can reduce the labor intensity and improve the productivity of workers.

#### 4. Adjustable fixture

Adjustable fixture, a combination of modular fixture and special fixture, can effectively overcome the disadvantages of the two kinds of fixture above, which can not only meet the requirements of machining accuracy, but also has some flexibility. The main difference between adjustable fixture and modular fixture is that it has a series of fixture bodies with good overall rigidity. In addition, T-grooves, stepped unthreaded hole and screw holes with multi-functions of positioning and clamping are installed on the fixture body, and various clamping and positioning elements are equipped.

#### 5. Multi-station fixture

Multi-position jig can clamp more than one workpiece at the same time, reduce the number of tool changing, and is easy to process and load or unload workpieces concurrently, which can shorten the auxiliary time, improve the productivity, it is suitable for medium-batch production.

#### 6. Group fixture

Group fixture appears with the development of group processing technology. The basis for the use of group fixtures is the classification of parts. Through process analysis, all kinds of parts with similar shape and size are grouped into groups, and then the same or similar parts with the same positioning, clamping and processing methods are put together to consider the fixture design as a whole. For the parts with similar structural shape, group clamp is adopted, which has the characteristics of economy and high precision of clamping.

#### 7. Principle for the selection of NC milling fixtures

The selection of NC milling fixture should be considered comprehensively according to the accuracy grade of parts, structural characteristics, product batch and machine tool accuracy.

① Universal fixture, modular fixture and adjustable fixture shall be widely used in single-piece production or product development, and other fixtures will be considered only when universal fixture, modular fixture and adjustable fixture cannot solve the problem of clamping the workpiece.

② Simple special fixtures may be considered in small batches or in batch production.

③ Multi-station fixtures and high efficient pneumatic and hydraulic fixture can be considered when the production batch is large.

## Ⅵ. Cutting Tool Compensation Function

### 1. Cutting Tool radius compensation instruction (G41, G42, G40)

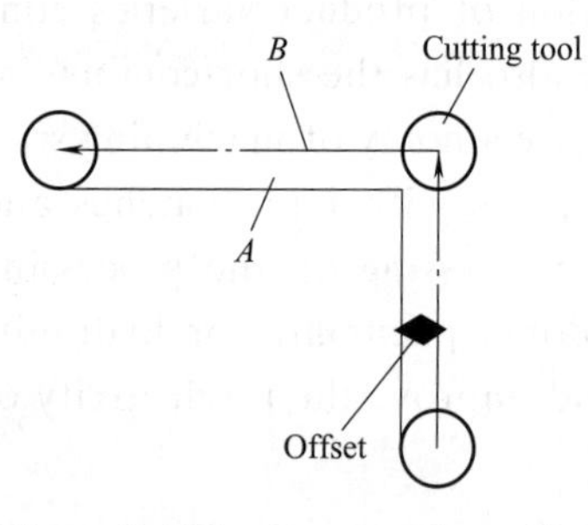


Figure 2-1-31 Cutter radius offset

(1) The concept of cutting tool radius compensation

The cutting tool radius compensation refers to calculating the cutting tool center track $B$ according to the workpiece contour $A$ and the tool offset, so the programmer can program on the basis of the workpiece contour $A$ or the size given on the drawing, and the cutter moves along the contour $B$ to produce the desired contour $A$, As shown in Figure 2-1-31.

(2) Working process of cutter radius compensation (see Figure 2-1-32)

① Establishment of cutter compensation It is a process of tool center transition from coincidence with programming trajectory to an offset with programming trajectory when the tool approaches the workpiece from its starting point. Parts should not be machined in this process.

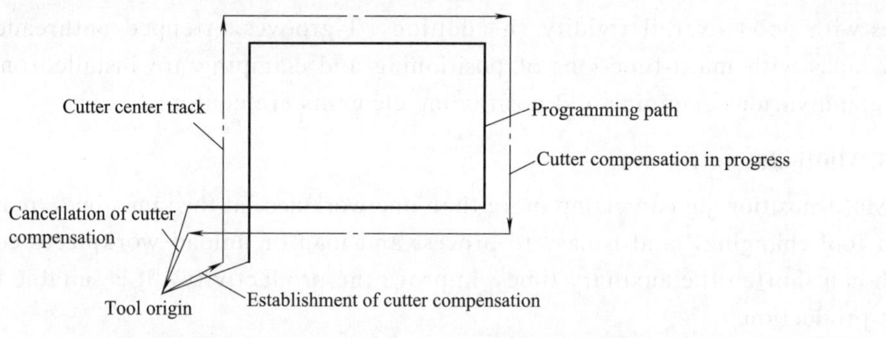


Figure 2-1-32 Working process of cutter radius compensation

② Execution of cutter compensation The center path of the tool is always a tool offset away from the programming path.

③ Cancellation of cutter compensation The tool withdraws from the workpiece, making the end point of the tool center track coincide with the end point of the programming path. It is an inverse process established by cutter compensation, and also parts should not be machined in this process.

Format G41 G00/G01X __ Y __ D __;

G42 G00/G01 X __ Y __ D __;

G40 G00/G01 X __ Y __ Z __;

G41 is the tool radius left compensation, looking forward along the direction of the tool movement, the tool is located on the left side of the part, as shown in Figure 2-1-33(a).

G42 is the tool radius right compensation, looking forward along the direction of the tool movement, the tool is located on the right side of the part, as shown in Figure 2-1-33(b).

The G40 is the instruction to revoke the cutter compensation.

D refers to the code that the control system stores the register unit of tool radius compensation (which is called the tool compensation No.).

G41, G42 and G40 are modal codes that can be written off from each other. G40 is a default value.

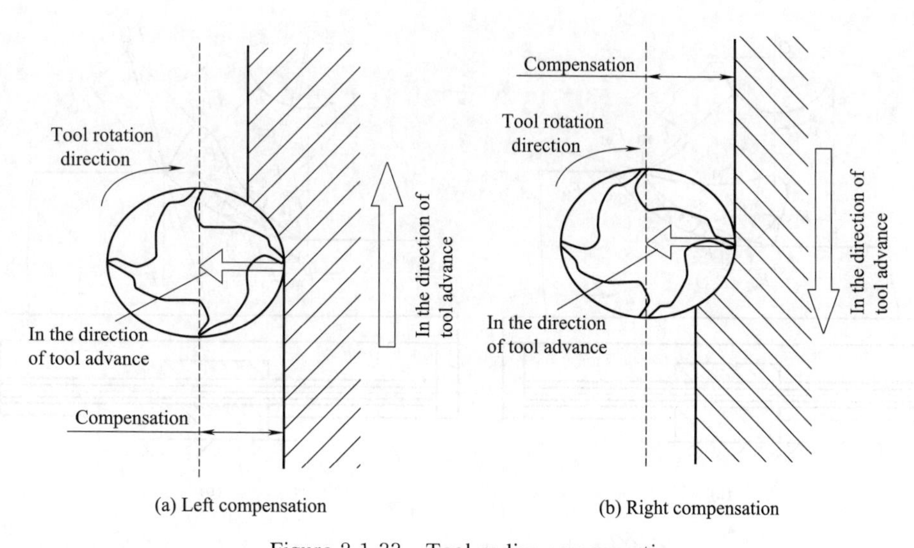

Figure 2-1-33 Tool radius compensation

Notes:

① The switch of tool radius compensation plane must be carried out in the mode of compensation cancellation.

② The tool radius compensation value is input by the operator to the tool compensation register.

③ The establishment and cancellation of tool radius compensation can only use G00 or G01 instruction, not G02 or G03 instruction. The so-called tool radius compensation establishment is the process that the cutter moves from no radius compensation to the desired starting point of cutter radius compensation, while the tool radius compensation cancellation is exactly the opposite.

**2. Climb milling and reverse milling**

(1) The concepts of climb milling and reverse milling

When the rotation direction of the milling cutter is opposite to the feed direction of the workpiece, it is called reverse milling, while that of the same is called climb milling.

(2) Characteristics of climb milling and reverse milling

As shown in Figure 2-1-34(a), when the reverse cutting is carried cut, the cutting tool is cut from the machined surface, the cutting thickness increases gradually from zero. The cutting edge of the cutter has a blunt circular radius $r_\beta$. When $r_\beta$ is greater than the instantaneous cutting thickness, the actual cutting rake angle is negative. The cutter teeth are extruded and slid on the machined surface, and can not cut scraps, which results in severe cold and hard layer on the surface. When the next cutter tooth is cut into, it is extruded and slid on the surface of the cold hard layer, the cutter tooth is easy to wear, and the roughness value of the surface of the workpiece is increased. Moreover, when the cutter teeth cut into the workpiece at the machined surface, due to the large deformation of the chip, the force exerted by the chip on the cutter makes the cutting depth increase in practice, which may result in a "digging cutter" type of too much cutting, resulting in insufficient machining margin, which is called "digging cutter" phenomenon. The phenomenon of "digging cutter" is extremely harmful to blanks of large and complex parts, which may cause parts to be scrapped in severe cases. At the same time, when the cutter teeth cut off

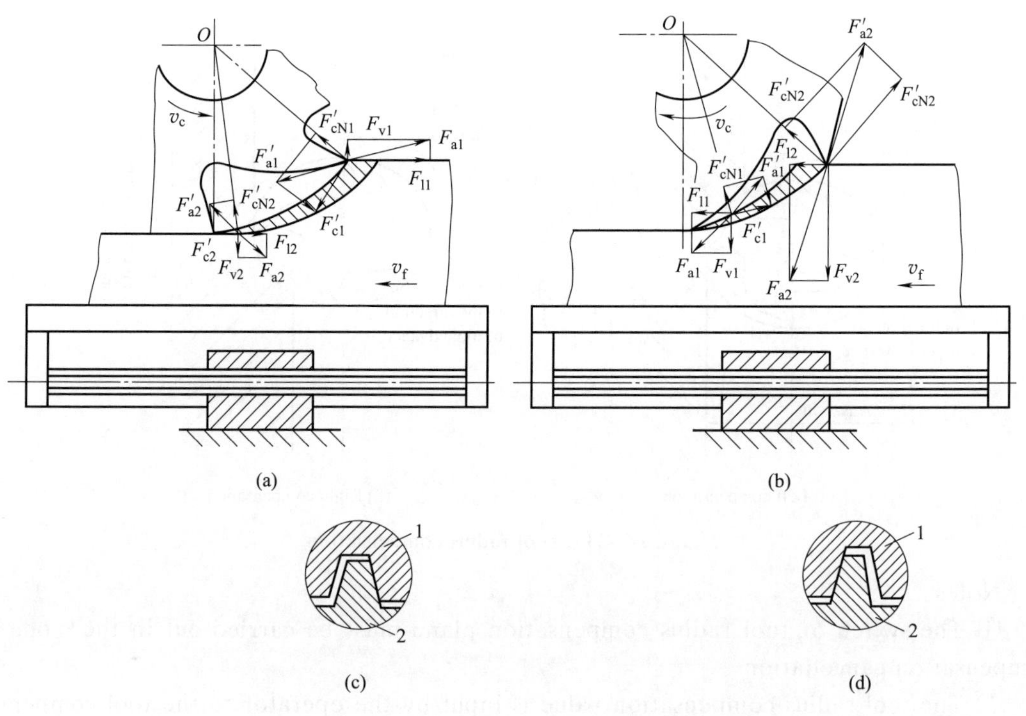


Figure 2-1-34 Reverse milling and climb milling

1—Nut; 2—Screw rod

the workpiece, the vertical component force $F_{v1}$ direction makes the workpiece off the worktable, which requires a larger clamping force. The advantage of reverse milling is that the cutter teeth cut from the machined surface, which will not cause the problem of cutting directly from the hard blank surface and hinder the cutter. As shown in Figure 2-1-34 (b), when climb milling is carried out, the cutter is cut from the surface to be machined, the cutting thickness of the cutter is gradually reduced to zero from the maximum, thus avoiding extrusion and sliding, the cutting force is very small when the chip is separated, the surface is smooth and there is no "digging cutter". Besides, the cutting component force $F_{v2}$ in the vertical direction is pressed to the workbench all the time, which reduces the upper and lower vibration of the workpiece. Therefore, the durability of milling cutter and the surface quality of workpiece can be improved.

The longitudinal feed movement of the milling machine table is usually realized by the screw rod and nut under the table, namely, the nut is fixed, and the screw rod rotates while driving the table to move. If there is a gap between the screw rod and the nut transmission, the climb milling shall be adopted. When the longitudinal component force $F_l$ increases gradually to exceed the table friction force, the table drives the screw rod to the left, and the gap between the screw rod and the right side of the nut transmission appears, causing the table to vibrate and feed unevenly, even making the cutter tip in serious case, as shown in Figure 2-1-34(c). In addition, the hard surface on the machined surface will accelerate the wear of the cutter teeth and even hinder the cutter in the climb milling. In reverse milling, the longitudinal component force $F_l$ is opposite to the longitudinal feed direction so that the transmission surface between the screw rod and the nut is always close to each other, as shown in Figure 2-1-34(d). As a result, the workbench will not occur movement and the milling is smooth.

(3) Determination of climb milling and reverse milling

According to the analysis above, when the surface of the workpiece has a hard cover and the feed mechanism of the machine tool has a gap, the reverse milling method should be used to arrange the feed route according to the reverse milling method. When the reverse milling is carried out, the cutter teeth cut from the machined surface, there will be no tipping, and the gap between the feed mechanism of the machine tool will not cause vibration and creep, which is in accordance with the requirements of rough milling. Hence, reverse milling should be used as far as possible in rough milling. When there is no hard cover on the surface of the workpiece and no clearance in the feed mechanism of the machine tool, the feeding route should be arranged according to the way of the climb milling. After being machined by climb milling, the parts machined have good surface quality and the cutter teeth have little wear , which is in accordance with the requirements of finish milling. Therefore, the climb milling should be adopted as far as possible in finish milling, especially when the parts are made of aluminum-magnesium alloy, titanium alloy or heat-resistance alloy.

Whereas, it should be emphasized that the NC equipment used in NC milling basically all adopt ball screw nut transmission, feed mechanism generally does not have the clearance, at this time, if the blank to be machined does not have high hardness, but has big size, complex shape and high cost, even in rough machining, the climb milling also should be adopted, which will be helpful to reduce the wear of the tool, avoid the shortage of margin caused by too much cutting of "digging cutter" in rough milling.

When the spindle rotates clockwise and the cutter is a right-handed milling cutter, the climb milling coincides with the left tool compensation (G41) and the reverse milling coincides with the right tool compensation (G42). Therefore, in general, finish milling uses G41 to establish the tool radius compensation, while rough milling uses G42 to establish the tool radius compensation.

**3. Tool length compensation instruction (G43, G44, G49)**

The format of tool length compensation is as follows:

```
G43 G00/G01 Z_ H_;
G44 Z_ H_;
G49 Z_;
```

G43 is a positive compensation of the tool length, that is, the coordinate value of $Z$ coordinate actual movement is the amount by adding $Z$-coordinate dimension word and tool length compensation value, as shown in Figure 2-1-35(a). When executing G43: Z actual value = Z instruction value + (H × ×).

G44 is a negative compensation of the tool length, that is, the coordinate value of $Z$ coordinate actual movement is the amount by subtracting the tool length compensation value from the $Z$-coordinate size word, as shown in Figure 2-1-35(b). When executing G44: Z actual value = Z instruction value − (H × ×).

H is the code that the control system stores the register unit of tool length compensation.

G43, G44 and G49 are modal codes that can be deleted from each other. G49 is a default value.

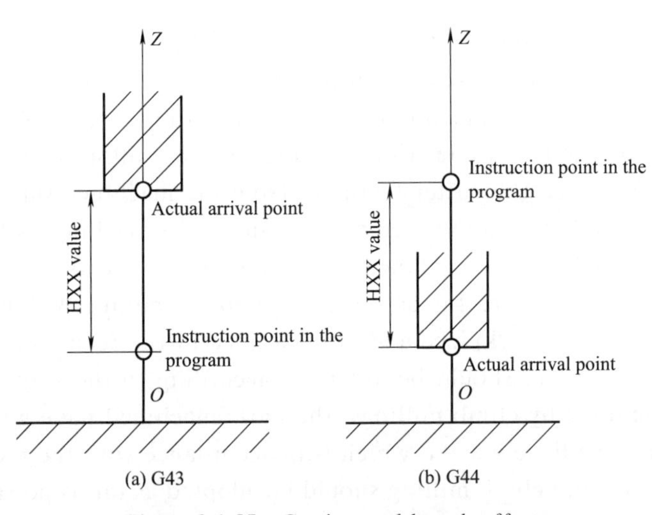


Figure 2-1-35 Cutting tool length offset

**Exercises**

1. What is the classification of CNC milling machine?
2. What is the main function of CNC milling machine?
3. What are the common CNC milling cutters?
4. How to choose the blade grade?
5. Please make a brief introduction to the processing range of CNC milling machine.
6. What is the difference between an absolute value instruction and a relative value instruction?
7. What are the types of common jigs?
8. What are the tool radius compensation instructions?
9. What is the difference between climb milling and reverse milling?
10. What are the tool length compensation instructions?

# Task Ⅱ Machining Baseplate

## Ⅰ. Drawing and Technical Requirements

Shown in Figure 2-2-1, Figure 2-2-2, Table 2-2-1, Table 2-2-2.

Figure 2-2-1 Baseplate physical drawing

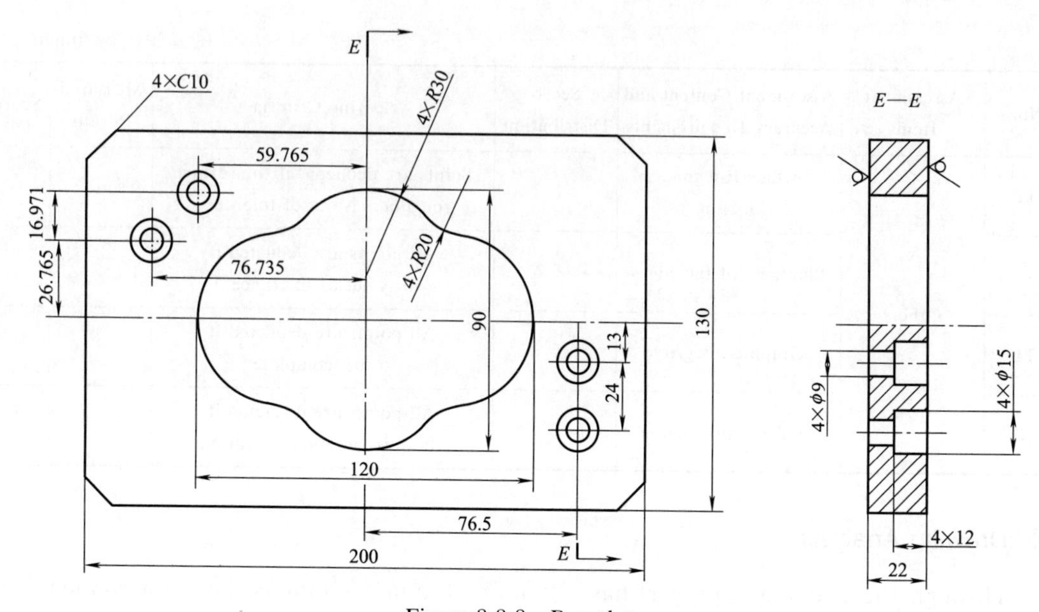


Figure 2-2-2 Baseplate

**Table 2-2-1 List of training equipment**

| No. | Type | Name | Specification | Quantity | Note |
|---|---|---|---|---|---|
| 1 | Material | 6061 | 210mm×140mm×30mm | 1 | |
| 2 | Cutting Tool | High-speed steel end mill | φ12mm, φ6mm | 1 for each | |
| | | Twist drill | φ9mm | 1 | |
| | | Centre drill | A3 | 1 | |
| 3 | Fixture | Precision plain vice | 0-300mm | 1 set | |
| 4 | Tool | Milling chuck | 0-13mm | 1 | |
| | | Clamping bush | φ12mm, φ6mm | 1 for each | Matching with cutting tools |
| | | Parallel clamp | | 1 pair | |
| | | Oil stone | | 1 | |

**Table 2-2-2 Part inspection items and score sheet (100 points)** Score ______

| No. | Assessment Items | Assessment Content and Accuracy Requirements | Score Distribution | Scoring Criteria | Measured Results | Score |
|---|---|---|---|---|---|---|
| 1 | Over all dimensions | $200_{-0.05}^{0}$ | 7 | No score when it is out of tolerance | | |
| 2 | | $130_{-0.05}^{0}$ | 7 | No score when it is out of tolerance | | |
| 3 | | 22±0.02 | 7 | No score when it is out of tolerance | | |
| 4 | | $120_{0}^{+0.05}$ | 7 | No score when it is out of tolerance | | |
| 5 | | $90_{0}^{+0.05}$ | 6 | No score when it is out of tolerance | | |
| 6 | | 76.5±0.02 | 6 | No score when it is out of tolerance | | |
| 7 | | φ15(4) | 10 | No score when it is out of tolerance | | |
| 8 | | φ9(4) | 10 | No score when it is out of tolerance | | |
| 9 | | 12±0.05(4) | 7 | No score when it is out of tolerance | | |
| 10 | | 59.7±0.02 | 7 | No score when it is out of tolerance | | |
| 11 | | 76.7±0.02 | 6 | No score when it is out of tolerance | | |

Continued

| No. | Assessment Items | Assessment Content and Accuracy Requirements | Score Distribution | Scoring Criteria | Measured Results | Score |
|---|---|---|---|---|---|---|
| 12 | Others | Surface Roughness 3. 2μm | 5 | Points are deducted if the surface roughness is out of tolerance | | |
| 13 | | Bluntness of the edges | 5 | All points are deducted if it is out of tolerance | | |
| 14 | | Graphic integrity | 5 | All points are deducted if it is incomplete | | |
| 15 | | Civilized production | 5 | All points are deducted if there is violation operation | | |

## Ⅱ. Drawing Analysis

Through the drawing, the part has 210mm×140mm×30mm as the main boundary dimension, the middle arc is inner contour; The bottom has four counter bores, the hole $\phi$15 has a depth of 12mm, and the through hole of $\phi$9mm has a depth of 10mm.

The part's shape is simple with a maximum tolerance requirement of 0. 05mm.

### 1. Selection of cutting tools

The choice and rational application of cutting tools' materials is very important, at present, the cutting tool materials commonly used in cutting are mainly high-speed steel and cemented carbide. The workpiece material in this example is hard aluminum, and the cutting tool selects high-speed steel solid end mill with sharp edge, good straightness accuracy and high precision.

According to the drawing, considering the structure of the part, the cutting tools are selected, see Table 2-2-3.

**Table 2-2-3 Card of cutting tools**

| Name of Cutting Tools | Specifications of Cutting Tools | Material | Quantity | use of Cutting Tools | Note |
|---|---|---|---|---|---|
| End Mill | $\phi$12mm | High-speed | 1 | Plane processing, rough and finish machining of contour | |
| End Mill | $\phi$6mm | High-speed Steel | 1 | Countersink part rough and finish machining | |
| Center Drill | $\phi$3mm | High-speed | 1 | Drilling center holes | |
| Twist Drill | $\phi$9mm | High-speed | 1 | Drilling | |

### 2. Selection of cutting parameters

According to the materials of the manufacturing objects, the materials and specifications of cutting tools, and by reading the book of metal cutting parameters to find the tool's cutting speed and per tooth's feed rate, the speed and the feed speed of the tool selected are determined, reference cutting parameters see Table 2-2-4.

**Table 2-2-4 Card of cutting parameters**

| Cutting Tool | Cutting Speed $v$/(m/min) | Feed Rate per Cut $f$/(mm/cut) | Speed of Spindle $S$/(r/min) | Feed rate $F$/(mm/min) | Note |
|---|---|---|---|---|---|
| $\phi$12m End Mill | 50 | 0.04 | 1300 | 200 | Rough Machining |
| | 80 | 0.03 | 2100 | 240 | Finish Machining |
| $\phi$6m End Mill | 50 | 0.04 | 2000 | 200 | Rough Machining |
| | 80 | 0.03 | 2500 | 240 | Finish Machining |
| $\phi$3mm Centre Drill | 30 | 0.03 | 3200 | 200 | |
| $\phi$9mm Drill | 25 | 0.05 | 1400 | 140 | |

### 3. Depth of cutting $a_p$

The depth of cutting is mainly limited by machine and tool rigidify during rough machining, and in general, the cutting depth is taken (0.6~0.8) $D_{cut}$ when the radial cutting volume is large. Otherwise the depth of cutting can be larger.

The part's contour processing volume is not large, processing depth for each contour can be processed according to the drawing dimension, no layered processing is needed.

### 4. Process procedure planning

The each contouring process of the part is planned as follows.

**Table 2-2-5 Card of part working process**

| Unit | | Product Name and Model | Part Name | Part Chart Number |
|---|---|---|---|---|
| | | | Simple Parts | |
| Process | Working procedure number | Fixture name | The Device Used | Materials for workpiece |
| 1 | O0001 | Precision plain vice | VMC850 | LY12 |
| Step | Content | Number of cutting tool | Tool and cutting volume | Note |
| 1 | Milling the upper surface | T01 | $\phi$12mm End Mill<br>$n$ = 2100r/min<br>$F$ = 240mm/min<br>$a_p$ = 0.3mm | Milling by finish machining |
| 2 | The processing origin is set at the center of the upper surface of the workpiece | T01 | | Using the test cutting method |
| 3 | Roughing 210×130 as the outer profile of the main size, leaving a margin of 0.2mm | T01 | $\phi$12mm End Mill<br>$n$ = 1300r/min<br>$F$ = 200mm/min<br>$a_p$ = 6.8mm | |
| 4 | Roughing the inner profile of the arc, leaving a margin of 0.2mm | T01 | $\phi$12mm End Mill<br>$n$ = 1300r/min<br>$F$ = 200mm/min<br>$a_p$ = 14.8mm | |
| 5 | Finishing 210×130 as the outer profile of the man size to the specified dimension | T01 | $\phi$12mm End Mill<br>$n$ = 1300r/min<br>$F$ = 200mm/min<br>$a_p$ = 4.8mm | |

Continued

| Step | Content | Number of cutting tool | Tool and cutting volume | Note |
|---|---|---|---|---|
| 6 | Finishing the inner profile of the arc to the specified dimension | T01 | $\phi$12mm End Mill<br>$n$ = 2100r/min<br>$F$ = 240mm/min<br>$a_p$ = 7mm | |
| 7 | Drilling four positioning center holes of $\phi$9mm | T02 | $\phi$3mm Centre Drill<br>$n$ = 3200r/min<br>$F$ = 200mm/min<br>$a_p$ = 3mm | |
| 8 | Drilling three through holes of $\phi$9mm | T03 | $\phi$5. 8mm Drill<br>$n$ = 1400r/min<br>$F$ = 140mm/min<br>$a_p$ = 2mm | |
| 9 | Roughing the part of counter bore, leaving a margin of 0. 2mm | T04 | $\phi$6mm End Mill<br>$n$ = 2000r/min<br>$F$ = 200mm/min<br>$a_p$ = 14. 8mm | |
| 10 | Finishing the part of counter bore to the specified dimension | T04 | $\phi$6mm End Mill<br>$n$ = 2000r/min<br>$F$ = 200mm/min<br>$a_p$ = 14. 8mm | |

## 5. Processing procedures

```
O0001(Main program)
G69  G40;
G28  G91  Z0;
G54  G90  G00  X0  Y0;
Z100;
M03  S1500;
XO  Y0;
Z5;
G01  Z-5  F100;
G01  G42  D01  X-60  Y0;
G02  X-33. 63  Y29. 77  R30;
G03  X-27. 12  Y31. 74  R20;
G02  X27. 12  Y31. 74  R30;
G03  X33. 63  Y29. 77  R20;
G02  X33. 63  Y-29. 77  R20;
G03  X27. 12  Y-31. 74  R20;
G02  X27. 12  Y-31. 74  R30;
G03  X-33. 63  Y-29. 77  R20;
G02  X-60  Y0  R30;
G01  G40  X0 Y0;
M98  P0002  L3;
G01  Z-22  F100;
G01  G42  D01  X-60  Y0;
G02  X-33. 63  Y29. 77  R30;
```

```
G03  X-27.12  Y31.74  R20;
G02  X27.12  Y31.74  R30;
G03  X33.63  Y29.77  R20;
G02  X33.63  Y-29.77  R20;
G03  X27.12  Y-31.74  R20;
G02  X27.12  Y-31.74  R30;
G03  X-33.63  Y-29.77  R20;
G02  X-60  Y0  R30;
G01  G40  X0  Y0;
G00  Z100;
M05;
M30;
O0002(Subprogram)
Z5;
G01  Z-5  F100;
G01  G42  D01  X-60  Y0;
G02  X-33.63  Y29.77  R30;
G03  X-27.12  Y31.74  R20;
G02  X27.12  Y31.74  R30;
G03  X33.63  Y29.77  R20;
G02  X33.63  Y-29.77  R20;
G03  X27.12  Y-31.74  R20;
G02  X27.12  Y-31.74  R30;
G03  X-33.63  Y-29.77  R20;
G02  X-60  Y0  R30
G01  G40  X0  Y0
Z5;
M99;
```

## Ⅲ. Analysis on Machining Process of NC Milling

The program of NC milling machine is the instruction document of NC milling machine. NC milling machine is controlled by program instruction, and the whole process of machining is carried out automatically according to program instruction. Therefore, the NC milling machine processing program is quite different from that of ordinary milling machine, and involves a wide range of contents. The NC milling process includes not only the process of the parts, but also the cutting parameters, the path of the cutter feed, the size of the tool and the movement of the milling machine. It is necessary for programmers to be quite familiar with the performance, characteristics, movement mode, tool system, cutting specification and workpiece mounting methods of NC milling machines. The quality of the process not only affects the efficiency of the milling machine, but also directly affect the processed quality of the parts.

### 1. The main contents of NC milling process

(1) NC milling processing technology

① Select parts suitable for NC milling machine and determine the process content.

② Analyze the drawings of the parts under processing, and clarify the processing contents and technical requirements.

③ Determine the processing plan of the parts and work out the NC processing technology route.

④ Design of processing procedure. Such as the selection of parts positioning criteria, fixture scheme determination, work-step division, cutting tool selection and determination of cutting requirements.

⑤ The adjustment of NC machining program. For example, selecting the point of tool setting and point of tool changing, determining the tool compensation and the machining route, etc.

(2) Select and determine the machining parts and contents of NC milling

The following aspects are suitable for NC milling.

① Internal and external contours composed of straight lines, circular arcs, non-circular curves and tabulated curve.

② Spatial curves or curved surfaces.

③ Parts with simple shape, but complicated size and difficult to de detected.

④ Inner cavities and inner boxes, which are difficult to be observed, controlled and detected when machining with ordinary machine tool.

⑤ Holes or planes with strict position and size requirements.

⑥ Simple surfaces or shapes that can be machined in a single clamp.

⑦ General processing content that NC milling can effectively improve productivity and reduce the labor intensity.

**2. Process analysis of NC milling machining parts**

(1) Part diagram and its structure process analysis

① Analyze the shape, structure and size of the part, determine whether there is any position on the part that hinders the movement of the tool, whether there is any area that will produce machining interference or that can not be machined, whether the maximum shape and size of the part exceed the maximum travel of the machine, whether the rigidity of the part changes too much with the progress of the machine, etc.

② Check parts processing requirements, for instance, whether dimensional machining accuracy, geometric tolerance and surface roughness can be guaranteed under existing processing conditions, whether there are more economical processing methods or schemes.

③ Whether there is a limit part and size requirements to the shape and size of the tool on the parts, such as fillet radius, chamfer angle, slot width and so on, whether these dimensions are too messy, whether they can be unified. Minimum cutting tools are used as far as possible to reduce tool specifications, tool changing and tool setting times and time, thus shortening the total processing time.

④ The technical datum used in the machining of parts should be taken into consideration, which not only determines the sequence of each machining process, but also has a direct effect on the position accuracy of each machining surface after each process is processed. It is necessary to analyze whether there are available process benchmarks on the parts. For general machining accuracy requirements, some datum or reference holes existing on the parts can be used, or the process benchmarks can be made specially on the parts. When the machining precision of the parts is extraordinary high, the advanced uni-

form reference positioning and clamping system must be adopted to ensure the processing requirements.

⑤ Only by analyzing the types, grades and heat treatment requirements of part materials and understanding the cutting performance of part materials, can tool materials and cutting parameters be reasonably selected.

⑥ The conditions (such as tangent, intersection, vertical and parallel, etc.) for the geometric elements (points, lines and surfaces) constituting part contour are the important basis for NC programming. Therefore, when analyzing the parts drawing, it is necessary to analyze whether the given conditions of geometric elements are sufficient so as to find out the problem in time and consult with the designer.

Making a process plan will affect the machining quality of the parts. Improving the structure of the parts to improve the process performance can effectively solve some technological problems . There are several typical methods to improve the process as shown in Table 2-2-6 below.

**Table 2-2-6　Improvement of parts structure and technology**

| Ways to improve processing | Structure | | Result |
|---|---|---|---|
| | Before improvement | After improvement | |
| Milling | | | |
| Improve inner-wall shape | 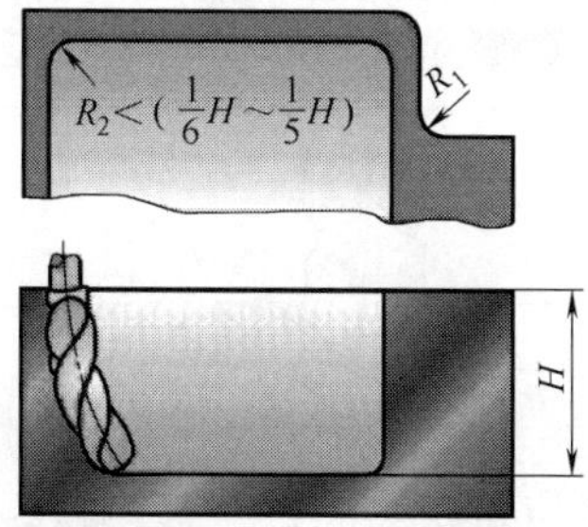 $R_2<(\frac{1}{6}H\sim\frac{1}{5}H)$, $R_1$, $H$ | 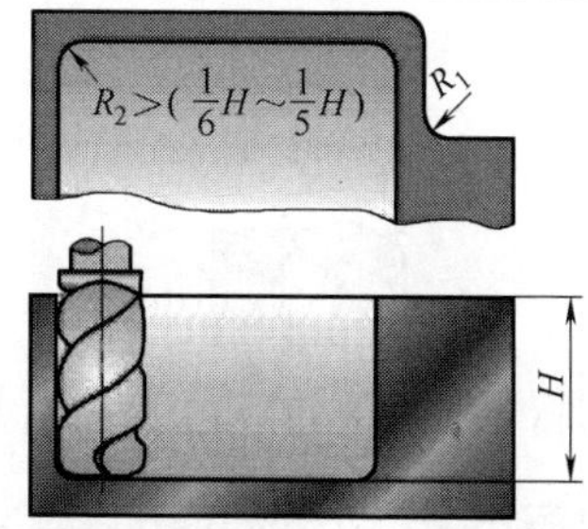 $R_2>(\frac{1}{6}H\sim\frac{1}{5}H)$, $R_1$, $H$ | Higher rigid tool can be used |
| Unify circular arc size | 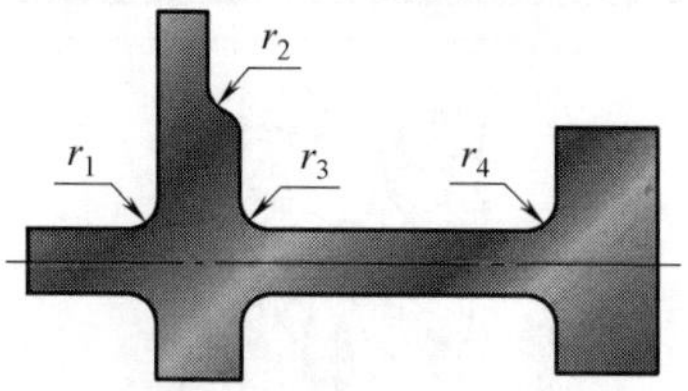 $r_1$, $r_2$, $r_3$, $r_4$ | 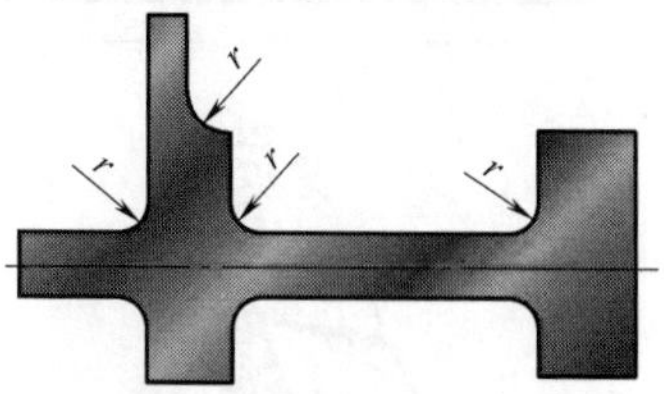 $r$, $r$, $r$, $r$ | Reduce the number of tools and the number of tool changing, reduce the auxiliary time |
| Select the appropriate circular radius $R$ and $r$ | 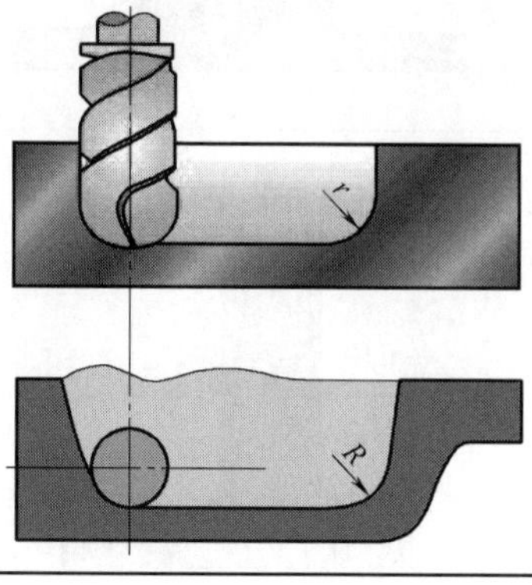 $r$, $R$ | 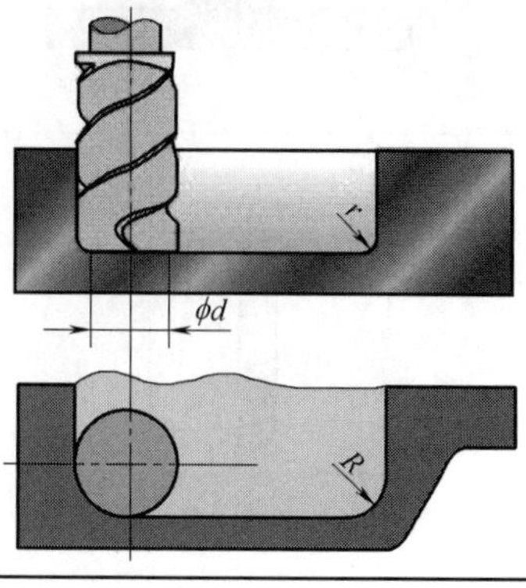 $r$, $\phi d$, $R$ | Improve production efficiency |

Continued

| Ways to improve processing | Structure | | Result |
|---|---|---|---|
| | Before improvement | After improvement | |
| Milling | | | |
| Apply bilateral symmetry structure | 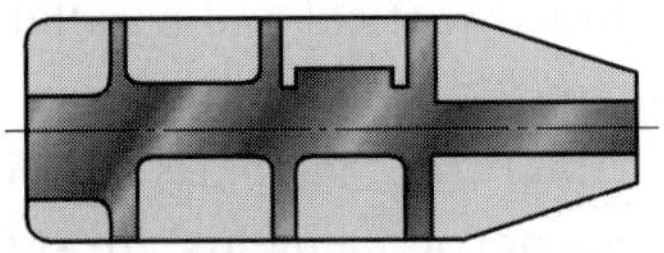 | 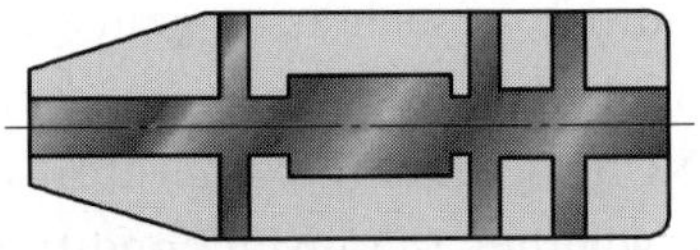 | Reduce programming time, simplify programming |
| Improve the distribution of boss reasonably | 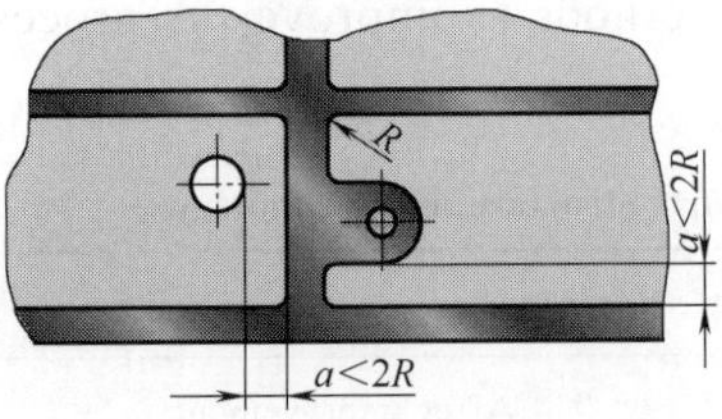 $a<2R$ $a<2R$ $R$ | 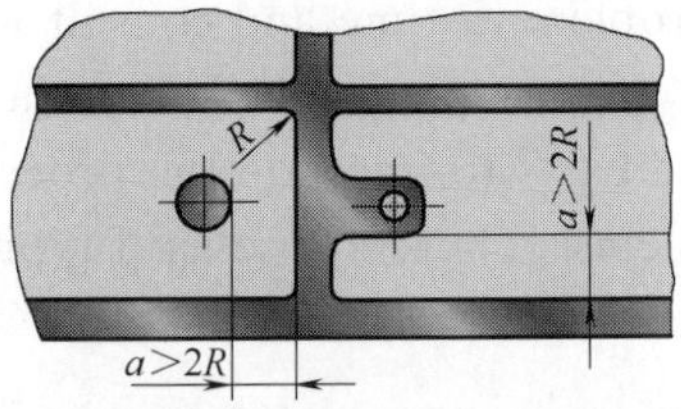 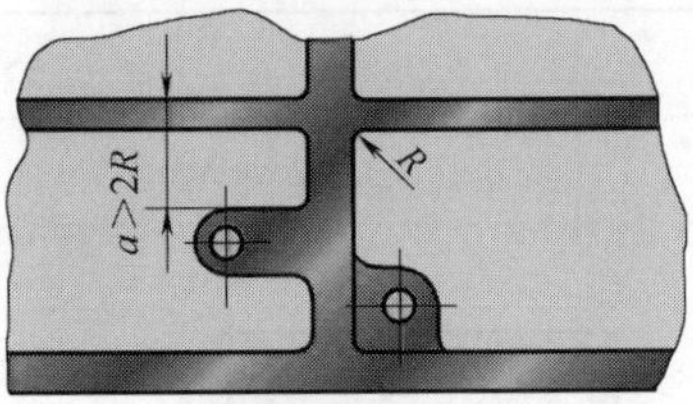 $a>2R$ $a>2R$ $R$ $a>2R$ $R$ | Reduce the amount of processing labor |
| Improve structural shape | 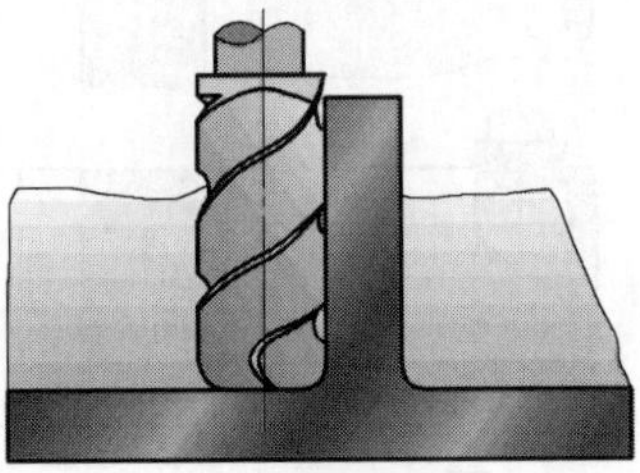 | 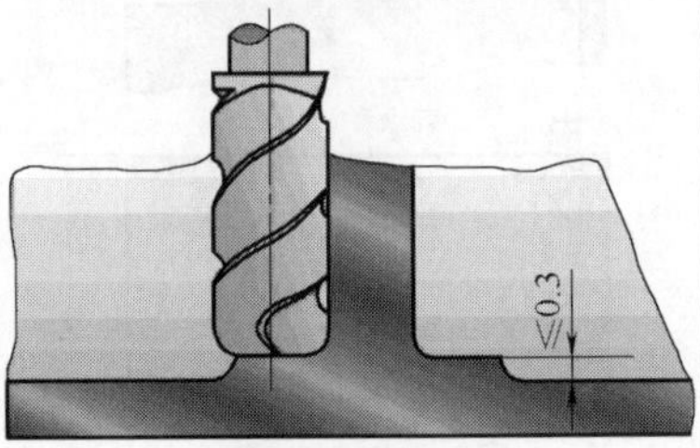 $\leq 0.3$ | Reduce the amount of processing labor |
| | 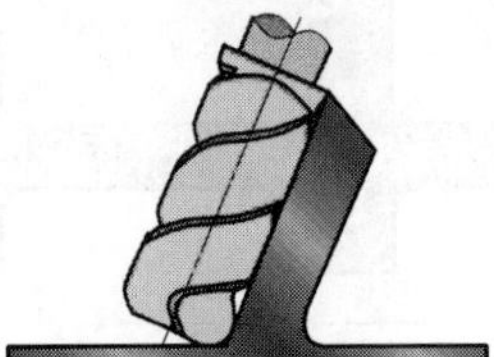 | 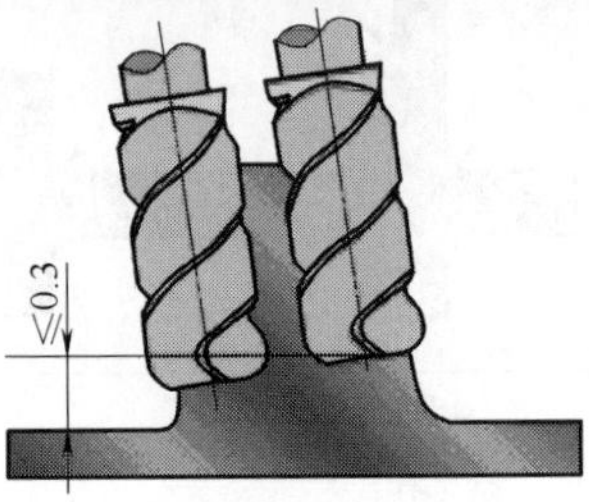 $\leq 0.3$ | |
| Improve size proportion | 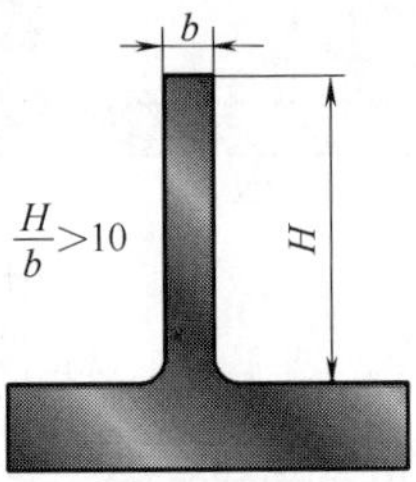 $\frac{H}{b}>10$ $H$ $b$ | 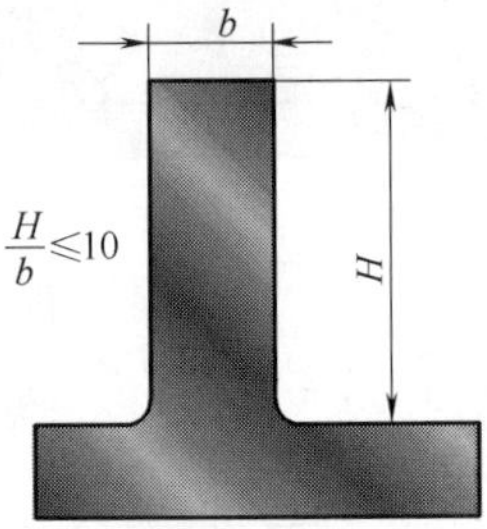 $\frac{H}{b}\leq 10$ $H$ $b$ | High rigidity cutting tools can be used to improve the productivity |

Continued

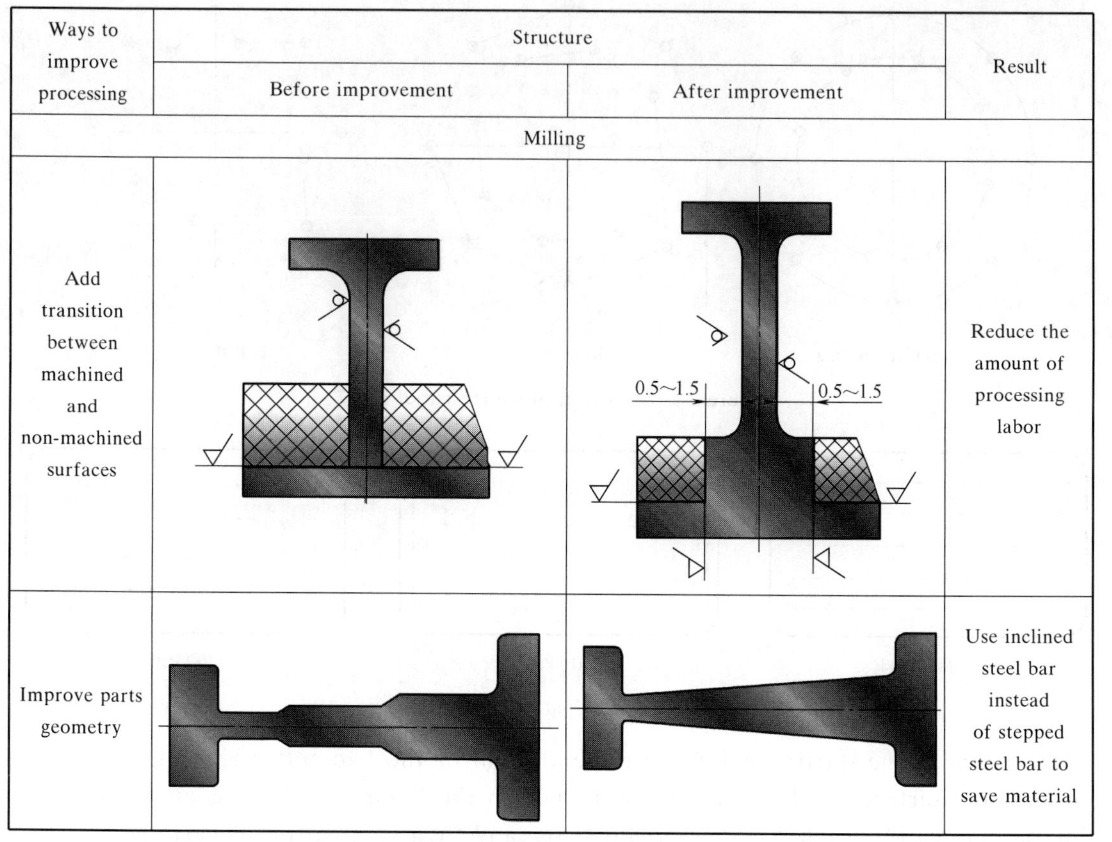


| Ways to improve processing | Structure | | Result |
|---|---|---|---|
| | Before improvement | After improvement | |
| Milling | | | |
| Add transition between machined and non-machined surfaces | | | Reduce the amount of processing labor |
| Improve parts geometry | | | Use inclined steel bar instead of stepped steel bar to save material |

(2) Process analysis of semi-finished parts

① The semi-finished parts should have sufficient and stable machining allowance.

② Analyze the clamping adaptability of semi-finished parts.

③ Analyze the residual size and uniformity of blank.

**3. Determine the route of the cutter and arrange the processing sequence**

Tool routing refers to the movement track of the tool in the whole processing process, which not only includes the content of steps, but also reflects the sequence of steps. It is also one of the bases for programming. The following points should be noted when determining the path of the cutter:

(1) Find the shortest machining route

For example, the hole system on the part is shown in processing Figure 2-2-3(a). The tool path of Figure 2-2-3(b) is to process the outer ring hole first and then the inner ring hole. If the tool path in the Figure 2-2-3(c) is adopted to reduce the idle run time, the positioning time can be saved nearly twice, thus improving the processing efficiency.

(2) The final contour should be completed by once tool feed

In order to ensure the roughness requirement after machining the contour surface of the workpiece, the final contour should be continuously machined in the last cutting.

Figure 2-2-4(a) shows a tool path for machining the inner cavity with a direction parallel cut, which can remove all the excess in the inner cavity, leave no dead angle or do not hurt the contour. However, the direction parallel cutting method will leave the residual

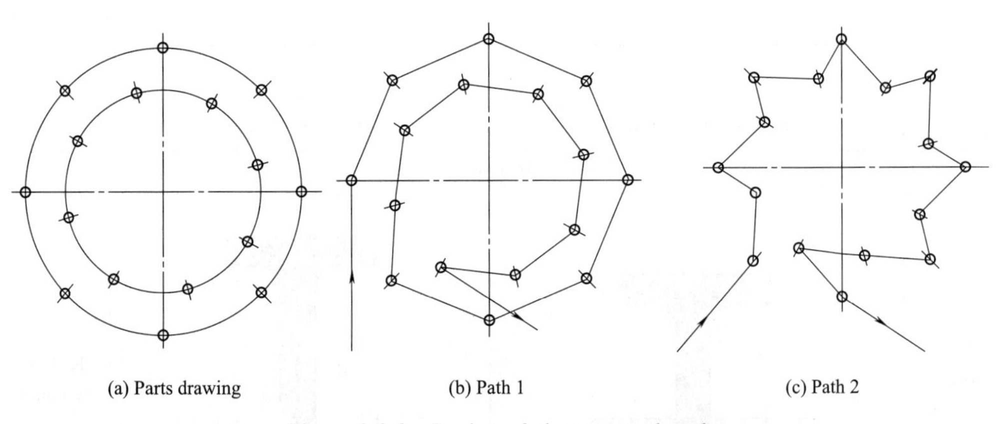

(a) Parts drawing (b) Path 1 (c) Path 2

Figure 2-2-3 Design of shortest tool path

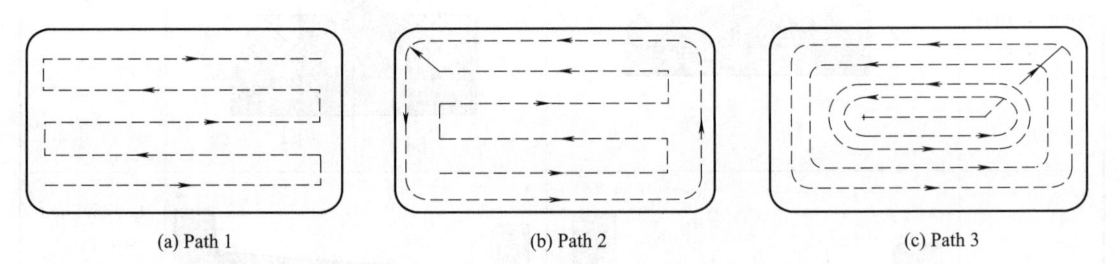

(a) Path 1 (b) Path 2 (c) Path 3

Figure 2-2-4 Three tool paths for milling inner cavity

height between the starting point and the end point of the two cutter feed and not reach the required surface roughness. So if the method in the Figure 2-2-4(b) is applied, direction parallel cutting should be performed first, and then ring cutting is carried out along circumferential direction to finish the contour surface so as to get outstanding performance. The way shown in Figure 2-2-4(c) is also a better tool path.

(3) Choose the direction of cutting in and cutting out

When considering the feeding and withdrawing (cutting in and cutting out) routes of the cutter, the cutting in and cutting out points of the cutter should be along the tangent line of the contour of the part to ensure the smooth contour of the workpiece. It is necessary to avoid scratching the workpiece surface by vertically feeding and withdrawing the cutter on the workpiece contour surface, and to minimize the pause (elastic deformation caused by sudden change of cutting force) in the cutting process of the contour, so as to avoid leaving tool marks, as shown in Figure 2-2-5.

Y
Starting point
Ending point
X

Figure 2-2-5 Extension of cutting-in and cutting-out of cutting tools

(4) Choose the route that can make the workpiece deform less after processing

Slender or thin plate parts with small cross-section area should be machined to the final size by several cutters feeding or should adopt symmetrical removal margin method to arrange the path of cutting tool. When working steps are arranged, the steps with less damage to the rigidity of the workpiece should be arranged first.

## Ⅳ. Basic Instructions for NC Milling

Take the FANUC 0i system as an example to introduce the G instruction words in the NC milling machine, as shown in Table 2-2-7, and the commonly used M instructions as shown in Table 2-2-8.

**Table 2-2-7 G instruction word and meanings in FANUC 0i**

| G code | Group | Function | Remarks |
|---|---|---|---|
| G00※ | 01 | Fast point positioning | |
| G01※ | | Line interpolation | |
| G02※ | | Clockwise circular interpolation | In G02 X Y I J or G02 X Y R,X、Y represents the end-point coordinates. I and J refer to the distance between the center of a circle and its starting point in the $X$ and $Y$ directions,and R represents the radius of a circle |
| G03※ | | Counterclockwise circular interpolation | |
| G04 | 00 | Pause(delay) | In G04 P,P denotes the duration of program stop |
| G17※ | 02 | $XY$ plane selection | |
| G18 | | $ZX$ plane selection | |
| G19 | | $ZY$ plane selection | |
| G20 | 06 | English measurement input | |
| G21 | | Metric system input | |
| G40※ | 07 | Cancel tool radius compensation | |
| G41※ | | Tool radius left compensation | Cutter compensation must be done in straight line section |
| G42※ | | Tool radius right compensation | |
| G43 | 08 | Tool length positive compensation | |
| G44 | | Tool length negative compensation | |
| G49※ | | Cancel tool length compensation | |
| G50 | 11 | | |
| G51 | | | In G51 X Y Z I J K,I,J,K represent $X$,$Y$,$Z$ axial scaling coefficients |
| G50.1 | | Cancel mirror image of coordinate system | |
| G51.1 | | Mirror image | G51.1 X is symmetrical with a line parallel to the $X$-axis<br>G51.1 Y is symmetrical with a line parallel to the $Y$ axis<br>G51.1Z takes($X$,$Y$)as its symmetric point |
| G53 | 00 | Set to machine tool coordinate mode | |
| G54-G59※ | 14 | Workpiece coordinate system | |
| G65 | 12 | Subprogram call | In G65 P L,P represents the subprogram number,and L indicates the number of calls |

Continued

| G code | Group | Function | Remarks |
|---|---|---|---|
| G68 | | Rotation of coordinate system | In G68 X Y R, X、Y represents the datum point, and R represents the rotation angle |
| G69 | | Cancel coordinate system rotation | |
| G70 | | Circumferential uniform spot drilling cycle | In I J L, I is the circular radius, J is the angle between the line from the starting point to the center of the circle and the $X$-axis, and L denotes the number of points uniformly distributed on a circle |
| G71 | | Circle uniform spot drilling cycle | In I J K L, I is the circular radius, J is the angle between the line from the starting point to the center of the circle and the $X$ axis, L is the number of points on the circle which is equally distributed, and K denotes the every aliquot angle |
| G72 | | Line uniform spot drilling cycle | In I J L, I represents the equidistant, J represents the angle between the straight line and the $X$ axis, and L represents point of the equidistant section |
| G80※ | 09 | Cancel a fixed drilling cycle | |
| G81※ | | Ordinary drilling cycle | In G81 X Y Z R F L, X、Y indicates $X$、$Y$ coordinates of the machining point, Z indicates drilling depth, R indicates reference plane position, F indicates cutting speed, and L indicates the number of repeated drillings |
| G82※ | | Drilling cycle(there is a stop at the bottom of the hole) | P indicates stop time in G82 X Y Z R F L P |
| G83※ | | Drilling cycle(clearance feed) | In G83 X Y Z R F L P Q I J K, Q indicates reduction in height per time, L indicates the first cutting depth, J indicates the reduction in quantity after each cutting, and K indicates the least cut |
| G84 | | Tapping cycle | G84 X Y Z R F L P |
| G85 | | Fine drilling cycle | G85 X Y Z R F L P |
| G86 | | Boring cycle | G86 X Y Z R F L P |
| G87 | | Reverse boring cycle | G87 X Y Z R F L P |
| G88 | | Reverse tapping cycle | G88 X Y Z R F L P |
| G90※ | 03 | Absolute value programming | |
| G91※ | | Relative value programming | |
| G92 | 00 | Coordinate system setting | |
| G94 | 05 | Feed per minute | |
| G95 | | Feed per turn | |
| G98 | 05 | Drilling cycle back to the initial point | |
| G99 | 10 | Ring back to R point | |

Note: The sign ※refers to commonly used instructions.

**Table 2-2-8 M instruction and meaning commonly used in FANUC 0i**

| M instruction | Function | Remarks |
|---|---|---|
| M00 | Program stop | Press the circular start button to restart |
| M01 | Optional stop | Whether the program stops depends on the jump switch on the machine operating panel |
| M02 | End of program | Do not return to the beginning of the program after the end of the program |
| M03※ | Clockwise rotation of spindle | When viewed from the tail end of the spindle to the front end of the spindle, it is clockwise |
| M04※ | Counterclockwise rotation of spindle | When viewed from the tail end of the spindle to the front end of the spindle, it is counterclockwise |
| M05※ | Spindle stop | |
| M06 | Tool switching | |
| M08 | Cutting fluid open | |
| M09 | Cutting fluid close | |
| M13 | Cutting fluid opening with clockwise spindle rotation | |
| M14 | Cutting fluid opening with counterclockwise spindle rotation | |
| M30※ | End of program | At the end of the program, automatically return to the location at the beginning of the program |
| M98 | Subprogram call | In M98 P L, P indicates program address, and L indicates number of calls |
| M99※ | Subprogram return | |

The auxiliary M instruction can be added to the program segment when needed, and the programmer only needs to remember its function and meaning, which makes it easy to use. The preparation function G instruction should be programmed according to the prescribed instruction format, which is relatively complicated. The definition and usage of the preparation function G instruction commonly used in NC milling machine will be introduced in the following.

(1) G00 quick point positioning instruction

The instruction format of the G00 quick point positioning is:

```
G00 X__Y__Z__;
```

Among them, X, Y, Z are the coordinate values which quickly locate to the end point. In G90 programming mode, the end point is the coordinates relative to the origin of the workpiece coordinate system. In G91 programming mode, the end point is the displacement relative to the starting point; G00 is the modal function instruction, which can be written off by G01, G02 or G03 function instructions.

Note: As the axis moves at its own speed, the composite path of the linkage linear axis is not necessarily a straight line when executing the G00 instruction.

As shown in Figure 2-2-6, if G00 programming is used, the tool is required to quickly locate from point $A$ to point $B$.

Absolute coordinate programming is G90 G00X90 Y45. 0;

Incremental coordinate programming is G91G00 X70Y30. 0;

To avoid a collision between the tool and the workpiece, it is common to move the Z axis to a safe altitude and then execute the G00 instruction, as shown in Figure 2-2-6.

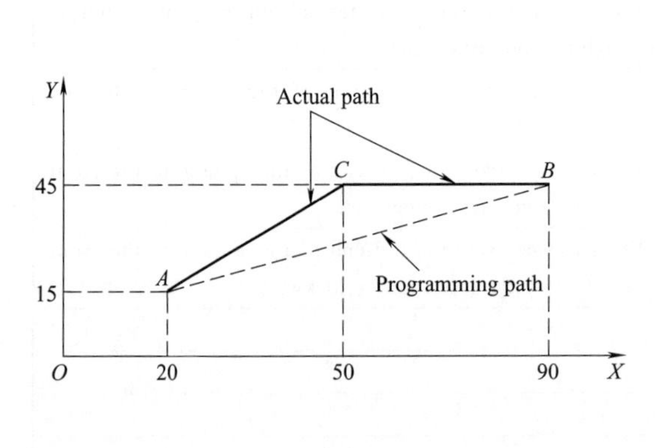


Figure 2-2-6　G00 programming

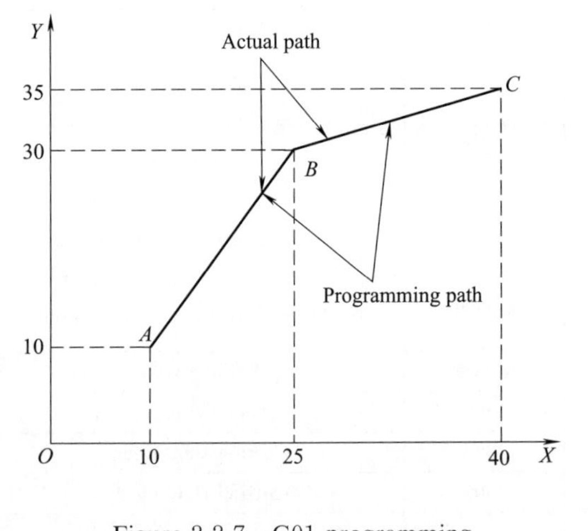


Figure 2-2-7　G01 programming

(2) G01 linear interpolation instruction

The instruction format of the G01 linear interpolation is as following:

```
G01 X__ Y__ Z__ F__;
```

Among them, X, Y and Z are the feeding end points of linear interpolation. In G90 programming mode, the end point is the coordinate relative to the origin of the workpiece coordinate system. In G91 programming mode, the end point is the displacement relative to the starting point; F is the synthetic feed speed, which is valid before there is a new F instruction. It is not necessary to write F instruction in each program segment.

G01 is a modal code instruction that can be canceled by G01, G02 or G03 functional instructions. G01 can instruct the tool to move from the current position to the end of the program section instruction in a linear way (the synthetic trajectory of the linkage linear axis is a straight line) according to the synthetic feed speed prescribed by F.

As shown in Figure 2-2-7, if G01 programming is used, the tool is required to feed linearly from point *A* through point *B* to point *C* (at this time, the feed route is a broken line from point *A* to point *B* to point *C*).

The absolute coordinates programming is shown as follow

```
G90 G01 X25. 0 Y30. 0 F100;
X40. 0 Y35. 0;
```

Incremental coordinates programming is shown as follow

```
G91 G01 X15. 0 Y20. 0 F100;
X15. 0 Y5. 0;
```

(3) G02/G03 circular interpolation instruction

G02/G03 circular interpolation instruction format:

$$\begin{Bmatrix} G17 \\ G18 \\ G19 \end{Bmatrix} \begin{Bmatrix} G02 \\ G03 \end{Bmatrix} \begin{Bmatrix} X\_\ Y\_ \\ X\_\ Z\_ \\ Y\_\ Z\_ \end{Bmatrix} \begin{Bmatrix} I\_\ J\_ \\ I\_\ K\_ \\ J\_\ K\_ \\ R\_ \end{Bmatrix} F\_$$

Among them, G02 is a clockwise arc interpolation; G03 is a counter-clockwise arc interpolation; X, Y, Z is the arc endpoint coordinates; I, J and K respectively represent the coordinates by deducting the coordinates of the origin of the arc from the coordinates of center of the axis of $X(U)$, $Y(V)$, $Z(W)$, as shown in Figure 2-2-8. The position of the circle center can also be expressed by the circular arc radius R (when the center angle of the circular arc is $\leqslant 180°$, R is positive; when the center angle of the circular arc is $>180°$, R is negative; and when the center angle of the circular arc is $= 360°$, R cannot be programmed, only I, J, K can be programmed).

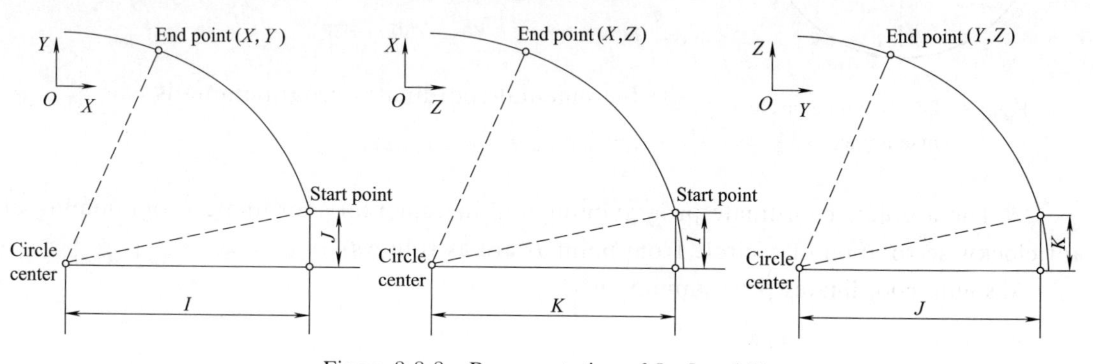


Figure 2-2-8 Representation of I, J and K

**【Example 2-2-1】** As shown in Figure 2-2-9, the minor arc *AB* and the major arc *BCA* complete their absolute and incremental programming respectively.

① Absolute coordinate programming and incremental coordinate programming for minor arc *AB* are as follows:

Absolute coordinates programming is shown as following

```
G90 G02 X0 Y30. 0 R30. 0 F80;
or G90 G02 X0 Y30. 0 I30. 0 F80;
```

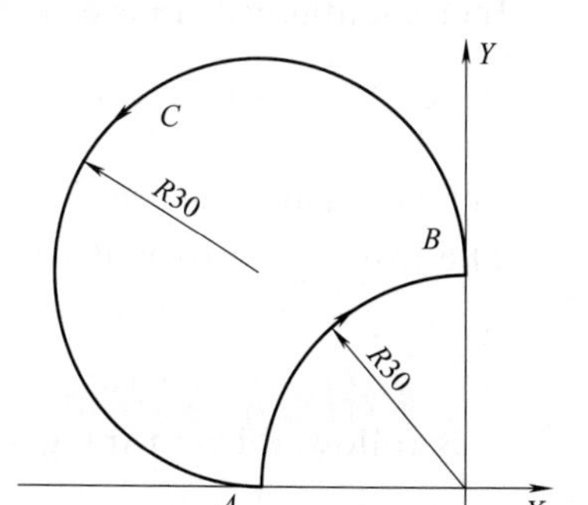


Figure 2-2-9 Minor arc *AB* and Major arc *BCA*

Incremental coordinates programming is shown as following

```
G91 G02 X30. 0 Y30. 0 R30. 0 F80;
or G91 G02 X30. 0 Y30. 0 I30. 0 F80;
```

② Absolute coordinate programming and incremental coordinate programming for major arc *BCA* are as follows:

Absolute coordinates programming is shown as following

```
G90 G03 X-30. 0 Y0 R-30. 0 F80;
or G90 G03 X-30. 0 Y0 J-30. 0 F80;
```

Incremental coordinates programming is shown as following

```
G91 G03 X-30. 0 Y-30. 0 R-30. 0 F80;
or G91 G03 X-30. 0 Y-30 J-30. 0 F80;
```

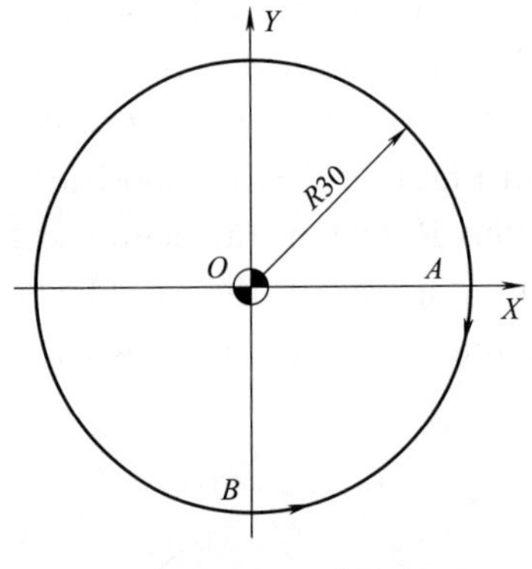

Figure 2-2-10 Whole circle machining

【**Example 2-2-2**】 The whole circle machining is shown in Figure 2-2-10. Complete the following absolute coordinate programming and incremental coordinate programming respectively.

① The absolute coordinate programming and incremental coordinate programming of clockwise rotation of a circle from point A are as follows:

Absolute coordinates programming is

```
G90 G02 X30. 0 Y0 I- 30. 0 F80;
```

Incremental coordinates programming is

```
G91 G02 X0 Y0 I- 30. 0 F80;
```

② The absolute coordinate programming and incremental coordinate programming of anticlockwise rotation of a circle from point *B* are as follows:

Absolute coordinates programming is

```
G90 G03 X0 Y- 30. 0 J30 F80;
```

Incremental coordinates programming is

```
G91 G03 X0 Y0 J30. 0 F80;
```

(4) G04 pause instruction

The instruction format for the G04 pause instruction is

```
G04 P__;
```

It is followed by an integer value, with the unit of ms.

```
or G04 X__;
```

The unit of number with decimal points behind it is s.

This command enables the tool to perform finishing cut in a short time without feeding, and is used for drilling blind via, turning slots, etc.

G04, a non-modal instruction, is only valid in the specified segment of the program.

## Ⅴ. Simplify Programming Instruction

(1) Call subprogram

M98 is used to call subprogram, while M99 represents the end of subprogram, which make control return to the main program.

The format of the subprogram is as follows:

```
OXXXX; subprogram number
...
...subprogram
...
M99; end of subprogram,return to main program
```

At the beginning of the subprogram, the subprogram number must be specified as the entry address of the call. M99 is used at the end of the subprogram to control the return of the main program after executing the subprogram.

The format for calling the subprogram is as follows:

```
M98 P__ L__
```

P is the subprogram number called; L is the number of repeated calls.

**【Example 2-2-3】** as shown in Figure 2-2-11, $Z$ starts at a height of 100mm and cuts at a depth of 20mm from the outside of the contour. Please try to compile the processing program.

Figure 2-2-11 Subprogram programming

Program

```
O6001;
N1;(Main Program)
G90 G54 G00 X0 Y0 S500 M03;
G00 Z100. 0;
M98 P100 L2;
G90 X120. 0;
M98 P100 L2;
G90 G00 X0 Y0 M05;
M30;
O0100;(Subprogram)
G91 G00 Z-95. 0;
G41 X20. 0 Y10. 0 D01;
G01 Z-25. 0 F50;
Y70. 0;
X20. 0;
Y-60. 0;
X-30. 0;
G00 G40 X-10. 0 Y-20. 0;
Z120. 0;
X40. 0;
M99;
```

(2) Mirror image function instructions (G51. 1, G50. 1)

```
Programming format:G51. 1 X__ Y__ Z__;
                   G50. 1 X__Y__Z__;
```

Among them, G51. 1 is the instruction to create the mirror image; and G50. 1 is the instruction to cancel the mirror image; X, Y and Z are mirror image positions (X0: $Y$ axial symmetry; Y0: $X$ axial symmetry; X0: origin symmetry).

G51. 1 and G50. 1 are modal instruction and can cancel each other.

**【Example 2-2-4】** Use mirror function to program (see Figure 2-2-12)

```
O6002;main program
N01 G92 X0 Y0 Z10;
N02 G91 G17 M03;
```

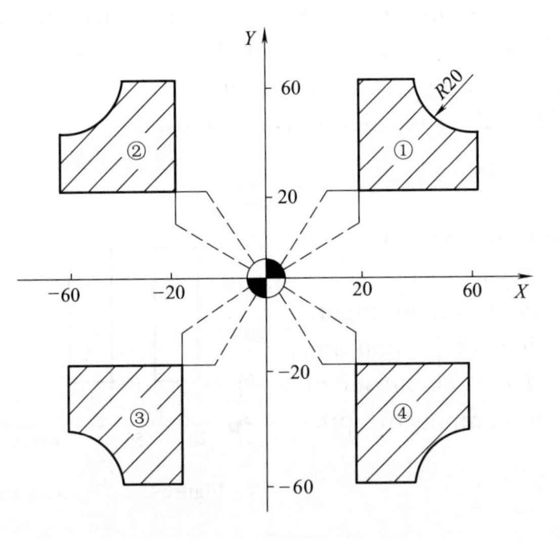


Figure 2-2-12 Mirror function

```
N03 M98 P100;            Processing ①
N04 G51. 1X0;            Mirror with Y axis
N05 M98 P100;            Processing ②
N06 G50. 1 X0;           Cancel Y -axis mirror
N07 G51. 1 X0 Y0;        With position point(0,0)
N08 M98 P100;            Processing ③
N09 G50. 1 X0 Y0;        Cancel point(0,0)mirror
N10 G51. 1 Y0;           Mirror with X -axis
N11 M98 P100;            Processing ④
N12 G50. 1 Y0;           Cancel X -axis mirror
N13 M05;
N14 M30;
% 100  subprogram
N01 G01 Z-5 F50;
N02 G00 G41 X20 Y10 D01;
N03 G01 Y60;
N04 X40;
N05 G03 X60 Y40 R20;
N06 Y20;
N07 X10;
N08 G00 X0 Y0
N09 Z10;
N10 M99;
```

(3) Scaling function instruction (G50, G51)

Programming format: G51X _ Y _ Z _ P _;

Among them, G51 is the command to set up zoom; G50 is the command to cancel zoom; X, Y and Z are the coordinate values of zoom center; P is the zoom multiple, which ranges from 0.001 to 999.999. If $0.001 < P < 1$ means reduction, $1 < P < 999.999$ means enlargement.

In the case of tool compensation, scaling should be done before tool radius compensation and tool length compensation.

G51 and G50 are modal instructions and can cancel each other.

(4) Rotation transformation instructions (G68, G69)

Programming format: G68X_ Y_ R_;

Among them, G68 is the instruction to establish rotation, G69 is the instruction to cancel rotation; X, Y, Z are the coordinate values of the rotation center; R is the rotation angle, with the unit of degree and the range from 0rom ran. Clockwise rotation is negative and counterclockwise rotation is positive.

In the case of tool compensation, tool compensation (tool radius compensation, length compensation) should be done after rotation, while in the case of scaling function, scaling should be done before rotation. G68 and G69 are modal instructions and can cancel mutually, G69 is a default value.

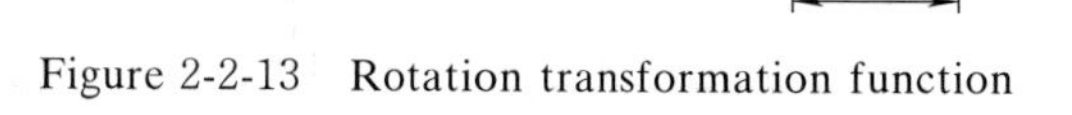


Figure 2-2-13 Rotation transformation function

**【Example 2-2-5】** The rotation transformation function program is shown in Figure 2-2-13.

```
O6003;              main program
N10 G90 G17 M03;
N20M98 P100;        Processing ①
N30 G68 X0 Y0 P45;  Rotation of 45°
N40 M98 P100;       Processing ②
N50 G69;            Cancel Rotation
N60 G68 X0 Y0 P90;  Rotation of 90°
N70 M98 P100;       Processing ③
N80 G69 M05 M30;    Cancel rotation
%100  subprogram(machining program for ①)
N100 G90 G01 X20 Y0 F100;
N110 G02 X30 Y0 I5;
N120 G03 X40 Y0 I5;
N130 X20 Y0 I 13;
N140 G00 X0 Y0;
N150 M99
```

**Exercises**

1. What is the process analysis of machining parts by CNC milling machine?
2. What is the process analysis of the rough parts?
3. How to determine the processing order of the tool path?
4. What is the difference between the instruction G00 and G01?
5. How to judge the clockwise circular interpolation instruction and counterclockwise circular interpolation instruction?
6. What are the instructions for simplifying programming?

7. What is the format for calling subprogram?

8. Please explain the meaning of the instructions in the rotation transformation instructions.

9. Semi-circle processing as shown in Figure 2-2-14. Please try to write a NC processing program.

10. Double-layer contour processing as shown in Figure 2-2-15. Please try to write the NC processing program.

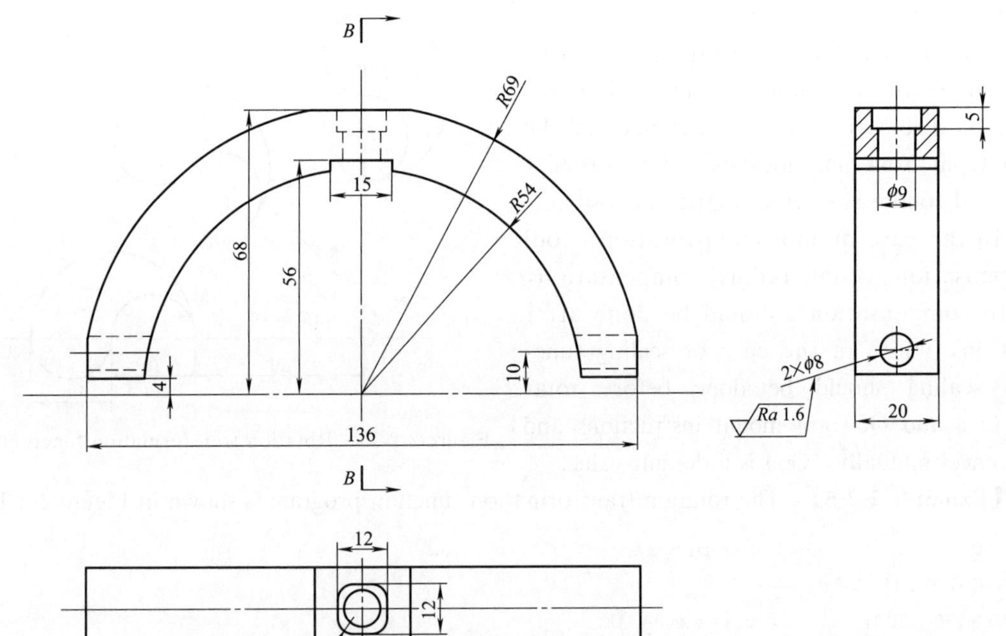


Figure 2-2-14 Semi-circle

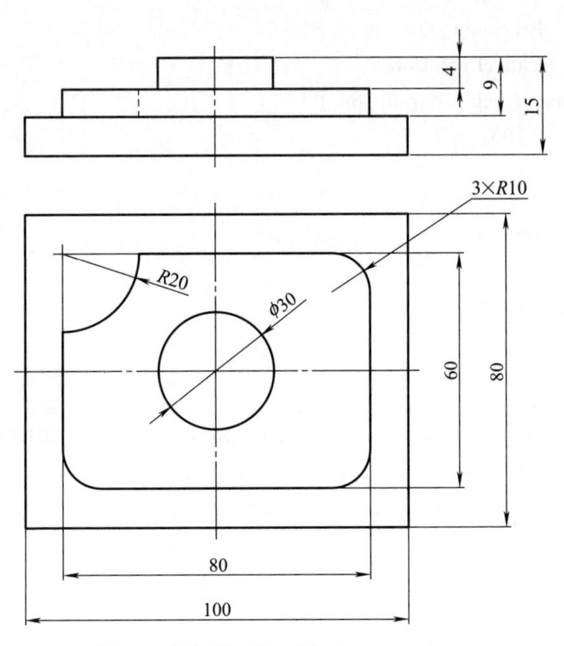


Figure 2-2-15 Double-layer contour

# Task Ⅲ Machining Contour and Hole Parts

## Ⅰ. Drawing and Technical Requirements

The contour and hole parts, as shown in Figure 2-3-1, are made of No. 45 steel, $\phi$60mm cylindrical rod, and is performed by normalizing treatment with hardness of HB200.

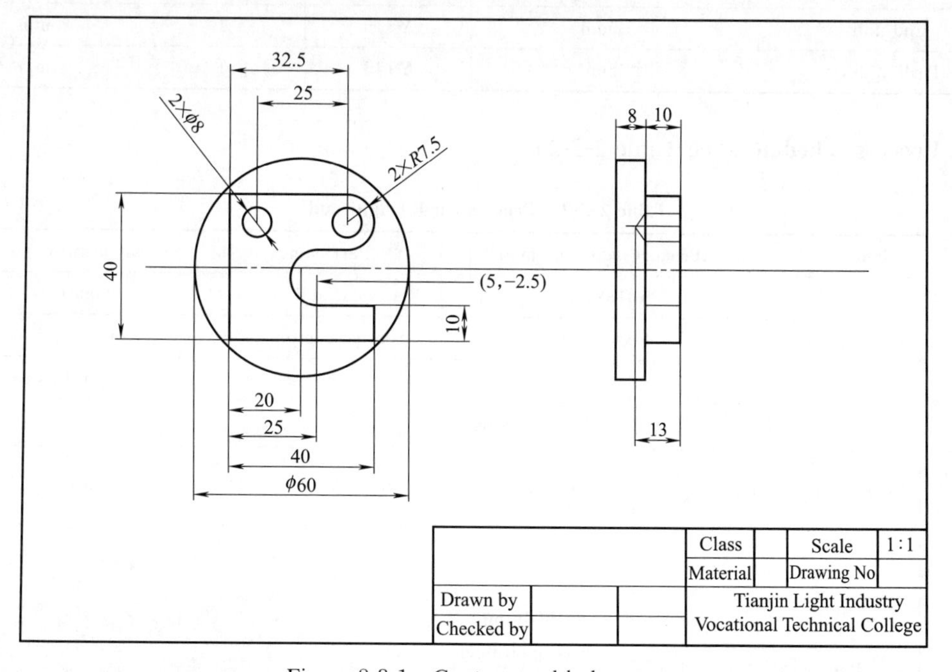


Figure 2-3-1 Contour and hole parts

## Ⅱ. Analysis of Drawings

Teaching strategy: group discussion, team reports, teacher summary

### 1. Parts diagram analysis

The outer contour of the part is a $\phi$60mm cylinder. It is composed of the outer contour and the holes from 2× $\phi$8, which belongs to the contour and the hole parts. Parts are designed according to the sequence of training and learning.

### 2. Process analysis

① Structure analysis: The parts are simple external contour and hole type parts.

② Accuracy analysis: The key dimension of the part is $2\times R7.5$ in two circular arcs, including one point coordinate of (5, − 2.5). When calculating the coordinates of the points, attention should be paid to that the outer contour depth is 10mm, and the hole depth is 13mm.

③ Positioning and clamping analysis: This part adopts bench vice for clamping. The clamping force of the workpiece should be moderate, which should is not only to prevents the deformation and scratching of the workpiece, but also to prevent the loosening of the workpiece in the process. When clamping, it is necessary to pad iron under the part, and

to hit and pad it tightly with wooden hammer.

④ Machining process analysis: After the analysis above, the overall arrangement sequence of the parts in this subject is to process the outer contour first and then punch.

### 3. Selection of main cutting tool (see Table 2-3-1)

**Table 2-3-1　Tool card**

| Tool name | Tool specification | Material | Quantity | Tool radius |
|---|---|---|---|---|
| End mill | $\phi$12mm | YT15 | 1 | 6mm |
| Drilling bit | $\phi$8mm | YT15 | 1 | 4mm |

### 4. Process schedule (see Table 2-3-2)

**Table 2-3-2　Process card (right end)**

| Unit | | Product name and model | | Part name | Part drawing number |
|---|---|---|---|---|---|
| | | Take Ⅲ | | Concur and hole | Figure 2-3-1 |
| Process | Program No. | Fixture name | | Service equipment | Workpiece material |
| 001 | O1301 | Bench vice | | NC milling machine | No. 45 steel |
| Part 1 | | | | | |
| Working step | Working step content | Tool No. | Cutting parameters | Remark | Working procedure sketch |
| 1 | Plunge cut | T01 | $n = 500$r/min (Feed multiplier switch is at the position ×10) | Automatic processing | |
| 2 | Cutting along the outer contour | T01 | $n = 500$r/min $f = 50$mm/min $a_p = 5.0$mm | Automatic processing | |
| 3 | Machining outer contour size and circular arc size | T01 | $n = 500$r/min $f = 50$mm/min $a_p = 5.0$mm | Automatic processing | |

Continued

| Part 1 | | | | | |
|---|---|---|---|---|---|
| Working step | Working step content | Tool No. | Cutting parameters | Remark | Working procedure sketch |
| 4 | Machining in the upper left hole | T02 | $n = 500r/min$<br>$f = 50mm/min$<br>$a_p = 13.0mm$ | Automatic processing | |
| 5 | Machining in the upper right hole | T02 | $n = 500r/min$<br>$f = 50mm/min$<br>$a_p = 13.0mm$ | Automatic processing | |
| 6 | Complete the whole part processing | T02 | $n = 500r/min$<br>$f = 50mm/min$<br>$a_p = 13.0mm$ | Automatic processing | |

## 5. Procedure

```
O1301;
G91 G28 Z0;
T1 M6;
G54 G90 G00 X0 Y-40 M03 S500;
G43 Z100. 0 H1;
Z10. 0;
G01 Z-5. 0 F50;
M98 P100 D1;
M98 P100 D2;
G01 Z-10. 0 F50;
G00 Z100. 0;
M05;
G91 G28 Z0;
T2 M6;
G90 G54 G00 X0 Y0 M03 S500;
G43 Z100. 0 H2;
G01 X-12. 5 Y12. 5;
G98 G83 Z-13 R4 Q3 F50;
```

```
X12. 5 Y12. 5;
G80 G00 Z100;
M05;
M30;
O100;
G41 G01 X20 Y-40 F120;
G03 X0 Y-20 R20;
G01 X-20 Y-20;
Y20;
X12. 5;
G02 X12. 5 Y5 R7. 5;
G01 X5 Y5;
G03 X5 Y-10 R7. 5;
G01 X20 Y-10;
Y-20;
X0;
G03 X-20 Y-40 R20;
G40 G01 X0 Y40 F120;
M99;
```

## Ⅲ. Technical Problems Related to NC milling

### 1. Concepts of program start point, return point and cut-in point (feed point), cut-out point (exit point)

① The program start point is the initial stop point of the tool tip (set the tool nose as the tool position point) at the beginning of the program.

② The return point of the program. The return point of the program refers to the stop point of the tip of the tool after it completes the program. It is generally the point to chang tools.

③ Cut-in point (feed point) is the contact point (also known as the cutting contact point) between the cutter and the curved surface at the initial cutting position of the curved surface.

④ Cut-out point (exit point) refers to the contact point between the cutter and the curved surface after cutting the curved surface.

### 2. Principles for determining the start point, return point, cut-in point and cut-out point of the program

① The principle of determining the start point and the return point. The start point and the return point are better to be the same in the same program. If the processing of a part needs several programs to complete, then the start points and the return points of these programs should be the same as far as possible, so as to avoid complex operation in processing. The coordinate values of the starting point and the return point should be set to zero for both $X$ and $Y$, which can make the operation convenient. The $Z$ coordinates of the start point and the return point shall be defined in a position where the tip of the tool is about 50-100mm above the highest point of the processed part, that is, it is the place where the start plane and the withdrawal plane are located, which is mainly due to the se-

curity of NC machining, the prevention of tool collision, and also considers the efficiency of NC machining, so that the non-cutting time is controlled in a certain range.

② The principle of selecting cut-in point. In the process of feeding or cutting curved surface, the cutter should be kept from damage. Generally speaking, for rough machining, the top corner point in the curved surface should be chosen as the cut-in point (initial cutting point) of the surface, because its cutting allowance (especially for the square profile and semi-finished forging with uneven allowance) is small and it is not easy to damage the tool when feeding. In terms of finish machining, a corner point with relatively flat curvature in the surface should be chosen as the cut-in point of the surface, because at that point, the allowance is uniform, and the bending moment of the cutter is small, which is not easy to break the cutter. In a word, the use of milling cutters as drills should be avoided, otherwise the cutter will be damaged because of the great force and the inconvenience of chip removal.

③ The principle of selecting cut-out point. When choosing the cutting point, the main consideration is that the surface can be processed continuously and completely, and the non-cutting time between the surfaces is as short as possible, so as to facilitate tool change and improve the effective working time of the machine tool. If the machined surface is an open surface, two corners of the surface can be used as cut-out points, and one of them is chosen as cut-out points according to the above principles. If the machined surface is a closed surface, only one corner of the surface is the cut-out point, which is automatically determined by the system in automatic programming.

**3. Concepts of start plane, return plane, feed plane, exit plane and safety plane**

① The start plane is the *Z* plane where the initial position of the tool tip is located at the beginning of the program, which is defined at a position of 50-100mm above the highest point of the machined part, generally higher than the safety plane. The corresponding height is called the start height. On this plane, the cutter travels at G00 speed.

② Return plane refers to the *Z* plane where the tool tip is located at the end of the program. It is also defined at a position 50-100mm above the highest point of the machined surface, which generally coincides with the start plane. Therefore, when the tool is on the return plane, it is safe. Its corresponding height is called return height. The cutter also travels at G00 speed on this plane.

③ Feed plane. At the beginning, the tool is cut at a high speed (G00). When the material is being cut, the tool is cut at the feed speed so as to avoid collision with the tool. The turning point of the speed is the feed plane, and its height is the feed height, which is also called the approach height. That is to say, the turning speed is called the feed speed or the approach speed. This height is generally between the working plane and the safe plane, 5-10mm from the working surface (the distance between the tool nose and the working surface), when the surface to be machined is the rough surface, the large value should be taken, while the small value should be taken when the surface to be machined is the machined surface. If this setting is not available, the NC milling machine will have a tool collision.

④ Exit plane. After machining (parts area), the cutter leaves the surface of the workpiece at a cutting feed speed to a certain distance (5-10mm) and then returns to the safety plane at a high speed (G00). The turning position is the exit plane and the height is the exit height.

⑤ Safety plane refers to the *Z* plane where the tool nose is located when the tool moves a distance along the tool axis after a surface has been cut. It is generally defined at a

position 10-50mm higher than the highest point of the machined part. When the tool is in the safety plane, it is safe and travels at G00 speed on this plane. The setting of the safety plane can not only prevent the tool from hurting the workpiece, but also control the non-cutting machining time in a certain range. The corresponding height is called the safe height. After the tool is processed in one position, it returns to a safe height and then moves to the next position along the safe height before the tool is machined on another surface.

**4. Determination of the direction of cutting and the way of cutting (feed)**

The direction of cutting refers to the motion direction of the tool during machining. The cutting method refers to the distribution of the tool motion path when the tool motion path is generated. These two concepts are very considerable in NC milling process analysis, whether the choice is reasonable or not will directly affect the machining precision and production cost of parts. The selection principle is as follows: according to the geometry shape of the surface of the parts to be machined, the machining time can be as short as possible on the premise of ensuring the machining accuracy. The selection of cutting mode and cutting direction in contour machining of 2D wire frame and area machining of 3D curved surface are discussed as follows respectively.

(1) Selection of cutting direction in two-dimensional wire frame contour machining

In the rough machining process of part contour, the reverse milling method should be adopted to reduce the vibration of machine tool on the basis of considering the large machining allowance of part surface. While in finish machining process of part contour, considering that the purpose of finishing is to ensure the processing accuracy and surface roughness of parts, the climb milling method should be adopted. In addition, attention should also be paid to prevent the tool from directly cutting into the workpiece surface, leaving a tool-holding trace, affecting the roughness of the machined surface, and cutting-in and cutting-out along the tangent direction of the part contour should be adopted.

(2) The selection of cutting direction and cutting way in the machining of three-dimensional curved surface area

There are three types of tool path generation techniques for 3D curved surface machining:

① Reciprocating type. Which is characterized by the alternation of climb milling and reverse milling during the cutting process. The surface quality is poor but the machining efficiency is high, as shown in Figure 2-3-2.

② One-way type. In the process of cutting, it can ensure the consistency of climb milling or reverse milling. Programmers can choose a tool-moving mode of climb milling or reverse milling according to the actual processing requirements. Due to the fact that the tool path adds a non-cutting motion track after completing a cutting path, the machining time of the machine tool is prolonged. The two methods shown in Figure 2-3-3 are also known as direction parallel cutting methods.

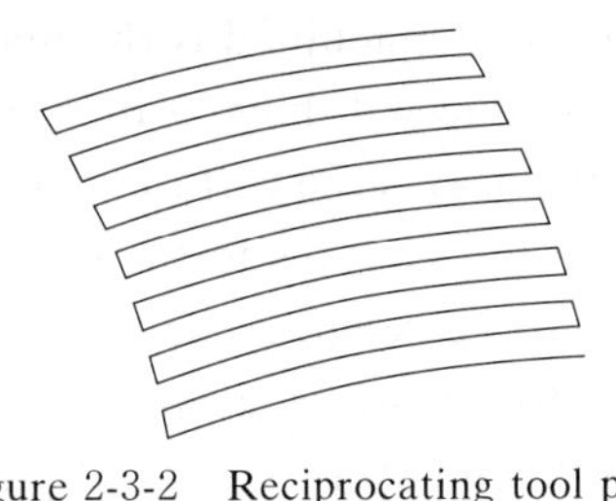

Figure 2-3-2 Reciprocating tool path

Figure 2-3-3 One-way tool path

③ The loop cutting path whose tool path is a set of equal-parameter closed curves of the machined surface, as shown in Figure 2-3-4. The loop cutting path is divided into equidistance loop cutting, contour loop cutting and helix loop cutting, which can be cut from the outside to the inside or from the inside to the outside. The loop cutting can not only ensure the consistency of climb milling or reverse milling, but also does not generate cutting motion track, with high machining efficiency and even machining path. Therefore, it is the main method to produce the moving path of closed ring curved tool.

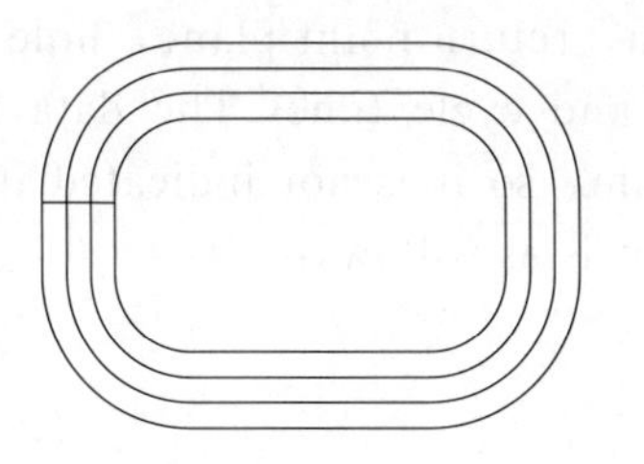

Figure 2-3-4 Loop Cutting Path

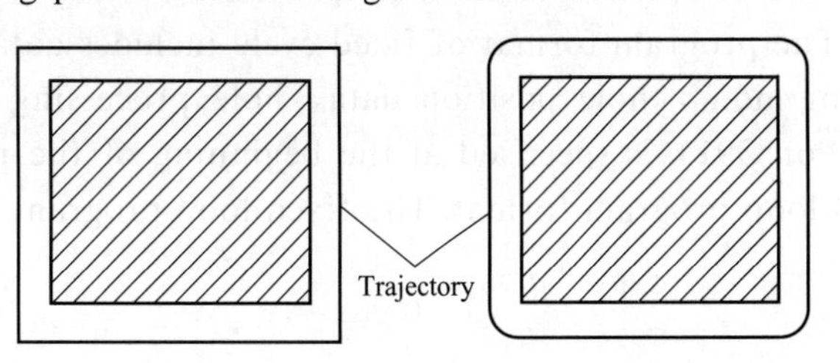


Figure 2-3-5 Transition Mode for Corner

Outer-contour corner transition refers to the transition mode when the outer-contour corner is met during the cutting process, which are generally the transition modes of the sharp corner and the circular arc, as shown in Figure 2-3-5.

The closed angle transition means that the tool transits in a straight line from one side of the contour to the other. The cutter of the way is prone to "over-range" phenomenon and the corner of the workpiece is sharp.

The circular arc transition means the transition of the tool from one side of the contour to the other side in the form of an arc. The cutter of the way is not easy to produce "over-range" phenomenon, but the corner of the workpiece is more smooth.

## Ⅳ. Hole Processing Instructions

In NC machining, some machining operation cycle has been typified. For example, the action of drilling and boring refer to the location of the hole plane, the rapid introduction, the work feed and the quick return, etc. The series of typical processing actions have been pre-programmed and stored in memory, which can be called by a program segment containing G code, thus simplifying the programming work. The G code, which contains a typical action loop, is called a loop instruction.

The works which can be finished by commonly used fixed loop instructions are: drilling, tapping thread and boring. These cycles usually include the following six basic operations (see Figure 2-3-6).

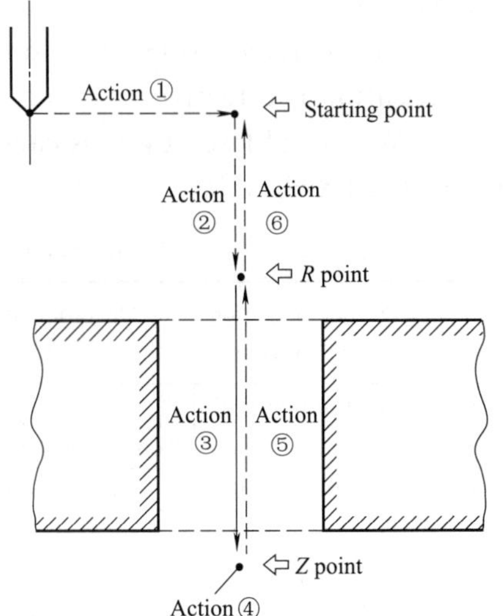


Figure 2-3-6 6 Actions for hole machining

Action 1——$X$ axis and $Y$ axis positioning: make the cutter quickly locate to the hole machining position.

Action 2——Fast feed to the $R$ point: tools can be quickly fed to R point from the starting point.

Action 3——Hole machining: the act of per-

forming hole machining by means of cutting feed ( $Z$ point).

Action 4——Hole bottom action: including pause, spindle accurate stop, tool movement and so on.

Action 5——Return to point $R$: move the tool safely while continuing to process other holes.

Action 6——Return to the starting point: it should normally return the starting point after the completion of the hole processing.

The program format of fixed cycle includes data form, return point plane, hole processing mode, hole position data, hole processing data and cycle times. The data form (G90 or G91) is specified at the beginning of the program, so it is not indicated in the fixed loop program format. The fixed loop program format is as follows:

$$\begin{Bmatrix} G90 \\ G91 \end{Bmatrix} \begin{Bmatrix} G98 \\ G99 \end{Bmatrix} G\square\square\ X\_\ Y\_\ Z\_\ R\_\ Q\_\ P\_\ F\_\ L\_;$$

Where:

G——Hole processing fixed cycle (G73-G89), see Table 2-3-3;

X, Y——Coordinate position of hole in $XY$ plane (absolute value or incremental value);

Z——$Z$ coordinate value of hole bottom (absolute or incremental value);

R——$Z$ coordinate value of point $R$ (absolute or incremental value);

Q——feed depth per time (G73, G83), or tool displacement (G76, G87);

P——pause time, ms;

F——Feed of cutting feed, mm/min;

L——Repeating times of the fixed cycle, L is not specified when loop is only performed once.

Note:

① G73-G89 are modal instructions. G01～G03 are instructions to cancel the loop.

② The parameters (Z, R, Q, P, F) in the fixed cycle are modal.

③ Make the spindle start before using the fixed loop instruction.

④ The fixed loop instruction cannot appear in the same segment with the post-instruction M code.

⑤ In the fixed cycle, the tool radius compensation is invalid while the tool length compensation is effective.

⑥ When the fixed cycle is canceled with G80, those interpolation modes are restored before the fixed cycle.

**Table 2-3-3 Common hole processing instructions**

| Code | Drilling operation (－$Z$ Direction) | Operation at the bottom of the hole | Exiting operation (＋$Z$ Direction) | Purpose |
|---|---|---|---|---|
| G73 | Intermittent feed | — | Rapid feed | High speed deep hole drilling cycle |
| G74 | Cutting feed | Pause →spindle forward rotation | Cutting feed | Countertapping |
| G76 | Cutting feed | Oriented stop of spindle | Rapid feed | Fine boring |
| G80 | — | — | — | Cancel fixed cycle |
| G81 | Cutting feed | — | Rapid feed | Drilling and counterboring |
| G82 | Cutting feed | Pause | Rapid feed | Drilling, stepped boring hole |
| G83 | Intermittent feed | — | Rapid feed | Deep hole drilling cycle |

Continued

| Code | Drilling operation ( - Z Direction) | Operation at the bottom of the hole | Exiting operation ( + Z Direction) | Purpose |
|---|---|---|---|---|
| G84 | Cutting feed | Pause →spindle reverse rotation | Cutting feed | Tapping |
| G85 | Cutting feed | — | Cutting feed | Boring |
| G86 | Cutting feed | Spindle stop | Rapid feed | Boring |
| G87 | Cutting feed | Spindle forward rotation | Rapid feed | Backcutting |
| G88 | Cutting feed | Pause →spindle stop | Manual feed | Boring |
| G89 | Cutting feed | Pause | Cutting feed | Boring |

(1) G81: Drilling cycle (fixed point drilling)(see Figure 2-3-7)

G98(G99)G81 X__Y__Z__R__F__K__

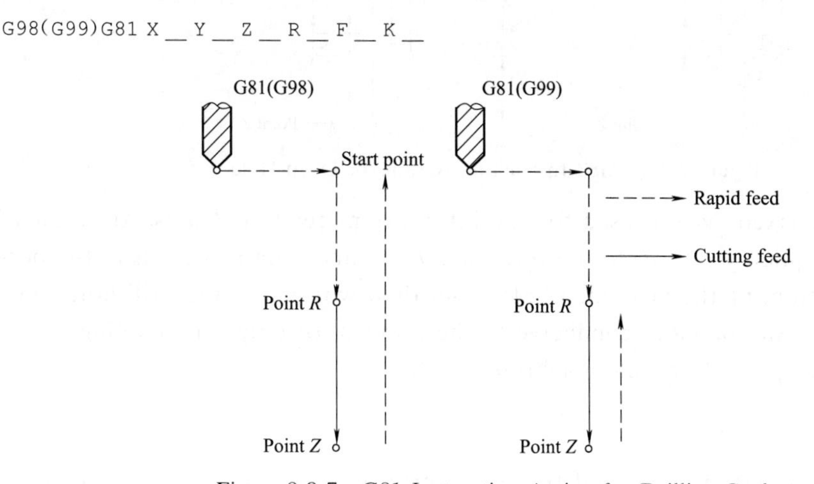


Figure 2-3-7 G81 Instruction Action for Drilling Cycle

(2) G82: drilling cycle with stops (see Figure 2-3-8)

G98(G99)G82 X__Y__Z__R__P__F__K__

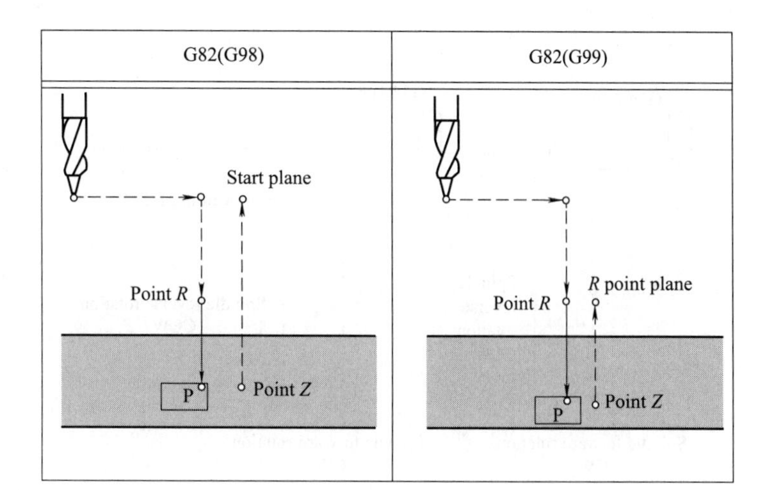


Figure 2-3-8 G82 Instruction action for drilling cycle

Function: This instruction is mainly used for machining counterbore and blind holes to improve the precision of hole depth. This instruction has the same action as G81 excluding pause at the bottom of the hole.

(3) G83: Deep hole machining cycle (see Figure 2-3-9)

```
G98(G99)G83 X__Y__Z__R__Q__F__K__
```

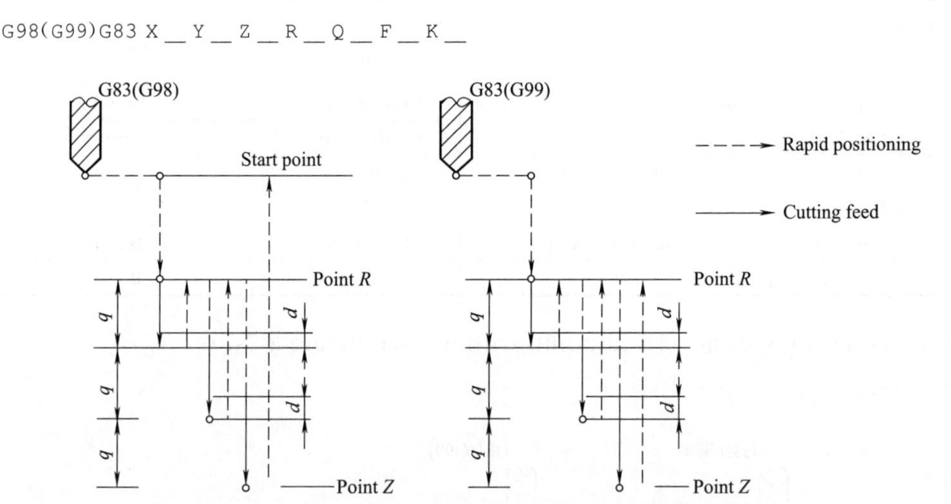


Figure 2-3-9 G83 Instruction Action for Deep Hole

Function: The fixed cycle is used for the intermittent feed of $Z$-axis. After each hole is drilled down, it quickly returns to the reference $R$ point, then fast feeds to the position of $K$ above the bottom of the machined hole, and then works into the drill hole, making the deep hole processing be more conducive to the removal of chips and cooling.

(4) G73: high-speed deep hole machining cycle

```
G98(G99)G73 X__Y__Z__R__Q__F__K__
```

Function: The fixed cycle is used for the intermittent feed of $Z$-axis, which makes the deep hole machining easy to remove chips, reduce the amount of backoff, and can be processed efficiently.

(5) G74: left thread tapping cycle instruction (see Figure 2-3-10)

```
G74  X__Y__  Z__  R__  F__  ;
```

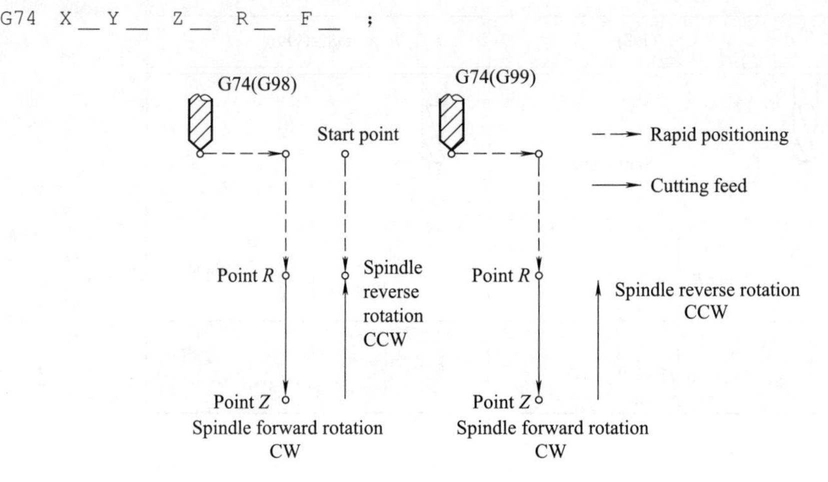


Figure 2-3-10 G74 instruction action for left thread tapping cycle

Where: F is feed speed of tapping thread (mm/min), F(mm/min) = thread lead $P$(mm) × the spindle speed $n$(r/min).

The G74 instruction is used for cutting left thread holes. The spindle cuts in reverse

and exits in forward direction, just opposite to the spindle rotation in G84 instruction. Other movements are the same as those in the G84 instruction.

(6) G76: fine boring cycle (see Figure 2-3-11)

```
G98(G99)G76 X_ Y_ Z_ Q_ F_ K_;
```

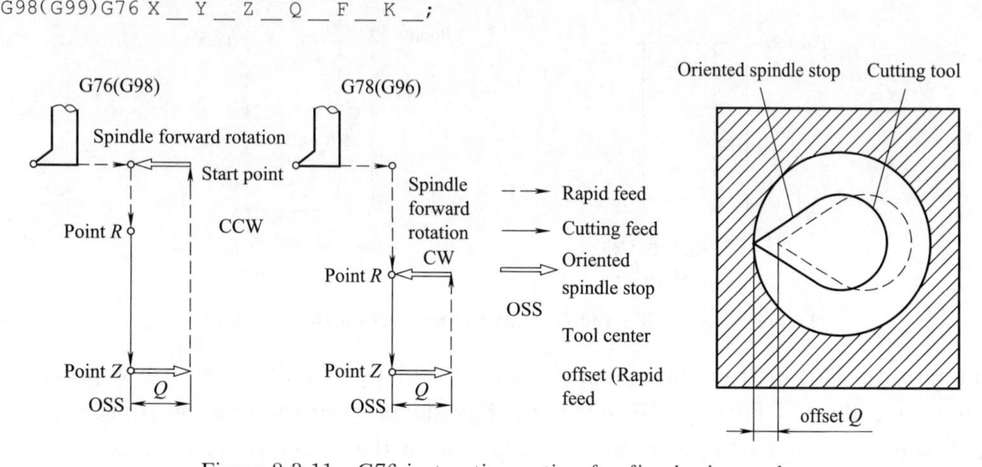


Figure 2-3-11 G76 instruction action for fine boring cycle

G76 instruction is used for fine boring machining. When boring to the bottom of the hole, the spindle stops at the directional position, namely oriented stop, then causes the tool nose to deviate from the surface, and then exit the tool. In this way, the hole machining can be done with high precision and high efficiency without damaging the machined surface of the workpiece.

In the program format, Q represents the offset of the tool nose, which is generally positive, and the direction of movement is set by machine parameters.

(7) G84: Right thread tapping cycle instruction (see Figure 2-3-12)

```
G98(G99)G84 X_ Y_ Z_ R_ P_ F_ K_
```

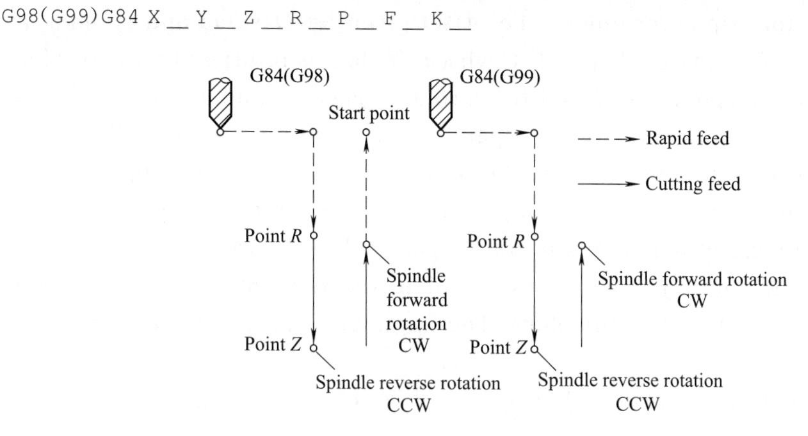


Figure 2-3-12 right-hand thread cycle G84 command action

The G84 instruction is used to cut the right thread hole. When cutting down, the spindle is turning forward, the movement of the bottom of the hole is changing from forward to reverse, and then exits. F indicates lead, the speed correction is invalid during G84 cutting thread, the movement will not stop halfway until the cycle is over.

(8) G85: Rough boring cycle [see Figure 2-3-13(a)]

```
G98(G99)G85 X_ Y_ Z_ R_ F_ K_
```

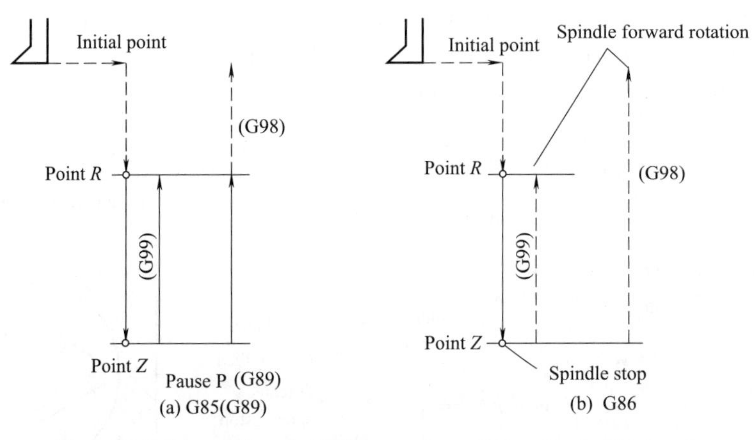


Figure 2-3-13　G85 and G86 instructions action for boring cycle

Function: This instruction is mainly used for boring machining with low precision. Its actions is as follows : boring at F speed cutting, delaying at the hole bottom, with drawing at F speed boring exiting, rotating the spindle in the whole process.

(9) G86: Boring cycle [see Figure 2-3-13(b)]

```
G98(G99)G86 X__Y__Z__R__F__K__
```

The difference between G86 and G85 is that when processing to the bottom of the hole, the spindle is stopped and the tool is quickly withdrawn.

(10) G87 reverse boring cycle instruction (see Figure 2-3-14)

```
G87 X__Y__Z__R__Q__F__K__;
```

After the $X$ and $Y$ axes are located, the spindle stops, the tool is offset in the direction opposite to the tool tip according to the offset given by $Q$ value and is quickly positioned at the bottom of the hole ($R$ point), where the tool is returned by the original offset ($Q$ value), then the spindle is turned forward and machined up along the $Z$ axis to $Z$ point. After the spindle is oriented to stop again at this position, the tool is moved back based on the original offset, then the spindle moves quickly along the upper part of the hole to the initial plane, and after returning to the original offset, the spindle rotates forward and continues to execute the next program segment. In this circulation mode, the tool can only be returned to the initial plane instead of to the $R$-point plane, because the $R$-point plane is lower than the $Z$ point plane. The parameters of this instruction are set in common with G76.

(11) G88: Boring cycle (manual boring)(see Figure 2-3-15)

```
G98(G99)G88 X__Y__Z__R__P__F__K__
```

When the tool reaches the bottom of the hole and delays, the spindle stops and the system enters the feed holding state. In this case, the manual operation can be performed, for the sake of safety, the cutter should be removed from the hole first. In order to start the machining again, the manual operation should be transferred to the tape mode or memory mode. Press the cycle start button, the tool can quickly return to the $R$ point (G99) or the initial point (G98), then the spindle begins to turn forward.

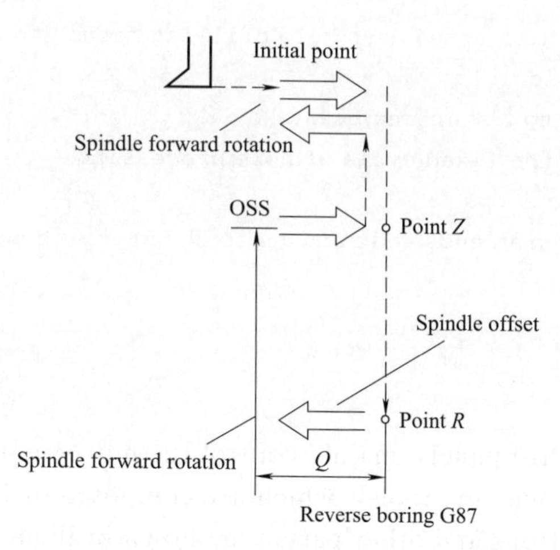

Figure 2-3-14 G87 instruction action for reverse boring cycle

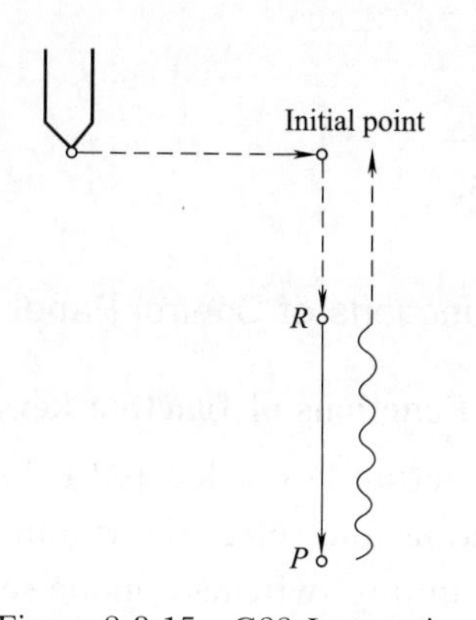

Figure 2-3-15 G88 Instruction Action for Boring Cycle

(12) G89: Counterboring cycle, stepped boring hole cycle

```
G98(G99)G89 X__Y__Z__R__P__F__K__
```

This instruction is the same as the G86 instruction, but has a pause at the bottom of the hole (hole bottom delays and spindle stops).

(13) G80

When G80 cancels the hole processing fixed cycle, the hole processing data in the fixed cycle instruction is also canceled. The interpolation modes before the fixed cycle will be restored.

For example, as shown in Figure 2-3-16, a hole with a depth of 50mm with $5\times\phi8$mm is to be machined. Obviously, this belongs to deep hole processing. G73 should be used for deep hole drilling and G80 instruction should be used. The program is as follows:

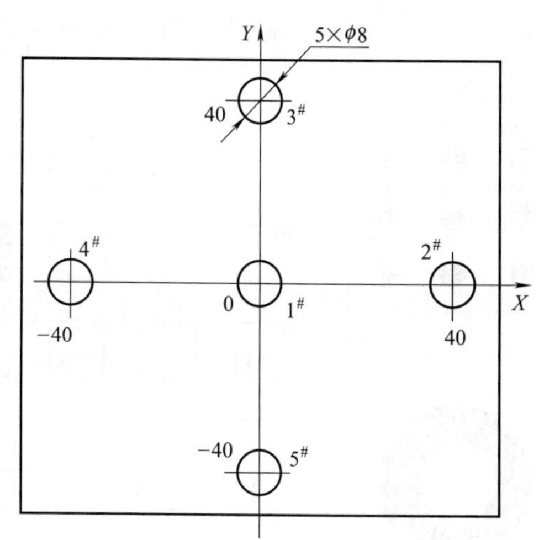

Figure 2-3-16 Application of G80 Instruction

```
O40
N10 G56 G90 G1 Z60 F2000            Select the processing coordinate system No. 2 to the Z
                                    starting point
N20 M03 S600                        Spindle start
N30 G98 G73 X0 Y0 Z-50 R30 Q5 F50   Select high speed deep hole drilling method to process
                                    No. 1 hole
N40 G73 X40 Y0 Z-50 R30 Q5 F50      Select high speed deep hole drilling method to process
                                    No. 2 hole
N50 G73 X0 Y40 Z-50 R30 Q5 F50      Select high speed deep hole drilling method to process
                                    No. 3 hole
N60 G73 X-40 Y0 Z-50 R30 Q5 F50     Select high speed deep hole drilling method to process
                                    No. 4 hole
```

| | |
|---|---|
| N70 G73 X0 Y-40 Z-50 R30 Q5 F50 | Select high speed deep hole drilling method to process No. 5 hole |
| N80 G01 Z60 F2000 | Return to Z starting point |
| N90 G80 | Cancel the fixed cycle of hole processing |
| N100 M05 | Spindle stops |
| N100 M30 | The program ends and returns to the starting point |

## V. Functions of Control Panel

### 1. Functions of function keys

Function keys, located below the control panel, mainly control the movement of machine tools and select the running state of machine tools, which are composed of NC program running switches, mode selection buttons and other parts, as shown in Figure 2-3-17 and Table 2-3-4.

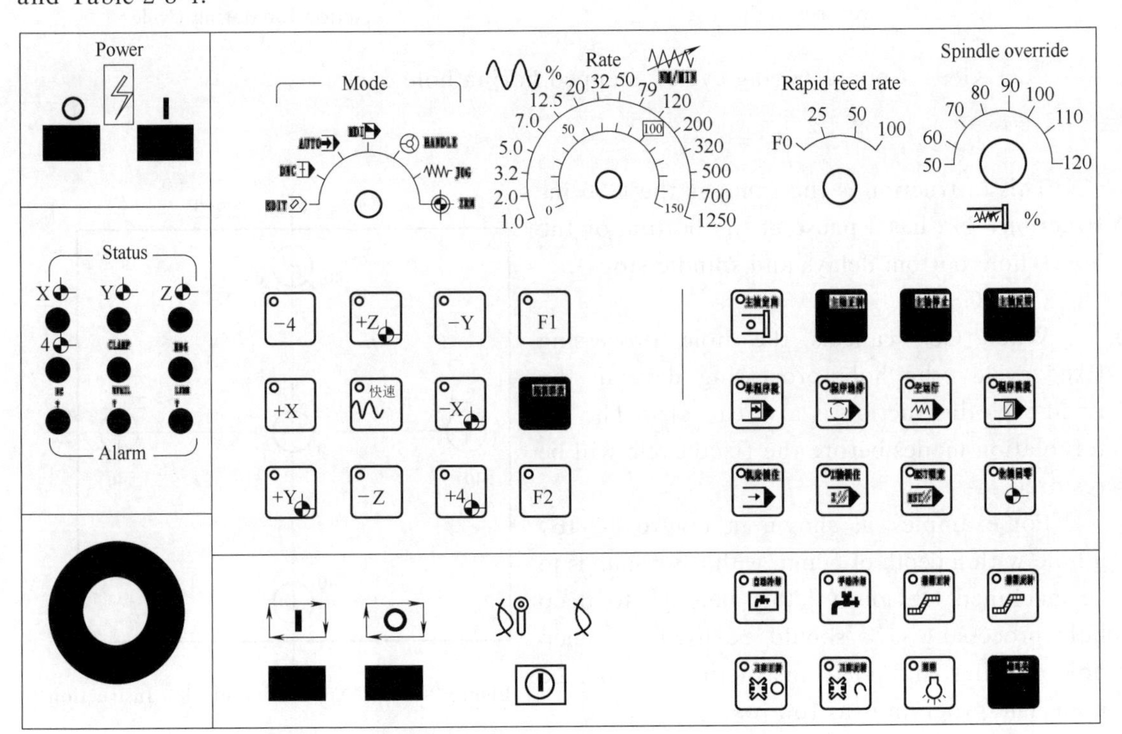


Figure 2-3-17 NC milling control panel

**Table 2-3-4 Functions of function keys**

| No. | Key Name | Icon | Function Description |
|---|---|---|---|
| 1 | Automatic process | | After pressing this key, operator can press the key of "Cycle Start" to complete the automatic process of the program |
| 2 | Editing mode | | When this key is pressed, the NC program can be input and edited |

Continued

| No. | Key Name | Icon | Function Description |
|---|---|---|---|
| 3 | MDI mode | | After pressing this key, operator can input data manually, operate the system panel and set the necessary parameters |
| 4 | DNC transmission | | After pressing this key, operator can control DNC remotely |
| 5 | Single-stage operation | | After pressing this key, the single-stage operation mode can be entered, and one line of NC instruction can be executed at a time |
| 6 | Function of skipping steps | | Pressing this key and adding a "/" sign before a single line to indicate the line is skipped, that is, the action of the line is not executed |
| 7 | M01 optional stop | | This key is M01 optional stop button. If this key is turned on, it means that the M01 instruction is running. If this key is turned off, the M01 instruction does not work |
| 8 | Homing | | When the program is in this position, if operator operates the corresponding keys such as [+X], [+Y], [+Z], he can make the machine tool return to the reference point |
| 9 | Manual feed | | Press this key to perform manually continuous feed or step-by-step feed operation |
| 10 | Hand-wheel | | After pressing this key, operator can move the machine precisely in three directions: $X$, $Y$ and $Z$ by operating the hand-wheel. This function is often used in tool setting |
| 11 | Feed holding | | The tool slows down and stops feeding when running automatically. Press the "Cycle Start" button again, and the machine will continue to feed |
| 12 | Cycle start | | Press this key, the automatic operation will be started and the program will be executed. In the automatic operation, the automatic operation indicator lamp lights up |

Continued

| No. | Key Name | Icon | Function Description |
|---|---|---|---|
| 13 | Display of homing | X 原点灯 Y 原点灯 Z 原点灯 | The lights will be on when the axles return to the home |
| 14 | Manual axle selection | X Y Z | When it is in manual mode, press the corresponding X/Y/Z key, then press the + / − key to manually control the movement of the machine tool along the corresponding coordinate axis |
| 15 | Multi-axis process | 4 5 6 | Reserve manual axle buttons for 4,5 and 6 axes |
| 16 | Rapid feed | + ∿ − | This key is pressed to match the positive and negative direction of the current axis, the tool will feed quickly |
| 17 | Spindle clockwise rotation/stop/anticlockwise rotation | | Press the corresponding key, and the spindle is in clockwise rotation/stop/anticlockwise rotation respectively |
| 18 | Emergency stop | EMERGENCY STOP | When pressing this button in case of emergency or critical situation, the machine tool is in an emergency stop state. After clearing the fault, it is necessary to rotate in the direction of the arrow on the button to reset the emergency button key. The reset of the emergency stop state also needs to press the "RESET" key on the MDI panel |
| 19 | Adjustment of feed rate | 0 1 2 4 6 8 10 15 20 30 40 50 60 70 80 90 95 100 105 110 120 (%) | The feed speed of NC program can be adjusted by choosing the ratio of feed speed of automatic and manual operations. The range of adjustment is 0-120% |
| 20 | Adjustment of spindle speed | 50 60 70 80 90 100 110 120 (%) | The feed speed of NC program can be adjusted by choosing the ratio of spindle speed of automatic and manual operations. The range of adjustment is 50%-120% |

Continued

| No. | Key Name | Icon | Function Description |
|---|---|---|---|
| 21 | Start |  启动 | Press this key to connect the power supply of CNC |
| 22 | Stop | 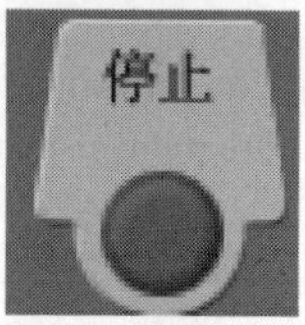 停止 | Press this key to cut off the power supply of CNC |
| 23 | Overrun unlocking |  超程/释放 | Press this key to unlock in case of overrun |
| 24 | Hand-wheel | 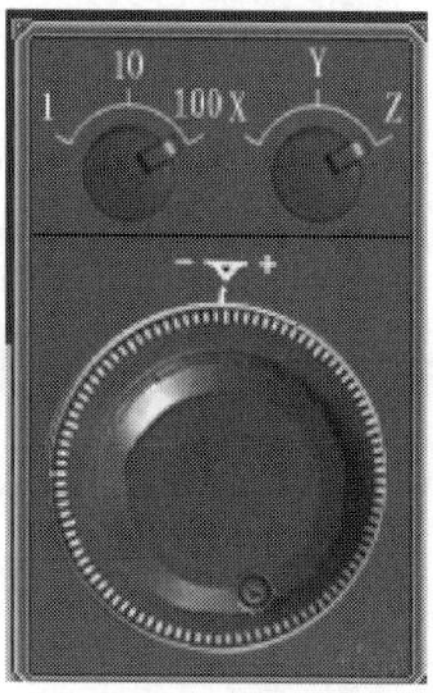 | The control key of the single step feed rate is used in conjunction with the hand-wheel to indicate the feed amount of each step on the working table, that is, × 1 is 0.001mm, × 10 is 0.01mm and × 100 is 0.1mm. The clockwise rotation of the hand-wheel indicates that the machine tool moves in the positive direction, and the counterclockwise rotation of the hand-wheel indicates that the machine tool moves in the negative direction. Each division on the hand-wheel dial scale represents a unit of moving rate. Each moving unit is coordinated by three gears of 0.1mm, 0.01mm and 0.001mm and three keys of speed changes |

**2. Functions of Input Area**

The address input key area can be seen in Figure 2-3-18.

Address input area is used to input data into the input field. The system can automatically distinguish whether a number or a letter is selected. For example, if the address key $G_R$ above is input directly, the letter G will be displayed on the CRT screen. If the SHIFT key is pressed first, then the $G_R$ key is pressed, the letter R will be displayed on the CRT screen. Similarly, numbers and letters, such as F and L, 9 and D, can be switched.

Figure 2-3-18 Address input key area

(1) The icons and functions of the address input key area ( shown in Table 2-3-5)

**Table 2-3-5 Icons and functions of address input key area**

| No. | Icon | Function Description |
|---|---|---|
| 1 | POS | The key displays the existing position of the machine tool and the position screen of coordinate. The position screen can be displayed by pressing the key, and there are three ways to display the position: relative coordinate, absolute coordinate and synthetic coordinate, among which the synthetic coordinate includes mechanical coordinate, relative coordinate, absolute coordinate and residual feed |
| 2 | PROG | The key must be used to edit and modify program by cooperating with EDIT mode. It is the key for program display and editing and can display the program in memory |
| 3 | OFFSET SETTING | This key is used for coordinate setting, displaying compensation values and macro program variables. Press the key for the first time to enter the setting page of coordinate system, press it for the second time to enter the tool compensation parameters page |
| 4 | SYSTEM | It is the key for setting system parameters. The system parameter screen and the self-diagnostic data and so on can be displayed after pressing the key |
| 5 | MESSAGE | It is the information page key, after pressing this key, information in the screen can be displayed, such as "alarm" information and so on |
| 6 | CUSTOM GRAPH | It is the graphic display page key, through which the completed graphics of user can be shown |

(2) The icons and functions of edit keys (shown in Table 2-3-6)

**Table 2-3-6 Icons and Functions of Edit Keys**

| No. | Icon | Function Description |
|---|---|---|
| 1 | SHIFT | It is the key to change gear. Keys in keyboard generally have two functions, pressing the "SHIFT" key can switch between the two functions |
| 2 | ALTER | It is the alternate key, which can be used to replace data in the cursor location with data in the input field |
| 3 | CAN | It is the cancel key that cancels the last character or data that has been entered into the buffer. If this key is pressed after inputting G02 X30 Z500, the final number 0 is deleted and it becomes G02 X30 Z50 |

Continued

| No. | Icon | Function Description |
|---|---|---|
| 4 | INSERT | It is the insert key, which inserts data from the input field behind the position of current cursor |
| 5 | INPUT | It is the input key, which inputs the data in the input field into the page of parameters or into an external NC program |
| 6 | DELETE | It is the delete key, which can delete the data where the cursor is located, or delete a or all programs |

(3) The icons and functions of page key (shown in Table 2-3-7)

**Table 2-3-7 Icons and functions of page keys**

| No. | Icon | Function Description |
|---|---|---|
| 1 | ↑ PAGE | This key is used to turn the page forward when the screen is displayed |
| 2 | PAGE ↓ | This key is used to turn the page backward when the screen is displayed |
| 3 | ↑ | This key is used to move the cursor up |
| 4 | ↓ | This key is used to move the cursor down |
| 5 | ← | This key is used to move the cursor to the left |
| 6 | → | This key is used to move the cursor to the right |

(4) Other icons and functions (shown in Table 2-3-8)

**Table 2-3-8　Other icons and functions**

| No. | Icon | Function Description |
|---|---|---|
| 1 | HELP | It is the system help key. When operator is unfamiliar with or doesn't understand the system control panel, the operator can press this key to get help |
| 2 | RESET | It is the reset key, which has the following functions: ①Press this key when the machine is running automatically, and all operations of the machine will stop; ②Cancel machine tool alarm; ③Make the machine tool reset and the cursor return to the beginning of the whole program; ④Clear the editing program in MDI mode |
| 3 | EOB E | This is the key to end the program segment, which ends the input of one line of program segment. When ";" appears at the end of one of the lines in the program, cursor will be changed to another line |

(5) Processing procedures (shown in Table 2-3-9)

**Table 2-3-9　Processing procedures**

| No. | Process Steps | Process Content |
|---|---|---|
| 1 | Step Ⅰ | Turn on the main switch of the power supply to power the machine tool and the NC system in the control panel |
| 2 | Step Ⅱ | Select the machine tool and operating system, and return the $X$ axis, $Y$ axis and $Z$ axis to the reference point respectively |
| 3 | Step Ⅲ | Enter, edit and check the processing program to ensure that the program and input are correct |
| 4 | Step Ⅳ | Choose the workpiece, clamp the workpiece and select various tools. Measure the compensation value of each tool by test-cut method and input the value into the tool compensation table. Attention should be paid to the positive and negative signs and decimal points |
| 5 | Step Ⅴ | Adjust the current processing procedures for automatic processing. Press the Cycle Start key to start processing. Use the function of single-stage program to check and process simultaneously, and at the same time, adjust the speed with appropriate feed rate so as to reduce the accidents caused by errors in program or tool setting |
| 6 | Step Ⅵ | After the first workpiece is processed, the size of each part should be measured. After modifying the compensation value of each tool, the second workpiece can be processed, and the workpiece can be produced in batches after the completion and determination of the correctness |

**Exercises**

1. What are the principles for determining the cutting-in point and cutting-out point?
2. What are the cutting (feeding) methods?
3. What are the actions of the hole machining instruction?
4. What is the difference between the instruction G81 and G82?
5. What are the instructions of deep hole machining, and what do they mean?
6. What is the function of the emergency stop in the control panel?
7. What does the "ALTER" key in the control panel mean and how to use it?
8. Please briefly describe processing steps of the CNC milling machine.

9. On the hole parts shown in Figure 2-3-19, drill 5 holes of $\phi10$. Choose the right tool and write the processing program.

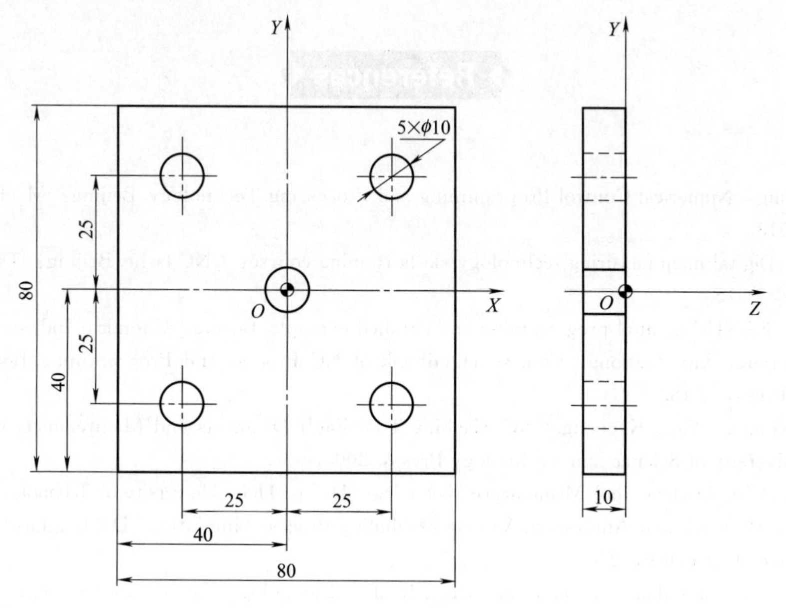


Figure 2-3-19　Hole Parts

10. Finish cover parts on NC milling, as shown in Figure 2-3-20. It is known that the blank is 120mm × 80mm × 30mm and the material is 45 steel.

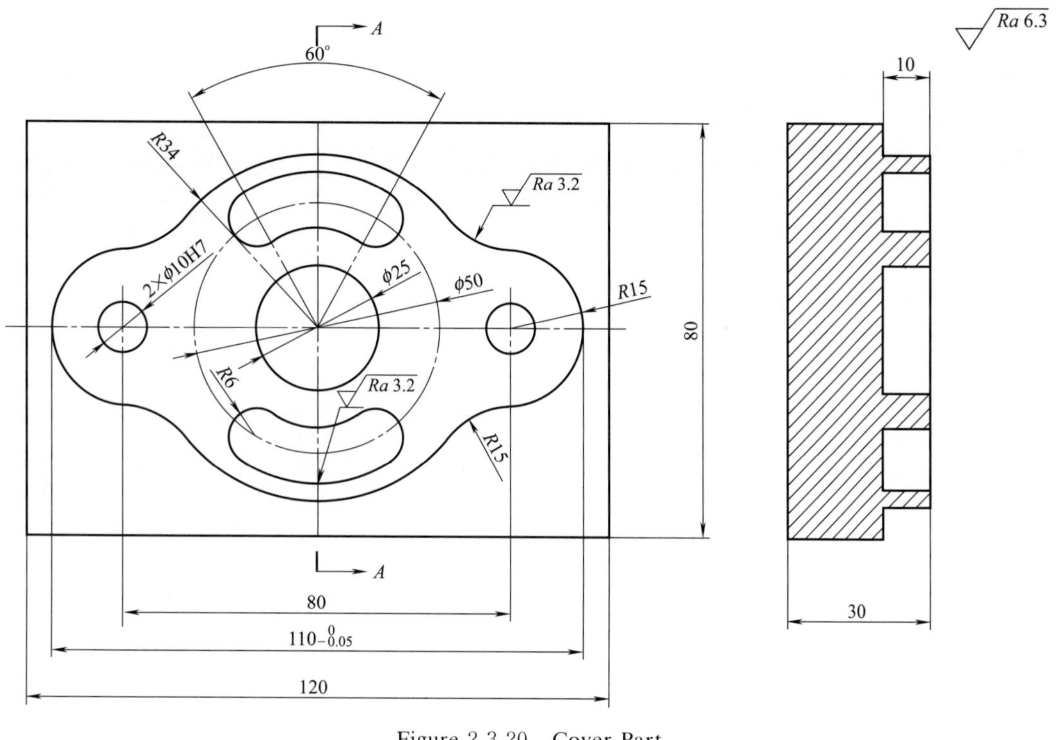


Figure 2-3-20　Cover Part

# References


[1] Wu Yizhong, Numerical Control Programming and Processing Technology. Beijing: Machanical Industry Press, 2018.

[2] Tan Bin, Digital manufacturing technology skills training course: CNC lathe. Beijing: Tsinghua University Press, 2015.

[3] Li Tiren, FANUC manual programming and detailed example. Beijing: Chemical Industry Press, 2018.

[4] Zhou Zhanxue, Liu Yuzhong, Concise Handbook of NC Process and Programming. Beijing: Chemical Industry Press, 2018.

[5] Zheng Xiaonian, Yang Kechong. CNC Machine Tool Fault Diagnosis and Maintenance. Wu Han: Huazhong University of Science and Technology Press, 2005.

[6] He Yinghe. CNC Machine Tool Maintenance Technology. Hefei: Hefei University of Technology Press, 2014.

[7] Liu Jinlei, Research and Analysis of Virtual Machining Process Simulation [D]. Beijing: North China Electric Power University, 2009.

[8] Liu Fengtian, Liu Yulan. Research on Virtral Reality Technology and Its Application in Education [J]. China Educational Technology Equipment, 2005.

[9] Wang An, NC Programming and Operation [D]. Beijing: China Light Industry Press, 2015.

[10] Li Yanxia, CNC Machine Tool Parts [D]. Xi'an: Xi'an Jiaotong University Press, 2016.